PEARSON
mybrady

MW00861646

Use the registration code below to gain access to this online st
all your student resources that support the textbook. By follow
will link to a wealth of opportunities to further your understanding of your course material. Use this
Website in conjunction with your textbook for a multidimensional learning experience. Enjoy!

STEP 1: Register

All you need to get started is a valid email address and the access code below. To register, simply:

1. Go to **www.bradybooks.com**, then click on **mybradykit**. Choose **myfirekit**.
2. Click "**Students**" under "**First-time users**."
3. Find the appropriate book cover. Cover must match the textbook edition being used for your class.
4. Click "**Register**" beside your book cover.
5. Read the **License Agreement** and **Private Policy**. If you accept, click "**I Accept**."
6. Leave "**No**" selected under "**Do you have a Pearson account?**"
7. Using a coin scratch off the silver coating below to reveal your access code. Do not use a knife or other sharp object, which can damage the code.
8. Enter your access code in lowercase or uppercase, without the dashes.
9. Follow the on-screen instructions to complete registration.

During registration, you will establish a personal login name and password to use for logging into the Website. You will also be sent a registration confirmation email that contains your login name and password. Be sure to save this email.

Your Access Code is:

Note: If there is no silver foil covering the access code, it may already have been redeemed, and therefore may no longer be valid. In that case, you can purchase access online using a major credit card. To do so, go to www.bradybooks.com. click "Students" under "First Time Users," find the cover of your textbook, then click "Buy Access," and follow the on-screen instructions.

STEP 2: Log in

1. Go to **www.bradybooks.com** and click on **mybradykit**. Choose **myfirekit**. Click on "**Students**" under "**Returning Users**."
2. Find the appropriate book cover. Click on "**Login**" next to your book cover.
3. Enter the login name and password that you created during registration. If unsure of this information, refer to your registration confirmation email.
4. Click "**Login**."

FIRE DYNAMICS

FIRE DYNAMICS

Gregory E. Gorbett

James L. Pharr

Pearson

Boston   Columbus   Indianapolis   New York   San Francisco   Upper Saddle River   Amsterdam
Cape Town   Dubai   London   Madrid   Milan   Munich   Paris   Montreal   Toronto   Delhi
Mexico City   São Paulo   Sydney   Hong Kong   Seoul   Singapore   Taipei   Tokyo

Publisher: Julie Alexander
Editor in Chief: Marlene McHugh Pratt
Senior Acquisitions Editor: Stephen Smith
Associate Editor: Monica Moosang
Development Editor: Anne Marie Masters
Editorial Assistant: Samantha Sheehan
Director of Marketing: Dave Gesell
Executive Marketing Manager: Katrin Beacom
Marketing Specialist: Michael Sirinides
Marketing Assistant: Judy Noh
Managing Production Editor: Patrick Walsh
Project Liaison: Julie Boddorf
Senior Media Editor: Amy Peltier
Media Project Manager: Lorena Cerisano
Production Editor: Karen Fortgang, bookworks publishing services
Manufacturing Manager: Ilene Sanford
Manufacturing Buyer: Pat Brown
Senior Design Coordinator: Christopher Weigand
Manager, Cover Visual Research & Permissions: Karen Sanatar
Creative Director: Jayne Conte
Cover Designer: Bruce Kenselaar
Cover Photo: James L. Pharr
Composition: Aptara®, Inc.
Printer/Binder: R.R. Donnelley
Cover Printer: Lehigh Phoenix Color

Credits and acknowledgments borrowed from other sources and reproduced, with permission, in this textbook appear on appropriate pages with text.

10 9 8 7 6 5 4 3 2

www.pearsonhighered.com

ISBN 13: 978-0-13-507588-3
ISBN 10:   0-13-507588-2

DEDICATION

This text is dedicated to those who have died and have been injured in fires. Through learning the lessons for which they paid at the greatest cost, we honor them and we hope to prevent others from repeating the actions that led to these devastating losses.

Carefully watching oxidation processes is very important for gaining a better understanding of what is occurring. Specialized equipment improves understanding by allowing the fire safety professional a way to examine factors that are not visible to the naked eye. The thermal imaging camera makes infrared energy visible so that we can study the dynamics that would otherwise be lost. This book examines fire from many different angles in an attempt to improve understanding.

CONTENTS

Chapter 10 Heat Transfer 181

Chapter 11 Ignition 205

Chapter 14 Extinguishment 284

No one is born with the knowledge and/or skills needed to perform jobs related to fire safety. Everyone must obtain this knowledge through practical experience, exploratory study, and theoretically based research. We recognize that students, especially students of fire, have different learning styles, so to be successful teachers, we try to introduce a variety of teaching methods to stimulate auditory, kinesthetic, and spatial learners. One of the difficulties with textbooks is trying to incorporate several methods so that all of these learning styles can be addressed, especially the learning styles of those students who learn better with hands-on exercises. We tried to supplement the text with many photographs and drawings to facilitate the visual learners. To facilitate the auditory and kinesthetic learners, we have created various videos to supplement this textbook; these videos can be found at http://firevideos.eku.edu/fire-science/about-eastern-kentucky-university-online/. You may also link to these videos via the myfirekit (www.myfirekit.com) created for this text. Here you will also find other review material. We encourage, challenge, and advise you to participate in such knowledge-gathering techniques to further develop your learning and understanding of this material.

One of the largest failures in the fire safety profession is technology transfer. Technology transfer is essentially the transferring of the knowledge held by few in our business and distributing it to many. Specifically, the knowledge is the experimentation and studies completed by those scientists who publish their findings in trade periodicals that typically do not appeal to the everyday fire safety professional. The studies and quantitative research are often groundbreaking, yet this information is not being transferred to those who may need it the most. Many of those scientific reports contain so much technical jargon and mathematical expressions that the important qualitative information is lost and not comprehensible to many. The problem of technology transfer is further exacerbated by the continued publishing of incorrect information in those trade journals that do influence fire safety professionals. This further propagates incorrect knowledge to another generation of firefighters and training officers because fire safety professionals cite those "reputable" magazines as the basis for their opinions. This is a principal reason that we chose to write this textbook. We believe that technology transfer must become a priority for all fire safety professions.

We do not believe that we have it all figured out. In fact, one of the most important things that we try to convey to our students is the concept of *intellectual humility*. This is especially true in the relatively young science of fire. We do believe, however, in providing what information is available and allowing students to make more informed decisions. We truly appreciate feedback. This textbook is an attempt to discuss fire dynamics in a format that is comprehensible to a wider audience. In our attempt to provide examples at a more comprehensible level, we may have lost something in translation and so we appreciate feedback to make this book better. It is our mission to make this book a conduit for technology transfer in an attempt to bridge the gap between practitioners and scientists.

Along the way, we have provided empirical correlations, physical laws, and their subsequent mathematical relationships. We did not provide an in-depth discussion of the derivation of these relationships because we did not feel it was appropriate for the audience of this text. We did, however, provide some basic examples and practical applications of various mathematical relationships that, when used correctly, can be powerful tools. For those wishing to gain a more in-depth discussion of any specific topic, we tried to incorporate references, in each chapter, to other textbooks and articles that go into more depth.

An old adage cautions us that when we engage in battle, we should know the enemy. Make no mistake, the battle between humans and the control of fire is a long, arduous battle that will continue while fire devastates before we reach a full understanding of its effects. This text attempts to explain concepts that are currently understood, while challenging you to continue studies that add to the body of knowledge in this important subject.

Gregory E. Gorbett
James L. Pharr

PEARSON
myfirekit™

As an added bonus, *Fire Dynamics* features a **myfirekit**, which provides a one-stop shop for online chapter support materials and resources. You can prepare for class and exams with multiple-choice and matching questions, weblinks, study aids, and more! To access **myfirekit** for this text, please visit **www.bradybooks.com** and click on **mybradykit**.

ACKNOWLEDGMENTS

We would like to thank and acknowledge the following reviewers:

Thomas Blair
Fire Science Degree Instructor, Weatherford College
Chief, Emergency Services District Six
Parker County, TX

Bruce Evans, MPA
Assistant Chief Support Services, North Las Vegas Fire Department
Fire Programs Coordinator, College of Southern Nevada
Las Vegas, NV

Samuel A. Giordano, Jr.
Professor and Department Coordinator, Fire Science Technology
William Rainey Harper College
Palatine, IL

Henry Goodrow
Fire Service Instructor
Kentucky State Fire/Rescue Training Area 6
Kentucky Community and Technical College System
Shelbyville, KY

William D. Hicks
Assistant Professor
Eastern Kentucky University
Richmond, KY

Joel Journeay
Department of Fire Science Chair
College of Safety and Emergency Services
Columbia Southern University
Orange Beach, AL

Matthew W. Knott, M.S.
Rockford Fire Department
Rockford, IL

Pat McVicker
Fire Professor
Skagit Valley Community College
Mount Vernon, WA

Richard Moss
Project Manager
Energy Solutions LLC
Bethel, CT

Jerry A. Nulliner
Division Chief
Fishers Fire Department
Fishers, IN

Tim Peebles
EMS Coordinator
Hall County Fire Services
Gainesville, GA

Guy Peifer, BS, NREMT-P
Yonkers Fire Department
Borough of Manhattan Community College
Yonkers, NY

Walter H. Thieme III
Director of Fire Science
Del Mar College
Corpus Christi, TX

Robert K. Toth
IRIS Fire Investigations, Inc.
Parker, CO

Gail Warner
University of Alaska, Anchorage
Fire and Emergency Services Technology
Anchorage, AK

Donnie P. West, Jr. MS, EMT-P, EFO, CFO
Center Point Fire/Rescue
Assistant Fire Chief/Fire Marshal
Professor, Columbia Southern University
Center Point, AL

First, I want to thank my coauthor, Jim Pharr, for recognizing the need for such a book and ensuring that we kept moving forward. Most important, Jim's passion and interest for teaching and improving the application of science toward the profession is to be greatly respected.

I must acknowledge my four mentors, who have had a special impact on my professional career. In the process of teaching me many things, they also became friends:

- **Dan Churchward, Kodiak Fire and Safety Consulting,** took a risk and provided me my first opportunity in this field. He has never once wavered in his confidence in my abilities, which provided me the support I desperately needed to keep pushing. Dan was the source of the spark that lit "the fire in my belly."
- **Dennis Smith, Kodiak Fire and Safety Consulting,** showed me that hard work and attention to detail result in accurate and defensible conclusions. In addition, his application of logic to the fire investigation field is unparalleled, and he has taught me to look at problems in a different way.

- *Patrick Kennedy, John A. Kennedy and Associates,* helped me to obtain a much greater respect for our profession and a new appreciation for the practical application of science to fire investigations. I thank Pat most for his persistent reminders to keep working toward that "light at the end of the tunnel."
- *Ron Hopkins, Eastern Kentucky University,* showed me that a person could have passion for and an excitement about teaching this science stuff. Ron *is* the reason that I now teach.

Most important, I am incredibly grateful to my wife and son, Kelly and Matthew: Your love and support were instrumental in helping me to complete this project, as they are in all parts of my life. You are my source of strength in this and all that I do. Thank you for your constant encouragement and the sacrifices that you have made, which have allowed me to give the necessary attention to this project. In addition, I would like to give thanks to my parents, who have sacrificed so much to provide me with so many opportunities. My mother, Kelly, slapped my hands more than once for playing with fire, and my father, Russell, taught me to respect fire. (And in one extreme case, he reminded me that eyebrows *do* grow back!) Finally, I want to thank my brother, Chris, who keeps me interested in fire and persistently reminds me that there is a difference between the book and the real world.

—Greg Gorbett

Noting each person who influenced me in the fire service throughout the past four decades is all but impossible; however, a few must be acknowledged. From the first chief I served under, Walter Burr, who encouraged young firefighters to study relentlessly; to my co-author Greg Gorbett, who is dedicated and driven to gain understanding of fire; and to Wayne Flora and Chris Hooper, who led me to get an education and have mentored me throughout my career, a simple thank you. I can never truly express my appreciation. Along the path, I have been fortunate to work with many students of fire who challenged traditional thinking. For each of these influences, I am eternally grateful.

To my parents, J. Lee and Emmy, thank you for providing me with the opportunity to pursue a career. You may have questioned my choice, but you always supported me. To my daughters, Mandy and Cindy, who tolerated many hours of separation as I worked and studied, yet still managed to love me, I am so grateful. Last, but certainly not least, I thank God for allowing me to live the dream of being a firefighter.

—Jim Pharr

Together we thank Brady Fire/Pearson/Prentice Hall and their fantastic support personnel. We appreciate the opportunity to affect lives with the written word, and we would like to express our gratitude to Anne Marie Masters, Monica Moosang, Heather Luciano, and Stephen Smith for their patience and confidence.

We gratefully acknowledge and thank the scientists who have dedicated their lives to better understanding fire. These people include, among many others, Dr. Dougal Drysdale, Dr. Vytenis Barauskas, Dr. James Quintiere, Dr. James Milke, Dr. Robert Zalosh, Dr. Craig Beyler, Dr. Kozo Saito, Dr. Dan Gottuk, Dr. Phillip Thomas, Dr. Harold "Bud" Nelson, and Daniel Madrzykowski, P.E. Without their efforts, misunderstanding of fire phenomena would continue.

In addition, many others have inspired portions of this text, including Ed Hartin, Dr. Dave Icove of the University of Tennessee, Ryan Cox of Kodiak Fire and Safety Consulting, Kathryn Kennedy of John A. Kennedy and Associates, the National Association of Fire Investigators, the International Fire Service Training Association, Wayne Chapdelaine, and Eugene Meyers.

We thank Eastern Kentucky University, specifically the Fire and Safety Engineering Technology Program, for the chance to teach. We especially thank Dr. Larry Collins, Paul Grant, Bill Abney, Shane LaCount, Branden Sobaski, Corey Hanks, and all of the students who have had to endure our never-ending homework and *new* (sometimes referred to as "crazy") ideas. Finally, a special appreciation for Professors Bill Hicks and Andrew Tinsley from Eastern Kentucky University, for their direct involvement in reviewing the manuscript and providing needed feedback.

Greg Gorbett
Jim Pharr

ABOUT THE AUTHORS

Gregory E. Gorbett is currently an assistant professor in fire and safety engineering technology at Eastern Kentucky University. Professor Gorbett specializes in instruction and curriculum development for fire behavior and combustion, fire dynamics, fire investigation, and fire protection systems courses at Eastern Kentucky University (EKU). Gorbett holds a B.S. in forensic science from Trine (Tri-State) University; a B.S. in fire science from the University of Maryland, University College; an M.S. in fire protection engineering from Worcester Polytechnic Institute; and an M.S. in executive fire service leadership from Grand Canyon University; and is currently pursuing a Ph.D. in fire protection engineering from Worcester Polytechnic Institute. Prior to joining EKU, Professor Gorbett was a fire and explosion analyst and investigator. Gorbett is a certified fire and explosion investigator (CFEI), certified fire investigator (CFI), and certified fire protection specialist (CFPS) through the National Fire Protection Association (NFPA). Gorbett serves on numerous NFPA and International Fire Service Training Association (IFSTA) validation committees as well as the board of directors for the National Association of Fire Investigators and has been a presenter nationally at a number of fire-related conferences.

James L. Pharr is currently assistant professor in fire and safety engineering technology at Eastern Kentucky University (EKU). Professor Pharr specializes in fire dynamics, building and life safety, supervision, emergency scene operations, and hazardous materials response. Pharr received an A.S. in fire science technology from Rowan Technical Institute and a B.S. in fire and safety engineering technology from the School of Applied Science at the University of Cincinnati. Pharr holds an M.S. in executive fire service leadership from Grand Canyon University. Professor Pharr has also completed the Executive Fire Officer Program at the National Fire Academy in Emmitsburg, Maryland, where he is an adjunct instructor. Prior to joining EKU, Pharr was the emergency management director and fire marshal in Gaston County, North Carolina. Pharr is a member of the International Association of Arson Investigators and the International Association of Fire Chiefs. Pharr has published a number of journal articles and research papers.

The following grid outlines fire dynamics course requirements and the chapters where specific content can be located within this text:

Course Requirements	1	2	3	4	5	6	7	8	9	10	11	12	13	14
Analyze building structural components for fire endurance and fire resistance.												X	X	
Understand the flame spread and smoke production properties of building furnishings and materials.									X	X	X			
Identify the physical properties of the three states of matter.				X		X	X	X						
Categorize the components of fire.					X	X	X	X		X				
Explain the physical and chemical properties of fire.			X	X										
Describe and apply the process of burning.						X	X	X	X					
Define and use basic terms and concepts associated with the chemistry and dynamics of fire.			X	X					X		X			
Discuss various materials and their relationship to fires as fuel.											X			
Demonstrate knowledge of the characteristics of water as a fire suppression agent.														X
Discuss other suppression agents and strategies.														X
Compare other methods and techniques of fire extinguishment.												X		X

FIRE DYNAMICS

1

Introduction

KEY TERMS

code enforcement personnel, *p. 8*

fire investigators, *p. 9*

fire protection engineers, *p. 9*

fire suppression personnel, *p. 7*

heat release rate (HRR), *p. 4*

time-temperature curve, *p. 5*

OBJECTIVES

After reading this chapter, you should be able to:

■ Identify and define the various roles of fire safety professionals.
■ Describe the importance of understanding fire behavior and fire dynamics as it applies to all fire safety professions.
■ Summarize fire fatality statistics and be able to analyze these statistics.

PEARSON

myfirekit

For additional review and practice tests, visit **www.bradybooks.com** and click on MyBradyKit to access book-specific resources for this text!

Introduction

Review of devastation wrought by fire throughout the centuries reveals that fire is a serial killer. Humans repeatedly create conditions that expose themselves and others to the probability of threatening situations (see Table 1.1). Few study those conditions prior to the devastating fire, yet voice great concern following such incidents. Easily one can correlate conditions that existed in the Cocoanut Grove Night Club (1942) with those in the Station Night Club (2003) where large life loss fires occurred. Separated by more than 60 years, these fires had fuels with high heat release rates and flammable materials lining walls, which quickly burned and prevented patrons from escaping through restricted exits. These are two examples of threatening conditions that were repeated because we failed to apply the understanding that was gained, at a great price.

Though not cited in most texts or studies, fires involving smaller numbers of fatalities occur more frequently, resulting in a greater number of total deaths and injuries (see Figure 1.1). The United States Fire Administration (USFA) reported that, in 2007, 84 percent of the 3,430 civilian fatalities in the United States occurred in residential fires. Another 17,675 civilians were estimated to have been injured by fire. Understanding conditions that led to these fatalities and injuries are paramount in reducing fire casualties.

The number of firefighters dying in the line of duty has remained above 100 annually for all but a few years since the National Fire Protection Association (NFPA) began compiling statistics in the late 1970s (see Figure 1.2). Many of the firefighters in these statistics die because those involved failed to gain sufficient

TABLE 1.1 Large Loss Fires in the United States	NUMBER OF DEATHS
1. World Trade Center, New York, NY, September 11, 2001	2,666
2. S.S. Sultana steamship boiler explosion and fire, Mississippi River, April 27, 1865	1,547
3. Forest fire, Peshtigo, WI, and environs, October 8, 1871	1,152
4. General Slocum excursion steamship fire, New York, NY, June 15, 1904	1,030
5. Iroquois Theater, Chicago, IL, December 30, 1903	602
6. Forest fire, northern MN, October 12, 1918	559
7. Cocoanut Grove nightclub, Boston, MA, November 28, 1942	492
8. S.S. Grandcamp and Monsanto Chemical Company plant, Texas City, TX, April 16, 1947	468
9. Forest fire, Hinckley, MN, and environs, September 1, 1894	418
10. Monongah Mine coal mine explosion, Monongah, WV, December 6, 1907	361

Sources: NFPA's Fire Incident Data Organization and other NFPA fire incident records.

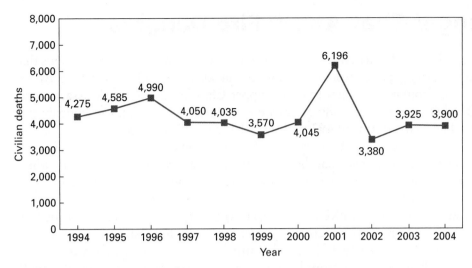

FIGURE 1.1 United States fire fatalities, 1994–2004.

understanding of conditions and dynamics involved in the chemical reaction they attempted to stop. Granted, some died during these years attempting to rescue human lives, and that is noble. Others, however, perished in situations where conditions precluded humans from living, yet risks were taken to attempt a "rescue." Fire behavior and the dynamics associated with fire development is an often simplified phase of study for those involved in protecting citizens and buildings from devastation resulting from flaming combustion. Research concerning fire phenomena has increased significantly in the latter decades of the 1990s, and that study continues today. Through increased understanding, increased efficiency in preventing and suppressing fires becomes possible, while at the same time decreasing exposure of humans to the adverse effects of those fires.

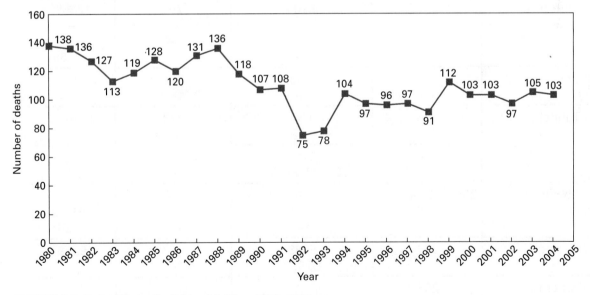

FIGURE 1.2 United States firefighter fatalities, 1980–2004.
Note: Excludes 330 firefighter deaths from 9/11/01.
Source: Adapted from data provided by US Fire Administration and NIOSH.

Changes That Affect Fire Dangers

Building and fire codes are largely promulgated based on engineering projections that respond to learning gained from fires and other incidents related to structures. An example comes from the Cocoanut Grove Supper Club fire in Boston, Massachusetts, in 1942, where inward swinging doors were identified as a prominent contributing factor leading to the deaths of 492 patrons. Change resulted in requirements for outward swinging egress doors in all facilities. Construction codes generally require outward swinging doors, but unauthorized changes can result in situations where improper doors are present. Fire code inspections easily identify these changes.

Testing of building components and assemblies has a long history. The American Society for Testing and Materials (ASTM) E119, Standard Test Methods for Fire Tests of Building Construction, fire growth rates for testing fire resistance, were developed in the early 1920s and continue as the standard used today (see Table 1.2). When these fire growth rates were developed, the concept of a fuel's **heat release rate (HRR)** was not considered; rather, a fire was modeled to match

heat release rate (HRR)
- The rate at which heat energy is generated by burning.

TABLE 1.2	Selected Standards for Fire Testing			
FIRE TEST STANDARDS	**NFPA**	**ASTM**	**UL**	**ISO**
Fire Resistance	251	E119	263	834[a]
	252	E2074 (replaced E152)	10B	3008[a]
	288	—	—	—
	257	E2010 (replaced E163)	9	3009[a]
	—	E814	1479[a]	—
	—	E1966	2079[a]	—
Exterior Fire Tests	256	E108	790	12468[a]
	268	—	—	—
	285	—	—	—
	—	—	—	13785
Ignitability	—	D1929	—	871
	—	—	—	5657
Steiner Tunnel Surface Flame Spread	255	E84	723	—
	262	—	910 (withdrawn)	—
Lateral Flame Spread	—	E162	—	—
	—	E1317, E1321	—	5658
Smoke Obscuration (Static)	258 (withdrawn)	E662	—	—
	270	E1995	—	5659-2
	—	—	—	5924
Floor Coverings	253	E648	—	9269
	—	D2859	—	—
Room-Corner Heat Release Tests	265	—	1715[a]	—
	286	—	—	—
	—	E2257	—	9705

TABLE 1.2	(continued)			
FIRE TEST STANDARDS	**NFPA**	**ASTM**	**UL**	**ISO**
Large-Scale Product Heat Release Tests	266 (withdrawn) 267 (withdrawn) 274 289 (draft[b]) — —	E1537 E1590 — — — E1822	1056 (withdrawn) 1895 (withdrawn) — — 1975 —	— — — — — —
Heat and Smoke Release (Bench Scale and Intermediate Scale)	271 263 (withdrawn) 272 — — — — 287 —	E1354 E906 E1474 E1740 D6113 F1550 E2102 E2058 E1623	2360 — — — — — — — —	5660 — — — — — 13927, 17554 — TR 14696
Potential Heat, Heat of Combustion, and Noncombustibility	259 — —	— E136 D5865	— — —	— 1182[a] 1716
Cigarette Ignition Tests	260 261 —	E1353 E1352 E2187	— — —	— — —
Smoke Toxicity	269	E1678	—	—
Smoke Corrosivity	— — —	D5485 — —	— — —	11907-4 11907-2 11907-3
Fabric Flammability	701 705	D568 (withdrawn) —	214 (withdrawn) —	— —
Plastics Flammability	— —	D2863 D635, D3801, D4804, D4986, D5048	— 94	4589 9772, 9773, 12992

a specific temperature increase occurring over a specified time. This model is known as the standard **time-temperature curve** (see Figure 1.3). The standard time-temperature curve was developed in the 1920s and is based on fuels of the day. It is still used today for fire resistance testing; however, the fuels in the 1920s did not include synthetic materials. Synthetics prevalent today typically have burning rates that are much higher than natural materials; consequently, an adverse impact on tenability and structural stability occurs much sooner than many burning rates based on historic fires and outdated fire resistance testing. This results in failure of protective barriers much sooner than their rating indicates; for example, an assembly rated for one half hour against the standard

time-temperature curve
■ A graph that depicts temperature at specific times during fire development. The standard time-temperature curve is used to evaluate components and assemblies to determine their performance during fire exposure.

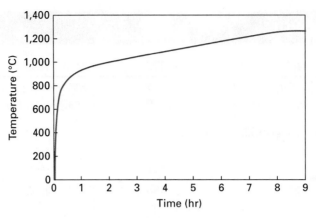

FIGURE 1.3 Standard time-temperature curve, circa 1928.
Source: Data derived from NFPA National Fire Experience Survey.

time-temperature curve may fail in fifteen minutes or less when subjected to a rapidly developing flame.

Even less was known about fuels involved in these fires. Greater documentation and research is now occurring on fuels and fuel arrangement within compartments to identify protective measures that will mitigate these hazards (see Figure 1.4). In the Cocoanut Grove Supper Club, flames from the combustible furnishings and lining materials contributed to rapid fire development, but only limited testing was possible for those fuels in 1942. Testing agencies and fire protection professionals knew the flame spread characteristics of soundproofing used in the Station Nightclub in West Warwick, Rhode Island, but evidence indicates flammability of that foam was not considered before installation or during fire code inspections.

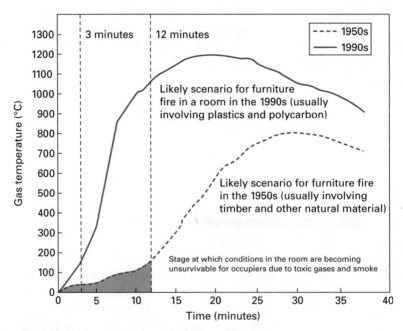

FIGURE 1.4 Time-temperature curve for a room fire.
Source: Way Forward New Zealand, Fire Commission.

Fire Dynamics: The Link to Collaborative Fire Protection

Fire dynamics refers to the interrelationship of fuel, heat, oxygen, chemical reactions, and the physical world. Study and control of these components provides safety for humans and structures. Understanding how heat is released from fire and how it affects nearby fuels provides insight into predicting fire intensity and development. Understanding how ventilation (infusion of fresh air and the outflow of fire gases) affects fire growth and tenability conditions within structures also facilitates critical insight into preventing life loss.

FIRE SUPPRESSION PERSONNEL

Fire suppression personnel have a vested interest in understanding fire dynamics as completely as possible. Conditions during emergencies preclude detailed study of exact dynamics; however, evaluations before emergencies (known as pre-fire surveys and more commonly called preplanning) yield information from more extensive review. Even in emergency conditions, understanding of fire dynamics allows the suppression forces to make critical observations so that they can develop better, safer strategies for resolving the situations. Firefighters who possess the knowledge and ability to "read" conditions can differentiate between those circumstances where successful attack is probable and those conditions that exceed human tolerance and thus will not allow a successful save. Proper and informed judgment saves firefighters' lives (see Figure 1.5).

fire suppression personnel
■ Persons who engage in operations to extinguish fires for public and private entities.

FIGURE 1.5 Fire suppression personnel. *Source: Photo by Katie Steenken.*

Significant increases in the volume of commodities stored in structures today, especially extremely combustible synthetic materials, results in rapid fire development that adversely affects assemblies that provide fire protection. Once the protective barrier is penetrated, structural members, especially lightweight members, fail quickly. In recent decades, failures of lightweight structural members have resulted in the deaths of many firefighters. Conditions that lead to collapse originate from the failure of the protective barriers due to high heat release rate (HRR) fires caused by modern fuels. Fire suppression forces have a significant vested interest in understanding dynamic conditions created by fire. Flashover, rollover, and backdraft conditions may cause injuries and deaths if fire conditions are not predicted and then handled properly. Interest is far more personal than other professions because fire suppression personnel place their lives in harm's way and too often lose that battle.

FIRE PROTECTION ENGINEERING AND CODE ENFORCEMENT PERSONNEL

code enforcement personnel
■ Persons who engage in inspections to ensure that conditions are within prescribed parameters and thus provide favorable fire protection conditions.

Fire protection professionals must fully explore fuels, fuel geometry, and compartment arrangements, then design equipment capable of detecting and suppressing fires in a manner that prevents conditions that may become life threatening. Educating building owners, occupants, and insurers of dangers associated with operating buildings beyond design considerations (that is, with higher heat release rate fuels or more dangerous fuel geometry, like that found with rack storage) is also critical to threat reduction. **Code enforcement personnel** involved in inspections are charged with identifying conditions that exceed design criteria

FIGURE 1.6 Fire protection engineering and code enforcement personnel.
Source: Photo by authors.

for fire protection systems, hinder active fire protection and detection elements from operating accurately, or adversely affect passive fire protection elements. In the event that any of these active or passive fire protection elements are compromised, fire prevention professionals must act to mitigate threats posed (see Figure 1.6).

Fire protection is becoming more performance-based. Thus, both design **fire protection engineers** and inspection personnel must become more proficient in understanding fire dynamics and how the by-products of combustion (i.e., heat, noxious gases) affect the people and property they are charged with protecting.

fire protection engineers
■ Persons who seek to use scientific methods to prevent or mitigate fires through study and testing.

FIRE INVESTIGATORS

When fires do occur, **fire investigators** must reverse-engineer conditions to determine origin, cause, spread, responsibility, and circumstances regarding fire origin and development. Only when fire dynamics are fully understood can investigators reliably identify a fire's origin. During investigations, hypotheses are developed regarding heat sources and fuels ignited. Understanding fire dynamics assists investigators in competently testing hypotheses. Once an origin is properly determined based on the dynamics of fire, investigators may then proceed to seek the fire's cause (see Figure 1.7).

fire investigators
■ Persons who work to determine the origin, cause, and circumstances of fires.

FIGURE 1.7 Fire investigation personnel.
Source: Photo by authors.

Outline of the Text

This text is dedicated to raising the understanding of fire and fire dynamics. The difference in knowledge and understanding is illustrated in Figure 1.8. The van owner in the figure experienced an engine fire. Life lessons had provided the owner with the knowledge that putting dirt or sand on the fire would extinguish the flames. The owner operating a small end loader dumped a bucket full of sand on the top of the van. The owner knew to use sand but did not understand the dynamics affecting heating of fuels to release gases that could then undergo combustion. Thus, the sand was not placed on the fuel source. Though the van owner's knowledge dictated what to do, the owner did not understand how or why. Had the owner understood the dynamics involved, a small shovel full of sand would have proven more effective.

This text guides readers through fundamental fire science (chemistry and physics), then moves to the dynamics involved in flaming combustion. With a better understanding of fire dynamics in conjunction with a cooperative approach to fire protection, we believe casualties and property loss will be reduced.

Chapters 2 through 4 in this text review math, chemistry, terminology, and physics, information that will be essential to gaining full understanding of fire dynamics. Life experience teaches us that many firefighters are reluctant to delve deeply into math, chemistry, and physics, but those concepts facilitate a much greater understanding of the phenomena involved. We recognize that nothing is more frustrating than performing tasks that are perceived to be irrelevant; consequently, the reviews have been directed to strengthen the reader's capabilities in

FIGURE 1.8 Improper extinguishment method.
Source: Photo by authors.

those specific areas that will enable the reader to become more proficient at performing tasks related to fire dynamics. Often students are attracted to technology programs because they are not comfortable with a mathematical derivation as the sole justification of a technical principle, as is customary in engineering science programs. On the other hand, once a principle is thoroughly understood in accurate qualitative terms, a mathematical expression of that principle does make sense. Consequently, all topics in this book are developed in a methodical, step-by-step, cause-and-effect approach geared to the need for technology students to be able to visualize why fire behaves as it does.

In this text, you will explore how matter exists in nature. Matter exists in three states: solid, liquid, and gas. However, combustion involves only fuels in a gaseous state. Study includes pyrolysis and vaporization, the processes whereby solids and liquids, respectively, are changed to the gaseous state by heating. Chapters are dedicated to the better understanding of fuels in all three states to enable better understanding of how fuels act in relationship to fire.

Chapter 5 details fire and terms associated with fire. Fire is a "rapid oxidation process, which is a chemical reaction, resulting in the evolution of heat and light in varying intensities" (NFPA 921, 2008). Concepts of the fire triangle and the fire tetrahedron are explained. Products of combustion, including fire gases, flame plumes, smoke, and other products, are explained. Through this text, you will gain understanding of heat transfer through conduction, convection, and radiation. Conduction is heat transfer through direct molecular contact (most prevalent in a solid medium), convection is heat transfer through fluid movement, and radiation is energy transfer without a medium. We discuss how the transfer of heat to fuels affects flame intensity and fire development within compartments (enclosures such as rooms). Substantial explanation is provided about how fire gases in compartments develop gas layers, and how heat transfer to other fuels within those enclosures affords fire professionals the ability to predict and counter the effects before conditions become untenable. This understanding is paramount in reducing fire deaths and injuries for both firefighters and civilians exposed to fire conditions.

Last, the reader will explore methods for extinguishing fires. The most common method of fire extinguishment is the application of water. Therefore, significant discussion is provided to enable a better understanding of how water affects fire and accomplishes extinguishment. Other extinguishing agents and methods are also explored. One of the greatest benefits of understanding fire extinguishment is gaining an appreciation for the effectiveness of applying extinguishing agents when flames are very small.

The theory and practical application of fire dynamics is still evolving, and this textbook will serve as a basic introduction to the many topics that are important to the general fire protection professional. Throughout the text, we will point out other texts and articles that have devoted more discussion about and research into the topic at hand for readers who are interested in further reading.

Summary

This text is dedicated to those who have died and been injured in fires. Through learning the lessons they paid at the greatest cost, we honor them and we hope to prevent others from repeating the actions that led to their devastating losses. We encourage you to learn from the past so that you can protect yourself in the future. By understanding fire, you will be able to understand the challenges posed by technological changes that are to come.

Case Studies

The National Institute of Occupational Safety and Health (NIOSH) studies selected firefighter fatalities within the United States and then publishes data regarding lessons learned. Reposts are accessed at http://www.cdc.gov/niosh/fire/. The following is a list of case-study recommended reading to increase awareness of how fire behavior affects safety within structures and to increase understanding of fire dynamics to better predict what will occur. Recommendations often relate to understanding risk versus gain in deciding tactical concerns. This understanding, combined with coordinated ventilation, is an indicator that personnel discussed in the case studies listed here did not fully account for fire dynamics in developing tactical priorities.

Case Study Number	Date	Description
F2008-34	Oct 29, 2008	Volunteer firefighter dies while lost in residential structure fire (Alabama)
F2008-09	Apr 08, 2008	A career captain and a part-time firefighter die in a residential floor collapse (Ohio)
F2008-07	Mar 07, 2008	Two career firefighters die and a captain are burned when trapped during fire suppression operations at a mill-work facility (North Carolina)

Case Study Number	Date	Description
F2007-28	Jul 21, 2007	A career captain and an engineer die while conducting a primary search at a residential structure fire (California)
F2007-18	Jun 18, 2007	Nine career firefighters die in rapid fire progression at commercial furniture showroom (South Carolina)
F2007-16	May 28, 2007	Career firefighter dies and a captain is injured during a civilian rescue attempt at a residential structure fire (Georgia)
F2007-12	Apr 16, 2007	Career firefighter dies in wind-driven residential structure fire (Virginia)
F2006-28	Oct 10, 2006	Career firefighter dies in residential row house structure fire (Maryland)
F2006-27	Aug 27, 2006	Floor collapse at commercial structure fire claims the lives of one career lieutenant and one career firefighter (New York)

Case Study Number	Date	Description
F2006-26	Aug 13, 2006	Career engineer dies and a firefighter is injured after falling through floor while conducting a primary search at a residential structure fire (Wisconsin)
F2006-24	Jun 25, 2006	Volunteer deputy fire chief dies after falling through floor hole in residential structure during fire attack (Indiana)

Case Study Number	Date	Description
F2002-11	Mar 04, 2002	One career firefighter dies and a captain is hospitalized after floor collapses in residential fire (North Carolina)
F32	Nov 06, 1998	Two volunteer firefighters are killed, and one firefighter and one civilian are injured during an interior fire attack in an auto salvage storage building (North Carolina)

Review Questions

1. Fire dynamics refers to:
   a. Fire and smoke development
   b. Study of how heat, fuel, and oxygen affect fire behavior
   c. Fluid mechanics
   d. None of the above
2. Fire suppression personnel are:
   a. Sprinkler system designers
   b. Alarm system designers
   c. Public and private responders to fire situations
   d. None of the above
3. Fire investigation personnel are:
   a. Fire suppression personnel
   b. Public and private personnel dedicated to studying the origin and cause of fires
   c. Laboratory scientists who study fire dynamics
   d. Engineers who study fire to develop building and fire safety codes
4. Fire protection engineering personnel are:
   a. Persons dedicated to mitigating fire effects through engineering practices
   b. Fire apparatus operators
   c. Professionals dedicated to ensuring that occupancies are situated around like occupancies
   d. None of the above

5. The standard time-temperature curve is relative to modern fire situations because:
   a. It outlines the actual fire conditions anticipated in modern buildings
   b. It offers an idea of what is anticipated in modern buildings
   c. It provides a standard to compare the relative resistance of building systems
   d. It has no validity in modern construction
6. Heat release rate (HRR) refers to the:
   a. Temperature from a fire
   b. Total energy released from a fully oxidized fuel package
   c. Amount of energy released when a specific amount of oxygen reacts
   d. Rate of energy release expressed in joules per second (watts) or British thermal units (BTUs) per second
7. The definition of the term *fire* is:
   a. Rapid oxidation process, which is a chemical reaction, resulting in the evolution of heat and light in varying intensities
   b. A chemical reaction
   c. Heat and light emitted from a chemical reaction
   d. None of the above

8. Understanding fire dynamics is important to:
   a. Fire suppression personnel
   b. Fire protection engineering personnel
   c. Fire investigations
   d. All of the above
9. Fire suppression personnel benefit from applying the principles of fire dynamics because they can then:
   a. Understand indicators of fire behavior to predict what is likely to occur in the near future
   b. Create fire dynamics models during fire suppression operations
   c. Understand indicators of fire behavior to determine what has already happened
   d. None of the above
10. The National Institute for Science and Technology (NIST) benefits those involved in fire protection by:
   a. Providing information on testing of specific fuel materials
   b. Providing information on testing of specific compartment arrangements
   c. Providing software that helps to model fires
   d. All of the above

PEARSON
myfirekit™

For additional review and practice tests, visit **www.bradybooks.com** and click on MyBradyKit to access book-specific resources for this text!

Register your access code from the front of your book by going to **www.bradybooks.com** and selecting the mykit links.

References

NFPA 921 (2008). *The Guide for Fire and Explosion Investigations*. Quincy, MA: National Fire Protection Association.

2
Math Review

OBJECTIVES

After reading this chapter, you should be able to:

- Understand conversions and formulas needed to understand fire from a mathematical concept.
- Recognize units of measure and convert between the systems of these units of measure.
- Demystify the concepts of math and feel confident about understanding fire through math.

For additional review and practice tests, visit **www.bradybooks.com** and click on MyBradyKit to access book-specific resources for this text!

Algebra

The term *algebra* inspires dread in many otherwise competent, confident people. Why? We suspect that rules applied to solving unknowns is somewhat confusing; thus, the fear relates to the unknown. For the purposes of this text, formulas are presented so there is no need to convert or factor to reach answers. Some of the formulas may be intimidating until you break them into smaller pieces for better understanding. To achieve this, let's examine some of algebra's more simple principles. If you need additional information on algebraic rules, please refer to *Brady's Mathematics and Problem Solving for Fire Service Personnel: A Worktext for Student Achievement* by Eugene Mahoney.

ALGEBRAIC EXPRESSIONS

Some may feel apprehension when they see letters or symbols where numbers are needed. Understanding what those letters and symbols represent is essential for solving the problem. Let's start with a simple expression:

$$LWH = V$$

V represents volume, L = length, W = width, and H = height. We must also understand that placing these symbols together indicates that we should multiply them (implied multiplication) so that $L * W * H = V$ is the represented equation. When we know the length, width, and height of any compartment, we can determine its volume.

Only multiplication is implied. The symbols for indicating multiplication and division come in various representations. Multiplication may be indicated by a simple $\times$ or $*$. Division may be shown as the traditional $\div$, or it may be shown with a solidus (x/y) or line showing a fraction x/y. Both are equivalent terms (that is, they indicate the same operation).

Exponents are used to indicate multiplication of the number by itself, for example 10^3 indicates that you should multiply $10 * 10 * 10$, which results in the product of 1,000. Exponents may be expressed as decimals or fractions; in either case, the easiest method for deriving the solution is using a calculator's exponent function. Fortunately, addition and subtraction are straightforward.

A simple algebraic expression is found in the combined gas law, which describes how gases react to changes in temperature, pressure, and/or volume. Here is the combined gas law expressed in algebraic terms:

$$\frac{P_1 V_1}{T_1} = \frac{P_2 V_2}{T_2}$$

This equation indicates that the relationship of pressure, temperature, and volume must remain equal. If one factor changes, others must change to balance the equation mathematically. If you know the initial and final temperatures, and the volume remains constant, the formula indicates what occurs with pressure. Let's examine that concept.

We have a compartment with 45 m³ of volume at a normal atmospheric pressure of 101.325 kilopascal (kpa) and ambient room temperature of 293°K. Fire within the compartment raises the temperature to 873°K (600°C) and all openings are closed. What is the pressure?

$$\frac{101.325 \text{ kpa} * 45 \text{ m}^3}{293 \text{ K}} = \frac{x \text{ kpa} * 45 \text{ m}^3}{873 \text{ K}}$$

To determine the answer, we must complete the equation so that we can get the unknown on one side of the equals sign. This is accomplished by moving the denominator from one side to the numerator on the other side of the equation. Also, we move the numerator of one side to the denominator of the other:

$$\frac{101.325 \text{ kpa} * 45 \text{ m}^3 * 873 \text{ K}}{293 \text{ K} * 45 \text{ m}^3} = x \text{ kpa}$$

Note that units cancel when they appear in both the numerator and denominator (like units are represented as the same color in the above equation and are therefore cancelled out):

$$\frac{101.325 \text{ kpa} * 45 \; \cancel{\text{m}^3} * 873 \; \cancel{\text{K}}}{293 \; \cancel{\text{K}} * 45 \; \cancel{\text{m}^3}} = x \text{ kpa}$$

$$= 301.9 \text{ kpa}$$

More simply stated, the pressure would be more than three times greater than normal ambient atmosphere. By plugging in known quantifiers and moving data, we were able to determine the answer.

ORDER OF OPERATIONS

Algebraic equations are written with symbols and letters indicating specific data points. How you process the data often determines results. The rules of algebra require you to multiply and divide from the left to right of a formula first. Then from left to right, complete addition and subtraction.

EXAMPLE 2.1

$$23 + 34 * 2 - 23/5 = x$$

First multiply 34 * 2 = 68.
Next divide 23 by 5 = 4.6.
Now add and subtract.

$$23 + 68 - 4.6 = 90.4$$

Data in parentheses and brackets are treated as separate formulas until their product is determined. Then they are factored by rules of multiplication/division and addition/subtraction order.

EXAMPLE 2.2

$$(23 + 34) * (2 - 23)/5 = x$$

First, factor the information in the parentheses.

$$23 + 34 = 57$$
$$2 - 23 = -21$$

Now, factor the multiplication and division.

$$57 * -21/5 = -239.4$$

Note that the examples include the same data. However, the different order means that much different answers resulted because of the difference in the order of operations.

When working formulas in fire dynamics, you will usually have some of the information needed to find an answer. Identify what information is needed, then plug it into the formula. Often, you may have to determine different data for specific attributes within formulas.

RATE

Rate refers to the quantity over time or another unit of measure. Common rates in fire dynamics are energy released over time (the heat release rate) expressed in watts (Joules per second) and heat flux, which is expressed as heat energy per unit of surface area, often described as watts per square meter.

SIGNIFICANT INTEGERS

Throughout this text, you will round numbers to significant integers. Typically three significant integers are sufficient because the first three integers of an answer are sufficient to yield acceptable results. Some retain numbers to two decimal places when whole numbers are associated; others use three decimal places. But rounding to whole numbers often occurs. The need for precision is the driving force in determining the level of detail, which in basic fire dynamics is two decimal places.

When rounding at the lowest significant integer, round up if the following integer is 5 or above and round down if the following integer is less than 5. This technique does open the door for some rounding error; however, that error is not significant enough to change basic understanding when dealing with fire development in compartments.

COORDINATE SYSTEM

coordinate system
- A system that uses incremental measures on three planes to designate the exact location of a point. Planes are commonly designated x, y, and z.

Inputs for many computerized fire modeling systems require accurate comparison of compartment dimensions with fuel package and ventilation locations within that compartment. A system of graphing coordinates is used to depict these points precisely. The **coordinate system** uses x to depict measurement points in one horizontal axis, y to depict horizontal measure in the plane perpendicular to the x axis, and a plane labeled z to depict measurements along the plane vertical to x and y axes (see Figure 2.1). Use of this system facilitates placement of specific points

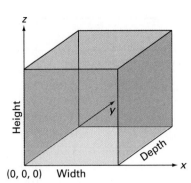

FIGURE 2.1 *x, y, z* coordinate system.

that can be plotted into three-dimensional representations of the compartment's features. The point labeled 0, 0, 0 serves as the starting point from which all axes emanate. Negative coordinates are permissible; they indicate simply that the axes extend in the opposite direction of positive coordinates.

EXAMPLE 2.3

A compartment measures 4 meters long (x), 3.5 meters wide (y), and 2.4 meters high (z). A door measuring 6′8″ (2.033 m) high and 3′0″ (.9144 m) wide is present in the x plane, 1 meter from the y coordinate.

The compartment is written as:

$$0, 4; 0, 3.5; 0, 2.4$$

The door's coordinates are written as:

$$0, 0; 1, 1.9144; 0, 2.033$$

Units of Measure

Many units of measure can be used to quantify length, **area**, volume, mass, and/or heat. Simply stated, units of measure involve standardization of increments to quantify the relative size of an item. This chapter will cover the **English system** and the International System of Units (abbreviated SI from the French Systeme International) **(SI)** because they are the primary systems used in the modern world. You may encounter other systems of measure; some can only be termed archaic (really old), and conversions can be obtained from tables and Internet conversion sites. In the English system, someone (probably a king) dictated the acceptable reference for a unit, whereas a committee of scientists specified references in SI. It is interesting that England does not use the "English" system; it adopted SI many years ago. Equally interesting is that the United States has resisted standardization to SI (also known as the **metric system**) many decades after Congress legislated change to the system, which is most commonly used in Europe and Asia. SI is used within the scientific community; thus, we urge adoption of this system for the purpose of understanding fire. Investigators in the United States will likely need to convert information to the English system when explaining scientific data to juries in the U.S. legal system.

area
- The quantity of space occupied in a two-dimensional plane (length × width).

English system
- A system for quantifying measurement that uses feet, gallons, pounds, and BTUs.

Standards International (SI)
- The system used for quantifying measurement that is accepted by the scientific community. Also known as *Scientific International*.

metric system
- A system for quantifying measurement that uses meters, liters, grams, and calories.

FIGURE 2.2 List of metric division designations.

FIRE DYNAMICS				
Prefix	**Symbol**	**Multiplication Factor**		
exa-	E	10^{18}	=	1 000 000 000 000 000 000
peta-	P	10^{15}	=	1 000 000 000 000 000
tera-	T	10^{12}	=	1 000 000 000 000
giga-	G	10^{9}	=	1 000 000 000
mega-	M	10^{6}	=	1 000 000
kilo-	k	10^{3}	=	1 000
hecto-	h	10^{2}	=	100
deca-	da	10^{1}	=	10
deci-	d	10^{-1}	=	0.1
centi-	c	10^{-2}	=	0.01
milli-	m	10^{-3}	=	0.001
micro-	μ	10^{-6}	=	0.000 001
nano-	n	10^{-9}	=	0.000 000 001
pico-	p	10^{-12}	=	0.000 000 000 001
femto-	f	10^{-15}	=	0.000 000 000 000 001
atto-	a	10^{-18}	=	0.000 000 000 000 000 001

LINEAR

Quantification of length is an essential tenet of quantifying space and energy transfer. The English system uses the *foot* as the base for other measures. Originally the length of a human foot, the measure has since been standardized at a length longer than most human feet. The foot is subdivided into increments of inches, 12 inches per foot. Inches are divided into fractional units (i.e., eighths and sixteenths). Longer distances are measured in miles, which consist of 5,280 feet. Other increments have been used in the English system (for example, rods, furlongs) but they are not prevalent in the United States.

Length is quantified by meters in the SI system. A meter is the distance that light travels through a vacuum in 1/299,792,458 of second. That random number was used to standardize the original measure thought to be 1/10,000,000 of the distance from Earth's equator to the North Pole. When it was learned that that distance was inaccurate, the light travel distance was used to standardize the measure. Increments of 10 are used to subdivide the meter, as is the case with other measures in the SI system (see Figure 2.2).

Conversion of Length Between Systems

A foot in the English system is equal to .3048 meters in SI; 39.37 inches (3.28 feet) equals 1 meter.

EXAMPLE 2.4

Convert 47 feet to meters.

Start by placing the known quantity (47 ft) on the left and the desired unit of measure on the right.

$$\underline{47 \text{ ft}} \qquad \underline{? \text{ m}}$$

Place the conversion factors, with the conversion factor for the desired unit on top: 1 foot is .3048 meter.

$$\frac{47 \text{ ft} * .3048 \text{ m}}{1 \text{ ft}}$$

Multiply the numerators and denominators, then divide the numerator by the denominator.

$$\frac{47 \text{ f\!t} * .3048 \text{ m}}{1 \text{ f\!t}} = \frac{14.3256 \text{ m}}{1} = 14.33 \text{ m}$$

By canceling units of measure by division, you can then reach the desired unit. Also note that the answer is rounded to two significant integers.

When converting English measure with inches and fractions of inches to metric, start with converting fractions of inches to decimal inches, then inches to decimal feet. Then conversion to metric involves multiplication or division by the conversion factor.

EXAMPLE 2.5

Convert 2 feet, 3⅝ inches to metric.

$$⅝ \text{ inch} = 5 \div 8 = .625 \text{ inch}$$

Then add the .625 to the inches (3) to derive 3.625 inches. That number divided by 12 (12 inches per foot) reveals the measurement of feet in decimal form.

$$3.625 \div 12 = .302 \text{ feet}$$

Adding this result to the whole-number total equals 2.302 feet; 2.302 feet times .3048 feet per meter equals .702 meter.

Occasionally you may need to convert from decimal to fractional measure. Fractional measure is somewhat less exact, which is important to remember as you round to the nearest increment. The fractional numerator is determined by multiplying the decimal by the denominator.

EXAMPLE 2.6

Convert 7.347 inches to whole inches and sixteenths (1/16).

1. The whole number of inches remains; that is, 7 inches.
2. Multiply the decimal (.347) by 16 = 5.552.
3. 5.552/16 rounds up to 6/16.
4. The answer is then 7⁶⁄₁₆.

FIGURE 2.3
Determining the area
of a rectangle.

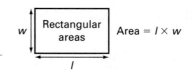

AREA

Area is a two-dimensional measurement to determine a quantity of flat space. Generally area is expressed as the square of a linear measure, for example, square feet or square meters. Determination of area follows rules of trigonometry. The most common is length by width ($L \times W$) or width times height ($W \times H$) (see Figure 2.3). Determination of area in fire dynamics involves identifying area of vents and area of a surface relative to heat transfer.

EXAMPLE 2.7

Determine the floor area of a house measuring 75 feet long and 32 feet wide.

1. 75 ft $\times$ 32 ft = 2,400 ft^2.
2. Converted to metric
   a. 75 ft * .3048 m/ft = 22.86 m.
   b. 32 ft * .3048 m/ft = 9.7536 m.
   c. 22.86 m * 9.7536 m = 222.97 m^2.
      or
3. 2,400 ft^2 * .0929 m^2/ft^2 = 222.96 m^2.

radius
- The measurement from the center of a circle to a point on its circumference.

Area in a circle is determined by multiplying the square of the circle's **radius** (one half the circle's diameter) by 3.1415 (π, or pi) expressed as πr^2, or A $= \dfrac{\pi D^2}{4}$ (see Figure 2.4).

EXAMPLE 2.8

Determine the area of a circle with a diameter of 5.54 m.

1. Determine the radius: 5.54 m/2 = 2.77 m.
2. Plug the radius into the formula πr^2.
3. Square the radius: 2.77 meters squared = 7.67 m^2.
4. 7.67 m^2 * 3.1415 = 24.1 m^2.

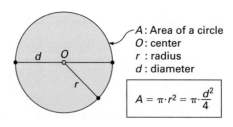

FIGURE 2.4
Determining the area
of a circle.

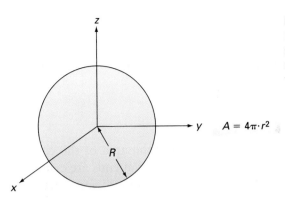

FIGURE 2.5
Determining the
surface area of a
sphere.

$A = 4\pi \cdot r^2$

Before we leave the discussion of measurements related to circles, let's discuss **circumference**, the distance around a circle. **Pi** (π) expresses the relationship between a circle's **diameter** and its circumference. By multiplying a circle's diameter by 3.1415, the circumference is derived.

A sphere is a perfectly round ball that contains volume. Determining the area of surface on the sphere is determined by the formula $4\pi r^2$ (see Figure 2.5). Radiant heat transfers equally around the flame in all directions, or on an imaginary sphere around the flame. Determination of heat transfer involves determining the area of a spherical area around that flame.

circumference
- The linear distance around the outside of a circle.

pi (π)
- The relationship of a circle's circumference to the diameter, which is 3.1415 to 1.

diameter
- The measure across the widest part of a circle.

EXAMPLE 2.9

Determine the area of a sphere with a diameter of 2.5 meters.

1. The radius is 2.5 m ÷ 2 = 1.25 m.
2. 1.25 m squared = 1.5625 m^2
3. 4 * 3.1415 * 1.5625 m = 19.63 m^2

VOLUME

Volume is a three-dimensional measure expressing total space occupied by an object or contained within a compartment and is expressed as a cubic measure. Volume is crucial in determining various unknowns in fire dynamics. In compartments, volume is a measure of multiplying length, width, and height ($L*W*H$) (see Figure 2.6).

volume
- The quantity of space occupied in three dimensions (length × width × height).

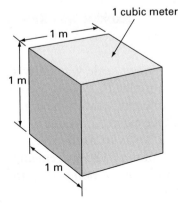

1 cubic meter

1 m

1 m

1 m

FIGURE 2.6 Volume: A cube that measures 1 meter on each edge. The volume occupied by this cube is 1 cubic meter (1 m^3), the approved SI unit of volume. *Source:* Meyer, *Chemistry of Hazardous Materials,* 5th edition, Brady, Figure 2.2, page 47.

EXAMPLE 2.10

Determine the volume of a compartment that measures 3.5 meters wide, 4 meters long, and 2.3 meters high.

$$3.5 \text{ m} * 4 \text{ m} * 2.3 \text{ m} = 32.2 \text{ m}^3$$

The volume of a cylinder is determined by factoring the area of the circular base with the height, which is expressed as $\pi r^2 h$. Tanks with flat ends are cylinders. If tanks have convex or concave ends, additional quantification is necessary.

EXAMPLE 2.11

Determine the volume of a cylinder 9 meters in diameter and 10 meters high.

1. Determine the area of the circle, which forms the top and the bottom of the cylinder.
   a. The radius is 9 m ÷ 2 = 4.5 m.
   b. The radius squared is 4.5 m squared = 20.25 m².
   c. The area is equal to π multiplied by 20.25 m² = 63.62 m²
2. The volume is area times height.
   a. 63.62 m² * 10 m = 636.2 m³

Fluid volume is measured in liters for the SI system and gallons in the English system. Each gallon equates to 3.8 liters.

EXAMPLE 2.12

A fire hose allows liquid to flow at 125 gallons per minute. Convert this flow rate to liters.

$$125 \text{ gallons per minute} * 3.8 \text{ liters per gallon} = 475 \text{ lpm}$$

TEMPERATURE

Temperature is the measurement of average kinetic energy. Later chapters of this text discuss heat energy in significant detail; thus, this section covers only the measurement of heat. Two systems are commonly used to measure the average kinetic energy: **Fahrenheit** and **Celsius**. Two other systems, **Rankine** and **Kelvin** correspond to the Fahrenheit and Celsius scales, but the difference is that they are absolute scales (see Figure 2.7).

Most commonly used in the United States, the Fahrenheit scale was derived by Dutch physicist Daniel Fahrenheit in about 1724. Fahrenheit's system used a mixture of salt, water, and ice to derive the lowest possible temperature he could achieve, which was designated as 0°. The point at which ice existed in water without changing the state of either the ice or the water (at 32°) was set as the freezing point. The temperature of a human was set at 96°. A total of 180 increments, commonly called degrees, separate the freezing point and boiling points of pure water at standard atmospheric pressure (Sizes, Inc., 2006).

---

temperature
■ Measurement of average kinetic heat energy.

Fahrenheit
■ A temperature measurement system that designates the freezing point of water at 32° and the boiling point of water at 212°. Absolute zero is −459.69°.

Celsius
■ A temperature measurement system that designates the freezing point of water at 0° and the boiling point of water at 100°. Absolute zero is −273.16°.

Rankine
■ A temperature measurement system that begins at absolute zero. The freezing point of water is 491.67°R, and the boiling point of water is 671.67°R.

Kelvin
■ A temperature measurement system that begins at absolute zero. The freezing point of water is 273.16°K, and the boiling point of water is 373.16°K.

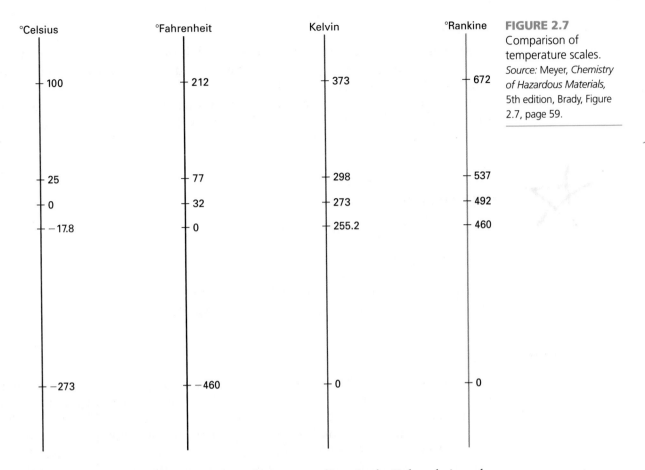

°Celsius   °Fahrenheit   Kelvin   °Rankine

°Celsius	°Fahrenheit	Kelvin	°Rankine
100	212	373	672
25	77	298	537
0	32	273	492
−17.8	0	255.2	460
−273	−460	0	0

FIGURE 2.7

Comparison of temperature scales. *Source:* Meyer, *Chemistry of Hazardous Materials,* 5th edition, Brady, Figure 2.7, page 59.

The Rankine scale of uses the same increments as those in the Fahrenheit scale, but it designates absolute zero (the temperature at which all molecular movement ceases) as the starting point. Water freezes at 491.67°R and boils at 671.67°R.

The Celsius, or centigrade, scale is named for Swedish astronomer Anders Celsius. Celsius worked in the 1740s to derive a system for measuring temperature based on two fixed points, one the freezing point of water and the second the boiling point of water. Celsius separated the two points on a thermometer by 100 increments, or degrees. Interestingly, Celsius indicated boiling as 0 degrees and freezing as 100 degrees. Other scientists of the period reversed the designation of the fixed points but retained the increments posed by Celsius water freezes at 0° and boils at 100° (Beckman, 2001).

The standard measurement for the SI is the Kelvin scale. With absolute zero as the starting point for Kelvin, the system uses identical increments, as does Celsius. Water freezes at 273.15°K and boils at 373.13°K.

Converting data in the Kelvin scale to Celsius involves adding 273.15. Conversely, change to the Celsius scale from Kelvin is accomplished by subtracting 273.15. Conversion between Fahrenheit and Rankine is equally simple, with the addition or subtraction of 459.67. Correlation between systems that use differing increments and a different basis for a freezing point requires a bit more work. One degree in the Celsius and Kelvin equates to 1.8° in Fahrenheit and Rankine. When converting from Fahrenheit to Celsius, subtract 32 from the temperature because 32°F is the freezing point, then divide by 1.8. When converting from Celsius to Fahrenheit, multiply by 1.8, then add 32.

ENERGY

The SI unit of measure for energy is Joule, which is the amount of energy required to move one kilogram (1 kg) one meter per second over a distance of one meter. One Joule per second is a Watt. Other measures of energy include the calorie, which is equal to the energy required to raise one gram of water 1°C, which in turn is equal to 4.18 Joules. Another measure is the British thermal unit (BTU), which is the energy required to raise one pound of water 1°F, which in turn equals 1,055 Joules and 252 calories.

EXAMPLE 2.13

Convert 2,408 BTU/second to watts.

1. 1 BTU equals 1055 joules.
2. 1 joule per second equals one watt.
3. Multiply 1055 J/BTU * 2408 BTU = 2,540,440 J.
4. 2,540,440 J/second = 2,540,440 watts = 2,540 kW or 2.5 MW

MASS

Kilogram is the unit of measure in the SI system; the English system uses pounds. Each kilogram contains 2.2 pounds; conversely, 1 pound is 0.454 kg. Fractions of pounds are expressed in ounces, with 16 ounces in a pound. There are 2,000 pounds per ton. Conversion between the systems involves multiplication or division in the correct order. When going from pounds to kilograms, the number of units will decrease. If your answer does not reflect this decrease, you probably multiplied or divided in the incorrect order.

EXAMPLE 2.14

Convert a person's weight of 180 pounds to kilograms.

$$180 \text{ pounds} \div 2.2 \text{ pounds per kilogram} = 81.82 \text{ kg}$$

PRESSURE

Pressure is a measure of force expressed in force per unit of area. In the English system, pressure is expressed in pounds per square inch (psi). Pressure in SI is expressed in kilograms per square meter (Pascal). Atmospheric pressure has the following units represented by different systems, but they are equivalent: 14.7 psi, 101.325 kps, 29.92 inches of mercury, and 33.9 feet of water (head pressure).

TIME

Fortunately, the measurement of time transcends other measurement systems. The rate of chemical reaction, release of heat, or application of extinguishing agent is based on the mass, volume, or heat measure per unit of time. Remember that each minute has 60 seconds and each hour has 60 minutes. Rarely will you have to convert beyond minutes and seconds.

TABLE 2.1 Table of Conversions

DIMENSION	METRIC	METRIC/ENGLISH
Acceleration	$1 \text{ m/s}^2 = 100 \text{ cm/s}^2$	$1 \text{ m/s}^2 = 3.2808 \text{ ft/s}^2$ $1 \text{ ft/s}^2 = 0.3048 \text{ m/s}^2$
Area	$1 \text{ m}^2 = 10^4 \text{cm}^2 = 10^6 \text{mm}^2$	$1 \text{ m}^2 = 1550 \text{ in}^2 = 10.764 \text{ ft}^2$; 1 acre = 43,560 ft^2; $1 \text{ yd}^2 = 0.836 \text{ m}^2$ $1 \text{ ft}^2 = 144 \text{ in}^2 = 0.0929 \text{ m}^2$; 1 acre = 4046.86 m^2
Density	$1 \text{ g/cm}^3 = 1 \text{ kg/L} = 1{,}000 \text{ kg/m}^3$	$1 \text{ g/cm}^3 = 62.428 \text{ lbm/ft}^3 = 0.036127 \text{ lbm/in}^3$ $1 \text{ lbm/in}^3 = 1728 \text{ lbm/ft}^3$ $1 \text{ kg/m}^3 = 0.06243 \text{ lbm/ft}^3$
Energy, heat, work, internal energy, enthalpy	$1 \text{ kJ} = 1{,}000 \text{ J} = 1{,}000 \text{ N} \cdot \text{m} = 1 \text{ kPa} \cdot \text{m}^3$ $1 \text{ kJ/kg} = 1{,}000 \text{ m}^2/\text{s}^2$ $1 \text{ kWh} = 3{,}600 \text{ kJ}$ $1 \text{ cal} = 4.184 \text{ J}$ $1 \text{ IT cal} = 4.1868 \text{ J}$ $1 \text{ kcal} = 4.1868 \text{ kJ}$	$1 \text{ kJ} = 0.94782 \text{ BTU}$ $1 \text{ BTU} = 1.05506 \text{ kJ} = 5.40395 \text{ psia} \cdot \text{ft}^3 = 778.169 \text{ lbf} \cdot \text{ft}$ $1 \text{ BTU/lbm} = 25.037 \text{ ft}^2/\text{s}^2 = 2.326 \text{ kJ/kg}$ $1 \text{ kJ/kg} = 0.430 \text{ BTU/lbm}$ $1 \text{ kWh} = 3412.14 \text{ BTU}$ $1 \text{ therm} = 10^5 \text{ BTU} = 1.055 \times 10^5 \text{ kJ}$ (natural gas)
Force	$1 \text{ N} = 1 \text{ kg} \cdot \text{m/s}^2 = 10^5 \text{ dyne}$ $1 \text{ kgf} = 9.80665 \text{ N}$	$1 \text{ N} = 0.22481 \text{ lbf}$ $1 \text{ lbf} = 32.174 \text{ lbm} \cdot \text{ft/s}^2 = 4.44822 \text{ N}$
Length	$1 \text{ m} = 100 \text{ cm} = 1{,}000 \text{ mm} = 10^6 \text{ um}$ $1 \text{ km} = 1{,}000 \text{ m}$	$1 \text{ m} = 39.370 \text{ in} = 3.2808 \text{ ft} = 1.0926 \text{ yd}$ $1 \text{ ft} = 12 \text{ in} = 0.3048 \text{ m}$ $1 \text{ mile} = 5{,}280 \text{ ft} = 1.6093 \text{ km}$ $1 \text{ in} = 2.54 \text{ cm}$
Mass	$1 \text{ kg} = 1{,}000 \text{ g}$ $1 \text{ metric ton} = 1{,}000 \text{ kg}$	$1 \text{ kg} = 2.2046226 \text{ lb}$ $1 \text{ lbm} = 0.45359237 \text{ kg}$ $1 \text{ ounce} = 28.3495 \text{ g}$ $1 \text{ slug} = 32.174 \text{ lbm} = 14.5939 \text{ kg}$
Power, heat transfer rate	$1 \text{ W} = 1 \text{ J/s}$ $1 \text{ kW} = 1{,}000 \text{ W} = 1.341 \text{ hp}$ $1 \text{ hp} = 745.7 \text{ W}$	$1 \text{ kW} = 3412.14 \text{ BTU/h} = 737.56 \text{ lbf} \cdot \text{ft/s}$ $1 \text{ hp} = 550 \text{ lbf} \cdot \text{ft/s} = 0.7068 \text{ BTU/s}$ $= 42.41 \text{ BTU/min} = 2{,}544.5 \text{ BTU/h}$ $= 0.7457 \text{ kW}$; $1 \text{ BTU/h} = 1.055056 \text{ kJ/h}$
Pressure	$1 \text{ Pa} = 1 \text{ N/m}^2$; $1 \text{ mmHg} = 0.1333 \text{ kPa}$ $1 \text{ kPa} = 10^3 \text{ Pa} = 10^{-3} \text{ Mpa}$ $1 \text{ atm} = 101.325 \text{ kPa}$ $= 1.01325 \text{ bars}$ $= 760 \text{ mm Hg at } 0°C$ $= 1.03323 \text{ kgf/cm}^2$	$1 \text{ Pa} = 1.4504 \times 10^{-4} \text{ psia}$ $= 0.020886 \text{ lbf/ft}^2$ $1 \text{ psi} = 144 \text{ lbf/ft}^2 = 6.894757 \text{ kPa}$ $1 \text{ atm} = 14.696 \text{ psia} = 29.92 \text{ in Hg at } 30°F$ $1 \text{ in Hg} = 3.387 \text{ kPa}$
Specific heat	$1 \text{ kJ/kg} \cdot °C = 1 \text{ kJ/kg} \cdot \text{K} = 1 \text{ J/g} \cdot °C$	$1 \text{ BTU/lbm} \cdot °F = 4.1868 \text{ kJ/kg} \cdot °C$ $1 \text{ BTU/lbmol} \cdot \text{R} = 4.868 \text{ kJ/kmol} \cdot \text{K}$ $1 \text{ kJ/kg} \cdot °C = 0.23885 \text{ BTU/lbm} \cdot °F = 0.23885 \text{ BTU/lbm} \cdot \text{R}$

(continued)

TABLE 2.1 Table of Conversions (*continued*)

DIMENSION	METRIC	METRIC/ENGLISH
Specific volume	$1\ m^3/kg = 1{,}000\ L/g = 1{,}000\ cm^3/g$	$1\ m^3/kg = 16.02\ ft^3/lbm$ $1\ ft^3/lbm = 0.062428\ m^3/kg$
Temperature	$T(K) = T(°C) + 273.15$ $\Delta T(K) = \Delta T(°C)$	$T(R) = T(°F) + 459.67 = 1.8T(K)$ $T(°F) = 1.8\ T(°C) + 32$ $\Delta T(°F) = \Delta T(R) = 1.8\Delta T(K)$
Velocity	$1\ m/s = 3.60\ km/h$	$1\ m/s = 3.2808\ ft/s = 2.237\ mi/h$ $1\ mi/h = 1.4467\ ft/s$ $1\ mi/h = 1.6093\ km/h$
Volume	$1\ m^3 = 1{,}000\ L = 10^6\ cm^3$ (cc) $1\ mL = 1\ cm^3$	$1\ m^3 = 6.1024 \times 10^4\ in^3 = 35.315\ ft^3$ $= 264.17\ gal$ (US)
Volume flow rate	$1\ m^3/s = 60{,}000\ L/min$ $= 10^6\ cm^3/s$	$1 m^3/s = 15{,}850\ gal/min$ (gpm) $= 35.315\ ft^3/s$ $= 2118.9\ ft^3/min$ (cfm)
Universal gas constant		$R_u = 8.31447\ kJ/kmol \cdot K = 8.31447\ kPa \cdot m^3/kmol \cdot K$ $= 0.0831447\ bar \cdot m^3/kmol \cdot K = 82.05\ L \cdot atm/kmol \cdot K$ $= 1.9858\ BTU/lbmol \cdot R = 1545.37\ ft \cdot lbf/\ lbmol \cdot R$ $= 10.73\ psia \cdot ft^3/lbmol \cdot R$
Standard acceleration of gravity		$g = 9.80665\ m/s^2 = 32.174\ ft/s^2$
Standard atmospheric pressure		$1\ atm = 101.325\ kPa = 1.01325\ bar = 14.696\ psia$ $= 760\ mmHg = 29.9213$

Source: Data derived from Cengel and Boles, *Thermodynamics: An Engineering Approach*, 4th Edition (2002).

Summary

The behavior of fire and combustion relates directly to fuel and compartment geometry. Fire behavior can be quantified and predicted with math. To get accurate results, however, inputs must be carefully and correctly extrapolated. This chapter reviewed how to determine area and to perform calculations commonly associated with basic fire behavior calculations (see Table 2.1).

Review Questions

1. Convert 12 feet 6 inches to meters.
2. Convert 235 gallons per minute to liters per minute.
3. Convert 54°C to °F.
4. What is the area of a compartment that measures 4 meters by 12 meters and is 3 meters high?
5. What is the volume of the compartment in Review Question 4?
6. What is the volume of a tank 12 meters in diameter and 34 meters high?
7. A sofa weighs 94 pounds. Convert that weight to kilograms.
8. How many degrees separate the temperatures at which water freezes and water boils on the Kelvin scale?
9. What is the area of a sphere that has a diameter of 2.4 meters?
10. Convert 750°K to °F.

PEARSON
myfirekit™

For additional review and practice tests, visit **www.bradybooks.com** and click on MyBradyKit to access book-specific resources for this text!

Register your access code from the front of your book by going to **www.bradybooks.com** and selecting the mykit links.

References

Beckman, O. (2001). *History of the Celsius Scale*. Retrieved June 16, 2009, from http://www.astro.uu.se/history/celsius_scale.html.

Sizes, Inc. (2006). *Fahrenheit Temperature Scale*. Retrieved June 16, 2009, from http://www.sizes.com/units/temperature_Fahrenheit.htm.

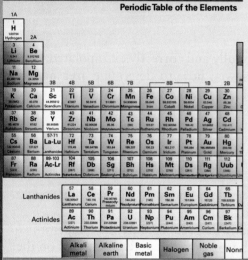

CHAPTER 3

Chemistry Review

Periodic Table of the Elements

KEY TERMS

alkanes, *p. 35*

alkenes, *p. 35*

alkynes, *p. 36*

atom, *p. 31*

atomic number, *p. 31*

atomic weight, *p. 31*

compound, *p. 32*

decomposition reaction, *p. 33*

double replacement reaction, *p. 34*

element, *p. 31*

hazard classification system, *p. 38*

ionic bonds, *p. 32*

isotope, *p. 31*

mixture, *p. 32*

mole, *p. 34*

monomer, *p. 37*

polymer, *p. 37*

single replacement reaction, *p. 34*

solution, *p. 32*

stoichiometric mixture, *p. 33*

synthesis reaction, *p. 33*

valence electrons, *p. 32*

OBJECTIVES

After reading this chapter, you should be able to:

- Comprehend chemical terms related to fire dynamics.
- Use the periodic table to describe characteristics of specific elements.
- Identify the chemical composition of various fuels.
- Differentiate among elements, compounds, mixtures, and solutions.
- Explain how the oxidation reaction occurs with specific chemicals.

Fire is a chemical reaction that results in changes explained by the laws of physics. This chapter reviews the basics of chemistry, but it does not delve deep into chemistry to provide the knowledge needed for fully understanding fire. Rather, this chapter should serve as a reminder and catalyst for seeking other information. Much of the information provided is a summation of that contained in the text *Chemistry of Hazardous Materials,* Fifth Edition, by Eugene Meyer, which is published by Prentice Hall (Meyer, 2010). Refer to Meyer's text for greater detail on specific chemicals and their reactions.

Atoms

Atoms are the building blocks of all matter. Atoms are the smallest particle of an **element**. Atoms are comprised of protons, neutrons, and electrons (see Figure 3.1). Protons have a positive electrical charge and can be found in the atom's center, or its nucleus. The number of protons defines the atom's **atomic number** and thus classifies which element that atom is.

Neutrons have no electrical charge and are also located within the atom's nucleus. The total number of protons and neutrons defines an atom's **atomic weight**. Usually the number of neutrons is identical to the number of protons, but this is not always the case. The term **isotope** defines those nuclei with differing numbers of neutrons. The atomic weight indicated on the periodic table often shows decimal points for the weight, which indicates the average atomic weight based on isotopes of that atom.

Electrons are negatively charged particles spinning around the atom's nucleus. Electrons move more quickly around the nucleus in layers known as valence orbits. Except in hydrogen and helium, where the valence orbit contains a maximum of two electrons, valence orbits contain up to eight electrons. When the valence orbit

atom
- The smallest particle of an element that can be identified with that element.

element
- A substance comprised of only one type of atom.

atomic number
- The number of protons possessed by an atomic nucleus; the number of atoms possessed by an atom.

atomic weight
- The mass of an atom compared to carbon-12, which is assigned the atomic weight of 12.

isotope
- Any of a group of nuclei having the same number of protons but a different number of neutrons.

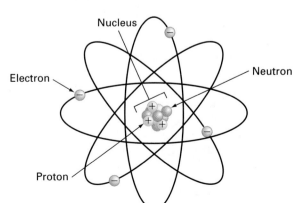

FIGURE 3.1
Depiction of an atom.

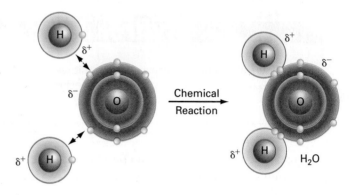

FIGURE 3.2 Covalent bond: water (left). Ionic bond: sodium chloride (right).

contains the maximum capacity of electrons, that is two or eight, it is said to be noble because that chemical cannot enter into an ionic or covalent bond with other chemicals. The electrons in the outer orbit are identified as the **valence electrons** because they dictate how that atom reacts with other atoms.

Compounds are masses comprised of two or more types of chemically connected atoms. Only a chemical reaction can separate elements that are contained within a compound. Within compounds, atoms are held tightly by *ionic* or *covalent* bonds. **Ionic bonds** involve transfer of electrons from the valence orbit of one atom to fill the valence orbit of another atom, rendering one atom to have a positive charge and the other a negative charge. These opposite charges bond together. Covalent bonding occurs when atoms share electrons in their valence orbits (see Figure 3.2).

Mixtures result when elements are combined but do not chemically react to form a compound. Chemicals in a mixture can be separated without a chemical reaction. **Solutions** are mixtures that involve a *solvent* (substance in which another is dissolved) and a *solute* (a substance that is dissolved into a solvent).

CHEMICAL REACTIONS AND EQUATIONS

Chemical reactions are depicted by equations that show reactants and products. Reactants are the chemicals that enter into the reaction; products are the chemicals that result from the reaction. Reactants and products are depicted in chemical equations with an arrow pointing from left to right to represent the chemical reaction.

A simple reaction is illustrated by the combination of sodium (Na) and chlorine (Cl) (see Figures 3.3 and 3.4). Sodium is a metal that produces a severe reaction with water. Chlorine is also a highly reactive chemical that produces hydrochloric acid when mixed with water, and it readily reacts with metals. Yet when combined, they become benign, or at least somewhat benign. The chemical reaction is written:

$$Na + Cl \rightarrow NaCl$$

FIGURE 3.3 Reaction of sodium and chlorine.

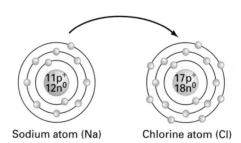

Sodium atom (Na)   Chlorine atom (Cl)

Periodic Table of the Elements

1A																	8A
1 **H** 1.00794 Hydrogen	2A											3A	4A	5A	6A	7A	**2** **He** 4.002602 Helium
3 **Li** 6.341 Lithium	**4** **Be** 9.012182 Beryllium											**5** **B** 10.811 Boron	**6** **C** 12.0107 Carbon	**7** **N** 14.0067 Nitrogen	**8** **O** 15.9994 Oxygen	**9** **F** 18.9984032 Fluorine	**10** **Ne** 20.1797 Neon
11 **Na** 22.989789 Sodium	**12** **Mg** 24.3050 Magnesium	3B	4B	5B	6B	7B	|—— 8B ——|			1B	2B	**13** **Al** 26.9815386 Aluminum	**14** **Si** 28.0855 Silicon	**15** **P** 30.973762 Phosphorus	**16** **S** 32.065 Sulfur	**17** **Cl** 35.453 Chlorine	**18** **Ar** 38.948 Argon
19 **K** 39.0983 Potassium	**20** **Ca** 40.078 Calcium	**21** **Sc** 44.955912 Scandium	**22** **Ti** 47.867 Titanium	**23** **V** 50.9415 Vanadium	**24** **Cr** 51.9961 Chromium	**25** **Mn** 54.938045 Manganese	**26** **Fe** 55.845 Iron	**27** **Co** 58.933195 Cobalt	**28** **Ni** 58.6934 Nickel	**29** **Cu** 63.546 Copper	**30** **Zn** 65.38 Zinc	**31** **Ga** 69.723 Gallium	**32** **Ge** 72.64 Germanium	**33** **As** 74.92160 Arsenic	**34** **Se** 78.96 Selenium	**35** **Br** 79.904 Bromine	**36** **Kr** 83.798 Krypton
37 **Rb** 85.4678 Rubidium	**38** **Sr** 87.62 Strontium	**39** **Y** 88.90585 Yttrium	**40** **Zr** 91.224 Zirconium	**41** **Nb** 92.90638 Niobium	**42** **Mo** 95.96 Molybdenum	**43** **Tc** [98] Technetium	**44** **Ru** 101.07 Ruthenium	**45** **Rh** 102.90550 Rhodium	**46** **Pd** 106.42 Palladium	**47** **Ag** 107.8682 Silver	**48** **Cd** 112.411 Cadmium	**49** **In** 114.818 Indium	**50** **Sn** 118.710 Tin	**51** **Sb** 121.760 Antimony	**52** **Te** 127.60 Tellurium	**53** **I** 126.90447 Iodine	**54** **Xe** 131.293 Xenon
55 **Cs** 132.90545 Cesium	**56** **Ba** 137.327 Barium	**57-71** **La-Lu** Lanthanides	**72** **Hf** 178.49 Hafnium	**73** **Ta** 180.94788 Tantalum	**74** **W** 183.84 Tungsten	**75** **Re** 186.207 Rhenium	**76** **Os** 190.23 Osmium	**77** **Ir** 192.217 Iridium	**78** **Pt** 195.084 Platinum	**79** **Au** 196.966569 Gold	**80** **Hg** 200.59 Mercury	**81** **Tl** 204.3833 Thallium	**82** **Pb** 207.2 Lead	**83** **Bi** 208.98040 Bismuth	**84** **Po** [209] Polonium	**85** **At** [210] Astatine	**86** **Rn** [222] Radon
87 **Fr** [223] Francium	**88** **Ra** [226] Radium	**89-103** **Ac-Lr** Actinides	**104** **Rf** [267] Rutherfordium	**105** **Db** [268] Dubnium	**106** **Sg** [271] Seaborgium	**107** **Bh** [272] Bohrium	**108** **Hs** [270] Hassium	**109** **Mt** [276] Meitnerium	**110** **Ds** [281] Darmstadtium	**111** **Rg** [280] Roentgenium	**112** **Uub** [285] Ununbium	**113** **Uut** [284] Ununtrium	**114** **Uuq** [289] Ununquadium	**115** **Uup** [288] Ununpentium	**116** **Uuh** [293] Ununhexium	**117** **Uus** [294] Ununseptium	**118** **Uuo** [294] Ununoctium

Lanthanides	**57** **La** 138.90547 Lanthanum	**58** **Ce** 140.116 Cerium	**59** **Pr** 140.90765 Praseody-mium	**60** **Nd** 144.242 Neodymium	**61** **Pm** [145] Promethium	**62** **Sm** 150.36 Samarium	**63** **Eu** 151.964 Europium	**64** **Gd** 157.25 Gadolinium	**65** **Tb** 158.92535 Terbium	**66** **Dy** 162.500 Dysprosium	**67** **Ho** 164.93032 Holmium	**68** **Er** 167.259 Erbium	**69** **Tm** 168.93421 Thulium	**70** **Yb** 173.054 Ytterbium	**71** **Lu** 174.9668 Lutetium
Actinides	**89** **Ac** [227] Actinium	**90** **Th** 232.03806 Thorium	**91** **Pa** 231.03588 Protactinium	**92** **U** 238.02891 Uranium	**93** **Np** [237] Neptunium	**94** **Pu** [244] Plutonium	**95** **Am** [243] Americium	**96** **Cm** [247] Curium	**97** **Bk** [247] Berkelium	**98** **Cf** [251] Californium	**99** **Es** [252] Einsteinium	**100** **Fm** [257] Fermium	**101** **Md** [258] Mendele-vium	**102** **No** [259] Nobelium	**103** **Lr** [262] Lawrencium

Alkali metal	Alkaline earth	Basic metal	Halogen	Noble gas	Nonmetal	Rare earth	Metalloid	Transition metal

FIGURE 3.4 Periodic table of elements.
Source: Courtesy Todd Helmenstine.

In the reaction, the atom of sodium, which has eleven electrons (two in the first valence ring, eight in the second, and one electron in the outer [third] valence ring), gives the outer electron to chlorine, which has a similar structure except that its outer (third) ring has seven electrons. When the electron moves, sodium assumes a positive charge and chlorine assumes a negative charge. Thus, the two atoms achieve an ionic bond. **When exact amounts of reactants are present to react fully into products, the mixture is said to be a stoichiometric mixture;** that is, no excess reactants remain after the chemical reaction. Stoichiometry is rare if not impossible because of the inability of ensuring exactly where each atom is within the mixture.

Chemical reactions are classified as follows:

Synthesis reaction—bonding of two or more elements into a compound.

$$Na + Cl \rightarrow NaCl$$
$$C + 2H_2 \rightarrow CH_4$$

Decomposition reaction—two or more elements separate from a compound. An example comes from the electrolysis of water.

$$2H_2O \rightarrow 2H_2 + O_2$$

stoichiometric mixture
- Chemical conditions where the proportion of reactants is such that there is no surplus of any reactant after the chemical reaction is completed. Also known as *stoichiometry*.

synthesis reaction
- The type of chemical reaction involving two or more substances that results in the formation of a single product.

decomposition reaction
- A chemical reaction involving the breakup of a compound into two or more elements or compounds.

single replacement reaction
■ The type of chemical reaction where an atom replaces another in a compound.

double replacement reaction
■ The type of chemical reaction in which two different compounds exchange their ions to form two new compounds.

Single replacement reaction—an element replaces another element in a compound, bonding the previously unattached element and releasing a previously bonded element.

$$Li + H_2O \rightarrow LiOH + H$$

Double replacement reaction—a reaction where elements transfer between compounds.

$$CH_4 + 2O_2 \rightarrow 2H_2O + CO_2$$

With relation to heat transfer, reactions are classified as either endothermic or exothermic. An *endothermic* reaction is a chemical reaction that absorbs heat energy. Conversely, *exothermic* reactions release heat energy. Fire is simultaneously an endothermic and an exothermic reaction. Fuel must be heated to the ignition temperature, which is the endothermic component; heat is released from the oxidation reaction, which is the exothermic component.

To simplify reactions, they are shown as equations. In the equations, the reactants are shown on the left and the products are shown on the right side of an arrow. Balanced equations indicate equal amounts of a chemical on either side of the reaction.

EXAMPLE 3.1

Simple Oxidation of Propane

$$C_3H_8 + 5O_2 \rightarrow 3CO_2 + 4H_2O$$

Though we use the simple reactant-to-product model to represent combustion reactions, oxidation actually involves a far greater number of elementary or intermediate steps. These steps represent the molecular level of the combustion reaction. Figure 5.5 in Chapter 5 represents the elementary or intermediate steps required to fully oxidize methane.

The traditional approach to balancing chemical equations required whole numbers because it was believed that partial atoms were impossible. More current thinking accepts that the numbers refer to the **moles** of that specific atom. In the early 1800s, Amedeo Avogadro posed that 6.022×10^{23} atoms of a chemical is equal to the mass, in grams, which is equivalent to the chemical's atomic number (carbon = 12). With this theory, the use of decimal numbers to balance equations became acceptable.

mole
■ Based on Avogadro's number: 6.022×10^{23} atoms of an element or molecules of a compound. Mass is the number of grams equal to the element's or compound's atomic weight.

EXAMPLE 3.2

By Atoms

$$2C_4H_{10} + 13O_2 \rightarrow 8CO_2 + 10H_2O$$

By Moles

$$C_4H_{10} + 6.5O_2 \rightarrow 4CO_2 + 5H_2O$$

The specific atoms within a compound, the number of those atoms, and their arrangement within the compound can alter that chemical's reactivity. In assessing fire and reaction danger, one must fully evaluate the chemicals available for reaction. The general rule for determining the number of carbon atoms within the compound uses the first three or four letters within the chemical designation:

EXAMPLE 3.3

meth-	1
eth-	2
prop-	3
but-	4
pent-	5
hex-	6
hept-	7
oct-	8
non-	9
dec-	10

The chemical bonds between the carbon atoms determine the ease with which the bonds can be broken. One may assume that multiple bonds would be stronger; however, the opposite is true: Single bonds are typically stronger than two or three. **Alkanes** are hydrocarbons that contain single bonds between carbon atoms and usually end with the letters *-ane*. An example of a simple formula for this is CH_4 (methane). Bonds between the carbon and hydrogen are shown as a dash, which represents a shared electron. The chemical arrangement for ethane is shown as:

alkanes
- Hydrocarbon compounds that have multiple carbons with a single bond connecting them with other carbons. Also known as *saturated hydrocarbons* because hydrogen atoms are present at every possible location.

EXAMPLE 3.4

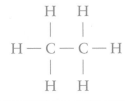

Alkenes are hydrocarbons arranged with a double bond between carbon atoms; they are usually designated by the ending *-ene*. The double bond means that two electrons are shared between the atoms. Note the shared electron depicted between the carbon atoms in ethylene:

alkenes
- Hydrocarbon compounds that have multiple carbons, where at least one carbon-to-carbon bond has a double bond.

EXAMPLE 3.5

```
        H   H
        |   |
        C = C
        |   |
        H   H
```

■ Hydrocarbon compounds that have multiple carbons, where at least one carbon-to-carbon bond has a triple bond.

Compounds that have triple bonds, or share three electrons between two carbon atoms, are called **alkynes**. These compounds are usually designated by the ending -*yne*. The formula for butyne has a triple bond between two carbons (Example 3.7).

Compounds in alignment are said to be straight chain. Butane is an example (Example 3.6).

EXAMPLE 3.6

Butane

$$\begin{array}{ccccccc}
 & H & H & H & H & \\
 & | & | & | & | & \\
H - & C & - C & - C & - C & - H \\
 & | & | & | & | & \\
 & H & H & H & H &
\end{array}$$

Isomers have the same chemical formula as straight-chain chemicals when it comes to atoms and the bonds between those atoms; however, the arrangement of those atoms differs from the straight chain. An example is isobutane (Example 3.8).

EXAMPLE 3.7

1-Butyne

$$\begin{array}{ccccc}
 & & H & H & \\
 & & | & | & \\
H - C & \equiv C & - C & - C & - H \\
 & & | & | & \\
 & & H & H &
\end{array}$$

Note the difference in the arrangement of atoms within the chemical, yet the same number of the same atoms are present. More than one *iso-* arrangement is possible. Thus, letters may be used to designate which arrangement is present.

EXAMPLE 3.8

Isobutane

$$\begin{array}{c}
H \quad\quad H \\
H \; | \; H \; | \; H \\
\diagdown C \; | \; C \diagup \\
H \diagup \; C \; \diagdown H \\
| \\
H \diagup C \diagdown H \\
| \\
H
\end{array}$$

Cycloalkanes are arranged so that fewer hydrogen atoms are present because the carbon atoms share electrons to complete valence. An example is cyclobutane (Example 3.9).

EXAMPLE 3.9

Cyclobutane

$$
\begin{array}{ccc}
& H & H \\
& | & | \\
H - & C - C & - H \\
& | & | \\
H - & C - C & - H \\
& | & | \\
& H & H
\end{array}
$$

Aromatic hydrocarbons exist in an arrangement wherein the six carbon atoms share six electrons in what is termed a benzyl ring. Benzene is the simplest type of this chemical structure, which is represented as C_6H_6. Each carbon atom completes its valence by sharing electrons with five other carbon and six hydrogen atoms. The benzyl ring with a methane atom attached to one carbon is toluene; the benzyl ring with methane atoms connected to two carbon atoms is xylene. Ethylbenzene is an ethane molecule chemically attached to a carbon within the benzyl ring.

POLYMERS

Monomers are single molecules of a chemical compound that stand alone. When these compounds chemically link together, they form **polymers**. When formed slowly, polymers result in matter that is useful in daily life; a prime example is cellulose. Other commonly encountered polymers result from polymerization of hydrocarbons to form plastics. An example is ethylene, a colorless gas that becomes polyethylene, a solid, when linked through polymerization:

monomer
- One or more single substances that combine to form polymers. Generally, hydrocarbons capable of linking to form polymers.

polymer
- High-molecular-weight substances produced by the linkage and cross-linkage of multiple subunits (monomers).

EXAMPLE 3.10

Ethylene monomer

$$
\begin{array}{cccc}
H & H & H & H \\
| & | & | & | \\
{\sim}{\sim}C = C & + & C = C{\sim}{\sim} \\
| & | & | & | \\
H & H & H & H
\end{array}
$$

EXAMPLE 3.11

Ethylene polymer

CARBOHYDRATES

Carbohydrates have a chemical composition similar to cellulose; however, they are not polymers. Carbohydrates (hydrated carbons) consist of carbons attached to water molecules, a structure that is represented as $(C\text{-}H_2O)_n$, with n equating to the number of hydrated carbons that are linked for the particular carbohydrate. Varying designations are attached to nutritional carbohydrates, but they have little bearing on fire dynamics. Fire professionals must understand that carbohydrates stored in food are also stored fuel for a fire. When such compounds are present in small high surface–to–mass ratio form, explosive conditions may exist. One example is the Imperial Sugar Refinery in Port Wentworth, Georgia. On February 7, 2008, sugar dust exploded, killing fourteen people. When in their more dense form, carbohydrates sustain continued combustion as would other fuels, but they are often not considered as fuels within structures.

Cellulose is a polymer comprised of carbohydrate monomers. Its chemical formulation is $C_6H_{10}O_5$. Cellulose is the major constituent of wood, paper, and cotton.

Fire Hazards Related to the U.S. Department of Transportation Hazard Classification System

The Fire and Emergency Services Higher Education (FESHE) program model indicates that the topic of hazards by classification should be discussed in the fire behavior and combustion curriculum. This section of the chapter covers fire-related hazards for the nine classifications indicated by the U.S. Department of Transportation, but it does not delve into all hazards related to each classification. That information is contained in the text *Chemistry of Hazardous Materials*, Fifth Edition, by Eugene Meyer, which is published by Prentice Hall. Reference for this discussion comes from Meyer and from 49 CFR 17, published by the U.S. Department of Transportation (USDOT). Summaries of the USDOT **hazard classification system** is discussed in the following subsections.

hazard classification system
■ A system of identifying substances by their hazard characteristics, including threat to humans, flammability, reactivity, and other threats.

CLASS 1—EXPLOSIVES

Explosives range from containers that experience explosive rupture due to over-pressurization, which in turn results from the expansion of their contents due to heating, to high-order detonations from chemicals that contain oxygen within their formulation, thus allowing flame development rapidly across the mass (atmospheric oxygen is less important for burning), in turn causing a *seated* explosion. Many explosions related to fire occur when fuels in the form of vapors or dusts ignite and burn, significantly raising pressure within confined areas, in a nonseated explosion. Backdraft is an example of fuels diffused in air and burning with rapid flame spread rates, which result in a pressure rise of explosive force. Explosions can occur with any fuel that chemically reacts quickly enough to raise pressure within a container where the oxidation reaction occurs. Dust from grain in a confined area and gasoline vapors dispersed through a structure are examples of common fuels that can become explosive.

Explosions are categorized into detonations and deflagrations; the categories indicate flame propagation speed through the burning mass. Detonations tend to shatter material with a shock wave associated with flame propagation above the speed of sound (3,300 fps to 1,005 mps). Explosions below the speed of sound tend to push more than shatter and are labeled as deflagrations.

CLASS 2—COMPRESSED GAS

Hazards associated with flammable gases are discussed in Chapter 6. Additional physical hazards are present when gases are stored in the compressed state. Heating of containers can result in a boiling liquid expanding vapor explosion (BLEVE) whether or not the gas is easily ignitable or inert. Pressure increase accompanied by compromise in the container can result in propulsion of the cylinder and/or cylinder parts in a manner that causes damage and threatens life.

Some chemicals that are considered compressed gases are so volatile that combustion can occur within the cylinder. Gases such as acetylene and ethylene oxide can burn inside their cylinders. This poses a hazard associated with overpressure from inside burning. Extreme care must be taken with such gases.

CLASS 3—FLAMMABLE AND COMBUSTIBLE LIQUIDS

The combustion hazards of liquids are discussed in Chapter 7.

CLASS 4—FLAMMABLE SOLIDS

The hazards and characteristics associated with the combustion of solids are covered in Chapter 8. Flammable solids described by the U.S. Department of Transportation pose greater hazard because of their flammability characteristics. These hazards include the following:

1. Explosives that have been desensitized but may still burn.
2. Reactive solids that experience exothermic reaction without atmospheric oxygen.
3. Chemicals that ignite from friction, such as the chemicals found on matchstick heads.
4. Pyrophoric solids, which are solids that ignite when in contact with oxygen in ambient air.
5. Solids that spontaneously heat in ambient air.
6. Materials that react with water to ignite or form flammable gases.

CLASS 5—OXIDIZING AGENTS

Hazards associated with this class of materials center on the materials' abilities to readily contribute oxygen to combustion. In and of themselves, many oxidizing agents are not ignitable. However, they enhance the propensity of ignitable fuels to ignite and increase their burning rate.

Organic peroxides are chemicals that contain carbon and oxygen (two oxygen molecules together) within their chemical composition. The atoms in peroxide molecules separate rapidly to combine chemically with other compounds in oxidation reactions. These reactions result in burning rates (flame propagation rates) that are sufficiently rapid to produce an explosive force.

CLASS 6—POISONS

Combustion of certain chemicals can result in poisonous gases (for example, cyanide [CN]), and some solid poisons are dissolved in ignitable liquids that facilitate even distribution. Other poisons are propelled by ignitable gases into the atmosphere. In these chemicals, ignitability is not the primary issue. The issue is the exposure of humans to poisons that are not neutralized by heat being transferred in the fire plume and/or the exposure of suppression personnel to poisons as they engage in mitigating damage.

CLASS 7—RADIOACTIVE MATERIALS

Fire hazards associated with radioactive materials result from fission and fusion reactions that produce heat at a rate sufficient to ignite surrounding combustibles. Nuclear reactors generate electrical power work on the principle of harnessing heat released from a controlled reaction, which is used in turn to convert water from liquid to steam. Pressure from the steam is then used to power turbines that turn generators, resulting in the generation of electrical current. If combustibles were in close proximity to the reactors, ignition could occur.

Thermonuclear devices (nuclear bombs) experience fusion; atomic bombs experience fission of mass. Both devices emit heat energy from their reaction that is sufficient to ignite combustibles for considerable distances; in addition, a significant shock wave results from rapid heating of gases in the area.

Nuclear reactions used for the purpose of energy development and bombs are beyond the scope of this text. Hazards associated with radioactive materials are not considered applicable for this text, even though they are realistic hazards associated with response to fires.

CLASS 8—CORROSIVES

Corrosives are chemicals with fewer electrons (acids) or surplus electrons (alkali) within the atomic structure. This adversely affects other substances as nature attempts to bring the substance back into balance (neutrality). Designation of a corrosive's strength is indicated on the pH scale, which goes from 0 to 14. The number 7 is designated as the midpoint, indicating neutral. Substances that have a pH lower than 7 are acidic; substances with a pH above 7 are alkalines. Corrosion is more intense as a substance moves away from the midpoint of 7.

Corrosives present a fire hazard because their reactions with other substances are often exothermic. The propensity of many reactions to produce hydrogen means that ignition from the heat of the reaction or other sources in the area is probable.

CLASS 9—MISCELLANEOUS HAZARDOUS MATERIALS

Miscellaneous hazardous materials typically present hazards associated with release during flight and/or marine transport; they do not involve fire-related issues and therefore are considered outside the scope of this text.

Summary

Fire is chemistry in action, or applied physics. Phenomena that result from oxidation reactions are predictable when reactants are known and understood. This chapter forms the basis for the discussion in later chapters of this text.

Review Questions

1. List the components of an atom.
2. Describe the difference between an element and a compound.
3. Describe the difference between a compound and a mixture.
4. What makes an isotope of an element?
5. What elements define hydrocarbons?
6. What is the difference between carbohydrates and cellulose?
7. Describe two hazards associated with corrosives.
8. Explain the difference between detonations and deflagrations.
9. As a prefix to an element, *pent-* indicates how many atoms of that element in a compound?
10. The name of a compound with a double bond in the chemical structure usually ends with what letters?

PEARSON

myfirekit™

For additional review and practice tests, visit **www.bradybooks.com** and click on MyBradyKit to access book-specific resources for this text!

Register your access code from the front of your book by going to **www.bradybooks.com** and selecting the mykit links.

References

Meyer, E. (2010). *Chemistry of Hazardous Materials,* 5th Edition. Upper Saddle River, NJ: Prentice Hall.

U.S. Department of Transportation (USDOT). (n.d.). 49 CFR 17.

4

Properties of Matter and Physics Review

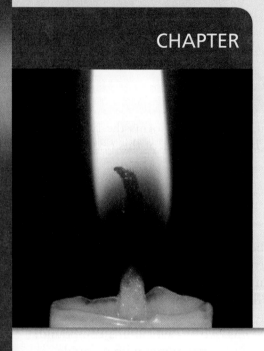

KEY TERMS

boiling point, *p. 52*

density, *p. 45*

energy, *p. 47*

expansion of matter, *p. 57*

gas, *p. 44*

heat, *p. 48*

heat capacity, *p. 51*

latent heat, *p. 52*

liquid, *p. 44*

mass, *p. 43*

matter, *p. 43*

pressure, *p. 52*

solid, *p. 43*

specific gravity, *p. 45*

temperature, *p. 50*

vapor, *p. 44*

vapor density, *p. 45*

vapor pressure, *p. 62*

OBJECTIVES

After reading this chapter, you should be able to:

- Identify the physical elements that affect the combustion process.
- Identify the properties that characterize the three physical states of matter.
- Understand the relationship of mass, volume, density, pressure, and the physical characteristics of the states of matter.
- Compare and contrast temperature and heat.
- Describe how thermodynamics, energy, and work play an important role in fire dynamics.
- Apply the fundamental gas laws to fire problems.

Fire, or combustion, involves matter entering into a chemical reaction. Anyone engaged in fire protection and investigation must understand the states of matter, especially what is required to change states of matter. The previous chapter dealt with how the basics of chemistry relate to fire science; this chapter will expand on the chemistry concepts and develop a bridge to the physics of fire science. Even though we separate the discussion between chemistry and physics, it is important to realize that, at the molecular level, these two sciences begin to become one. This chapter examines states of matter relative to fuels changing into a state where combustion can occur. It also examines how matter acts to extinguish fires.

Matter

Matter is defined as something that has **mass** and occupies space. It is distinguishable from empty space by its presence in it. Air is a form of matter and should not be confused with empty space. The *mass* (m) of a body is a quantity of matter that a body possesses. The term *weight* (W) is often incorrectly used to express mass. Unlike mass, w*eight* is a force. It is the gravitational force (g) applied to a body. In other words, weight is a measurement of the gravitational pull on that body of matter. Because weight is a measurement of force acting on a mass, it can be calculated using Newton's second law [Force = (mass)(acceleration)]:

matter
■ Anything that has mass and occupies space.

mass
■ A measure of matter's inertia, often referred to as weight; however, weight is matter's mass times gravitational pull.

Equation 4-1: Weight

$$W = m \times g$$

Where:

W = weight (N or kgf) [lbf]
m = mass (kg) [lbm]
g = gravity (m/s^2) [ft/s^2]

For instance, Earth has a distinct gravitational pull or force (9.81 m/s^2 or 32.2 ft/s^2 at sea level), and it is different than the gravitational pull or force found on Mars (3.73 m/s^2 or 12.24 ft/s^2). A body weighing 200 pounds (90.72 kg) on Earth would weigh 75.4 pounds (34.2 kg) on Mars. Thus, the weight of the matter would change from Earth to Mars; however, the mass would never change. The mass of a body remains constant regardless of its location in the universe. Another common unit used when discussing force is the Newton. One Newton is the force required to accelerate a one-kilogram mass at a rate of one meter per second per second, or:

$$1\,N = 1\frac{kg \cdot m}{s^2}$$

Matter exists in one of three states: solid, liquid, or gas. **Solids** are defined as matter that has a definite shape and volume. Molecules in solids have strong

solid
■ Matter that possesses definite shape and volume.

molecular bonds, which cause them to be closely packed and able to hold their shape regardless of the container.

Liquids are defined as matter that assumes the shape of containers. Liquids have weaker intermolecular forces than those found in solids, which cause them to lose shape unless contained. Liquids do not tend to disperse in air and generally do not compress. Liquids are heavier than air.

Gases (gaseous state) are defined as the state of matter wherein significant space separates molecules. They do not possess a characteristic shape. Gases have almost no intermolecular forces, which allows them to spread freely through the space, but the molecules collide frequently with each other and with the sides of the container. Gases are fluids that have neither a definite shape nor a definite volume, but they will spread throughout the shape of the container. The terms *vapor* and *gas* are often used interchangeably, even though there is an actual difference. The term *gas* is used for matter that exists in the gaseous state at standard temperature and pressure (STP: 1 atm at 0°C), while **vapors** are liquids or solids at standard temperature and pressure (STP).

Liquids and gases are considered fluids. Fluids assume the shape of their container or will freely seek levels consistent with their **specific gravity** and **vapor density**.

Density of Matter

Density describes the ratio of mass to volume (see Table 4.1). Densities of solids and liquids are often cited in the English system of units as pounds per gallon (lb/gal) or pounds per cubic feet (lb/ft^3), while gases and vapors are cited as pounds per cubic

TABLE 4.1	Density and Specific Gravity of Some Common Liquids and Solids*			
SUBSTANCE	KILOGRAMS/ CUBIC METER (kg/m^3) AT 20°C	GRAMS/ MILLILITER (g/mL) AT 20°C	POUNDS/ GALLON (lb/gal) AT 68°F	SPECIFIC GRAVITY AT 20°C
Acetone	792	0.792	6.6	0.792
Aluminum	2,700	2.7	22.5	2.7
Copper	8,940	8.94	74.61	8.94
Concrete	1,900–2,300	1.9–2.3	15.86–19.2	1.9–2.3
Gypsum plaster	1,440	1.44	12	1.44
Gasoline	660–690	0.66–0.69	5.5–5.7	0.66–0.69
Linseed oil	929.1	0.93	7.75	0.93
Kerosene	820	0.82	6.8	0.82
Mercury	13,546.2	13.55	113	13.55
Pine (yellow)	640	0.64	5.34	0.64
Polyurethane foam	20	0.02	0.17	0.02
Water	1,000	1.00	8.3	1.00

*Typical values; properties vary.

feet. In the metric system, solids and liquids are often cited as grams per milliliter (g/mL), while the densities of gases and vapors are cited as grams or kilograms per cubic meter (kg/m^3).

Equation 4-2: Density

$$\rho = \frac{m}{V}$$

Where: ρ = density (kg/m^3) [lb/gal or lb/ft^3]
                m = mass (kg) [lb]
                V = volume (m^3) [gal or ft^3]

In other words, a given mass that occupies more volume is less dense, while the same mass occupying less volume is denser.

EXAMPLE 4.1

Suppose we weigh exactly 75 pounds of water on a scale.

$$m = 75 \text{ lbs}$$

Then we place 75 lbs of water into 1-gal containers. The water entirely fills nine 1-gal containers at 20°F.

$$V = 9 \text{ gallons (gal)}$$

Knowing the amount of mass and the volume it occupies provides the density of water when Equation 4-2 is applied. The density of water is then computed as follows:

$$\rho = 75 \text{ lbs}/9 \text{ gal} = \textbf{8.33 lb/gal}$$

SPECIFIC GRAVITY

Specific gravity describes a comparison of a liquid or a solid substance's density with the density of water, which has a specific gravity of 1 (see Table 4.1).

Equation 4-3: Specific Gravity

$$\text{S.G.} = \frac{\rho_1}{\rho_{H_2O}}$$

Where: S.G. = specific gravity (dimensionless)
                ρ_1 = density of comparison substance (kg/m^3 or lb/gal)
                ρ_{H_2O} = density of water (1,000 kg/m^3 or 8.33 lb/gal)

Liquids with a specific gravity less than 1 are less dense than water and thus tend to rise to the surface of water (float). For example, most oils have densities less than water and tend to float. Conversely, liquids with a specific gravity greater than 1 are denser than water and thus tend to settle to the bottom of a body of water (sink). For example, elemental mercury is much more dense (about thirteen times) than water and sinks in water.

specific gravity
■ The ratio of the average molecular weight of a given volume of liquid or solid to the average molecular weight of an equal volume of water at the same temperature and pressure.

vapor density
■ The ratio of the average molecular weight of a given volume of gas or vapor to the average molecular weight of an equal volume of air at the same temperature and pressure. Also known as *specific gravity of gas*.

density
■ The property of a substance that measures its compactness. The mass of a substance divided by the volume it occupies.

EXAMPLE 4.2

What is the specific gravity of linseed oil? See Table 4.1 for the density of linseed oil.

Solution:

$$S.G. = \frac{\rho_1}{\rho_{H_2O}}$$

$$S.G. = 929.1 \text{ kg/m}^3/1,000 \text{ kg/m}^3 = 0.929$$

Because the specific gravity is less than 1, the linseed oil would float on water.

EXAMPLE 4.3

What is the specific gravity of mercury? See Table 4.1 for the density of mercury.

Solution:

$$S.G. = \frac{\rho_1}{\rho_{H_2O}}$$

$$S.G. = 13,546.2 \text{ kg/m}^3/1,000 \text{ kg/m}^3 = 13.546$$

Because the specific gravity of mercury is much greater than 1, it would definitely sink in water.

VAPOR DENSITY

The characteristic of gases that determines whether they rise or fall in air is vapor density (also known as gas specific gravity). Dry air is assigned a vapor density of 1. More dense gases (higher mass-to-volume ratio) have a vapor density greater than 1 and tend to displace the air and concentrate at lower levels (fall in air). For example, gasoline vapors have a vapor density of approximately 3 to 4 and tend to concentrate at lower levels. This is one of the reasons why codes require water heaters in garages or utility rooms to be placed 18 inches up off the floor. In the event of a gasoline spill, the vapors would otherwise concentrate near the floor and could possibly be ignited by the open pilot flame. Another common example of a gas that is heavier than air is propane, which has a vapor density of 1.5.

Conversely, less dense gases have vapor densities less than 1 and tend to rise in air. An example of this is methane, or natural gas, which has a vapor density of 0.553. There are only eight gases that are lighter than air: helium, ammonia, hydrogen, acetylene, methane, illuminating gas, carbon monoxide, and ethylene. The acronym commonly used to remember these eight gases is HAHAMICE.

The importance of understanding and recognizing vapor densities, especially the recognition of natural gas versus propane, will become important later in discussions regarding deflagrations and flash fires (see Table 4.2). One concept that can be emphasized now is the importance of venting gases for safety purposes; for example, outdoor barbeques and the relationship of the ventilation holes must be in coordination with the type of gas used, or a flash fire or explosion may occur.

Remember that all gases and vapors are miscible with air and other gases. In other words, unlike liquids, gases mix freely with other gases. The discussion regarding

TABLE 4.2	Vapor Density for Some Common Gases	
CHEMICAL	**VAPOR DENSITY**
Air	1
Methane	0.55
Ethane	1.04
Propane	1.50
Carbon monoxide	1.0
Carbon dioxide	1.53
Ethylene	0.97
Styrene	1.1
Toluene	3.14
Propylene	1.46
Ethylene oxide	1.49
Methanol	1.11
Ethanol	1.59
Butane	2.05

vapor densities and their general effects on fuel gases (i.e., propane gas, natural gas) in their relationship to rising or falling in a compartment (or container) must account for this diffusion of gases. When these gases start to mix freely with the surrounding air, the highest concentration of this gas will be based on its vapor density, but this does not mean all the gas will be located in this area. This mixing may allow the dissipation of ignitable mixtures, or it may even cause ignitable mixtures to form at ignition sources located in the center of the compartment. Vapors and gases are also greatly affected by turbulent mixing within the compartment or container (e.g., heating, ventilation, and air conditioning [HVAC]; ventilation openings; people shuffling through the room). Temperature changes and pressure differences also have a major impact on the behavior of these gases.

Energy, Work, and Thermodynamics

Energy can exist in many forms; some examples are thermal, mechanical, chemical, kinetic, potential, electric, magnetic, and nuclear. Energy is defined as the capacity to do work. It is important to point out that energy, as it relates to combustion, is present in all matter in the form of chemical energy. The chemical energy within matter relates to its chemical bonds and makeup. Work is associated with a force acting through a distance. An example of work is the pushing or dragging of a rock across a surface. A force must be applied to this mass to move it a distance. Work done over a period of time is termed *power*. The unit of power is kJ/s or kW.

energy
- The property of matter that enables it to do work.

Energy has a variety of units of measurement. In the United States, the British thermal unit (BTU) and the calorie are commonly used. One BTU represents the energy that must be supplied to raise 1 pound of water 1 degree Fahrenheit, from 63 to 64°F. One calorie represents the amount of energy required to raise the temperature of 1 gram of water 1 degree Celsius, from 14.5 to 15.5°C. The SI unit of energy is called the *Joule*. One joule is the quantity of energy done by a force of one Newton acting through a distance of one meter. At first glance, it does not seem like the Joule would relate to **heat** because it is measuring the energy done to move a weight over a distance. The Joule was related to heat by James Prescott Joule when he performed experiments to determine the amount of work involved in moving a standard weight over a standard distance and its equivalent change in the temperature of a body of water, which can be related to the calorie and BTU. Conversion between the systems is made possible by recognizing the following conversion factors:

heat
■ The form of energy transferred from one body to another because of temperature difference between them; energy arising from atomic or molecular motion.

$$1 \text{ BTU} = 252 \text{ cal}$$

$$1 \text{ BTU} = 1{,}054.8 \text{ Joules}$$

$$1 \text{ Watt} = 1 \text{ Joule/sec (Power)}$$

$$1 \text{ cal} = 4.184 \text{ J}$$

Energy within a system is often separated into three concepts: internal, kinetic, and potential. Potential energy is energy in a system based on the elevation of a mass against Earth's gravitational pull. Potential energy has the ability to do work once it is released. It is easy to visualize the potential energy in a boulder that is resting on a mountain and is ready to be pushed off. Energy is the ability to do work, and work is the product of force times distance and has units of foot-pounds; thus, potential energy is the potential or ability to move some unit of weight (force) some distance. The energy is typically expressed as a result of the elevation of the mass against the Earth's gravitational pull; it is expressed mathematically as:

Equation 4-4: Potential Energy

$$PE = mgh \quad or \quad PE = (W)(h)$$

Where:

m = mass (lbs or g)
g = gravity (ft/s^2 or m/s^2)
W = some unit of weight (lbs or g)
h = height (ft or m)

The kinetic energy in a system is a result of its motion. A vehicle increasing in velocity would be considered chemical energy (internal combustion) being transferred into kinetic energy. Mathematically, we can express kinetic energy as:

Equation 4-5: Kinetic Energy

$$KE = \frac{(m)(v)^2}{2} = \frac{1}{2}m(v)^2$$

Where:

KE = kinetic energy (Joules or BTUs or lb-force)
m = mass (g or lbs)
v = velocity (m/s or ft/s)

Microscopic forms of energy are related to the molecular structure of the system and the molecular activity. The internal energy is defined as the sum of all microscopic forms of energy in a system (Cengel and Boles, 2006). These microscopic forms of energy relate to the kinetic energy of the molecules and molecular bonds. Molecules for all matter have some movement or activity occurring at the molecular level, which is a result of the vibration and rotation of the molecules, neutrons, and electrons. As discussed in Chapter 3, atoms consist of neutrons and protons bound together by chemical bonds. The chemical energy within the material is related to these atomic bonds in a molecule. During a combustion reaction, which is a chemical reaction, these bonds are broken, which releases energy.

Thermodynamics can be defined as the science of energy (Cengel and Boles, 2006). Many laws and principles outlined in thermodynamics are useful to fire safety professionals. One of the fundamental laws of nature, and specifically thermodynamics, is known as the conservation of energy. This law simply states that the amount of energy remains constant. Energy can be converted from one form to another (potential energy can be converted to kinetic energy), but the total energy (TE) within the system remains fixed. Total energy is equal to the potential energy (PE), plus the kinetic energy (KE), and the internal energy (U). Therefore, by simple substitution, we can calculate the total energy.

Equation 4-6: Total Energy

$$TE = KE + PE + U$$

If we substitute the equations for kinetic energy and potential energy, we arrive at the following equation:

Equation 4-7: Total Energy, with Kinetic and Potential Energies Substituted

$$TE = \frac{(m)(v)^2}{2} + (m)(g)(h) + U$$

The best way to think of the difference among potential energy, kinetic energy, and the conservation of energy is by imagining a boulder at the top of a cliff. That boulder has potential energy because it has a mass and has been elevated from the ground against Earth's gravitational pull. Hypothetically, to give some numbers to the discussion, let's say that the boulder has 100 Joules of potential energy sitting on the cliff. Because the boulder is not moving, it has zero kinetic energy. Therefore, its total energy would be equal to the potential energy (100 Joules) plus kinetic energy (0 Joules) for a total energy of 100 Joules. If that boulder were to be pushed off the side of the cliff, it would begin to have kinetic energy as it fell. However, it would still have potential energy because it is still elevated off the ground; it just wouldn't be the same as it was when it was sitting on the cliff, up higher. Let's pause time at the point where the boulder has fallen only slightly from the maximum cliff height. The kinetic energy is now approximately 5 Joules. Therefore, the potential energy must be 95 Joules to maintain the total energy of 100 Joules. As the boulder reaches the midpoint from the top of the cliff and the ground, its potential energy is about half what it was at the start, so now it is 50 Joules. That means that the kinetic energy has increased accordingly to 50 Joules, still making the total energy of the system 100 Joules. Right before the boulder hits the ground, all of the potential energy will be converted into kinetic energy.

Heat and Temperature

temperature
■ Measurement of the average kinetic heat energy of an object, which results from molecular motion.

We need to make a distinction between energy and temperature. Remember that, at the molecular level, molecules in the material are in constant motion. As matter absorbs energy, the movement of these molecules begins to speed up, resulting in a temperature increase. A material's **temperature** is simply the measurement of the amount of motion that the molecules or atoms have. If the molecules within a material are moving very quickly, the temperature of this material will be high; slow motion equals a low temperature.

Several different scales have been developed to represent temperature (see Figure 4.1). The SI system includes the Celsius scale (°C) and the Kelvin scale (°K). In the Celsius scale, water freezes at 0°C and boils at 100°C, and there exists the possibility of negative temperatures. The Kelvin scale does not allow for negative temperatures and starts at the thermodynamic temperature of absolute zero, which equals −273.15°C. Absolute zero represents the theoretical lowest temperature, which is the theoretical point at which molecular activity stops. The English system uses the Fahrenheit scale (°F) and the Rankine (°R). The Fahrenheit scale is based on water freezing at a temperature of 32°F and boiling at a temperature of 212°F. The Rankine scale is based on an absolute temperature of −459.67°F.

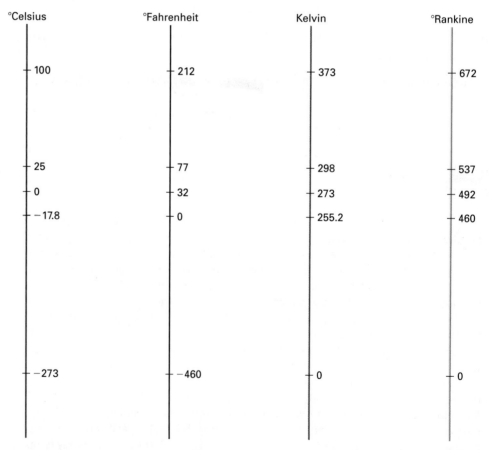

FIGURE 4.1 Interrelationship among the four temperature scales: Celsius (°C), Fahrenheit (°F), Kelvin (K), and Rankine (°R).
Source: Meyer, *Chemistry of Hazardous Materials,* Fifth Edition, Brady, Figure 2.7, page 59.

Conversion can occur between both systems by using the following conversion factors:

Metric

$T(K) = T(°C) + 273.15$

$\Delta T(K) = \Delta T(°C)$

Metric/English

$T(R) = T(°F) + 459.67 = 1.8T(K)$

$T(°F) = 1.8\ T(°C) + 32$

$\Delta T(°F) = \Delta T(R) = 1.8\Delta T(K)$

Heat is the amount of energy transferred from one object to another due to a difference in temperature. Heat is not transferred between two systems of the same temperature. For instance, when you get into a warm bath, you do not sense that it is warm because the water molecules are moving quickly; you feel that the water is hot because the molecules are colliding with your body and transferring some of their energy to your skin. Here is another good example of the difference between temperature and energy: Consider a single candle versus a small gasoline pool fire (2 ft²). The temperature of that candle will be approximately 1,000 to 1,200°C, with an approximate peak heat release rate of 70 kJ/s or 70 W (Hammins, Bundy, and Dillon, 2005). The temperature of the gasoline pool fire is approximately 1,000 to 1,200°C, but it has an approximate peak heat release rate of 400 kJ/s or kW. (The concepts of heat release rate and heat transfer will be covered more extensively in Chapters 9 and 10.)

Heat Capacity [C]

Heat capacity is the amount of heat required to change a unit of mass one degree in temperature. This concept also describes the ability of a material to store the energy when it is heated. In the United States, the most common measurement is the British thermal unit (BTU). A BTU is the amount of heat required to raise one pound of liquid water one degree Fahrenheit (1 lb H_2O ↑ 1°F) at one atmosphere. The SI unit of measure is the calorie, the energy required to raise one gram of water one degree centigrade (1 g H_2O ↑ 1°C) at one atmosphere. One calorie is equal to 4.186 Joules. Joules are the unit of energy that passes from an object of higher temperature to one of lower temperature. One could also define the heat capacity of liquid water as requiring 4.18 Joules to raise one gram of water one degree centigrade. The ratio of the heat capacity of a substance to the heat capacity of water (1.0) at the same temperature is a dimensionless number that is called the specific heat of a substance. Most fire behavior and dynamics studies utilize the SI system; thus, thorough knowledge and understanding of the SI system cannot be overemphasized.

Knowing the heat capacity of a liquid might seem sufficient to determine how much energy a unit of mass can absorb; however, an additional phenomenon occurs when a mass is in solid or gaseous form. When a mass is in solid form, its heat capacity is one half (.5) BTU (°F per pound). When a mass is in vapor form, its heat capacity is .48 BTU. By understanding the heat capacity of matter and by using the following formula, you can determine how much energy is required to change the temperature or how much energy can be removed from reactions such as fire.

heat capacity
- The amount of heat needed to raise one mass of a substance one degree. In the SI system, heat capacity refers to raising one gram 1°C. In the English system, it refers to raising one pound 1°F.

Equation 4-8: Energy Required to Change Temperature Given the Mass
of Material

$$Q = m \times C \times \Delta T$$

Where:
$$Q = \text{amount of heat energy (measured in Joules)}$$
$$m = \text{mass}$$
$$\Delta T = \text{change in temperature } (T_f - T_i)$$
$$T_f = \text{final temperature}$$
$$T_i = \text{initial temperature}$$

Change in States of Matter

Matter may exist in one of three states (solid, liquid, or gas). Temperature and **pressure** are the primary factors in determining when elements, compounds, and mixtures move from one state to another. To facilitate learning, we use water as an example. Each material can move among the states of matter; however, the points of change differ, often widely.

pressure
- Force applied per unit of area.

Most understand that water (H_2O) exists in these states: It is ice when it is solid, water when it is liquid, and vapor when it is a gas. Moving water from one state to another requires changes in heat or pressure. The most common way to change the state of water is by using heat.

The freezing point is the lowest temperature at which the mass can exist in liquid at normal pressure. The **boiling point** is the highest temperature at which the mass can exist in liquid form based on the pressure exerted by the atmosphere in which the liquid exists.

boiling point
- The temperature at which the vapor pressure of a substance equals the average atmospheric pressure.

latent heat
- The amount of heat required to change phase, from a solid to a liquid (latent heat of fusion) or from a liquid to a vapor (latent heat of vaporization).

Latent heat is the amount of energy required to change the state of matter; this is also called the psychophysical change (SFPE, 2009). Four latent heats are utilized in describing physical state change with matter:

Latent heat of vaporization —— liquid → gas
Latent heat of condensation —— gas → liquid
Latent heat of fusion —— solid → liquid
Latent heat of solidification —— liquid → solid

The *latent heat of vaporization* (L_v) is the amount of energy that must be applied to change matter, at the boiling point, from liquid to gas. Exactly the same energy, must be released from the mass to change it from a gas to a liquid at the boiling point; this energy is called the *latent heat of condensation* (L_C). For water, the value is 970 BTU, or 540 calories (2,260 J/g).

The *latent heat of fusion* (L_f) is the energy that must be absorbed to change a mass from a solid to a liquid. Conversely, the *latent heat of solidification* (L_S) is the amount of energy that must be released to change matter, at the freezing point, from a liquid to a solid. For water, this value is 143 BTU/lb, or 80 calories (334 J/g).

An equation can be used to assist in calculating the energy required to change the state of matter given a mass of a material:

Equation 4-9: Energy Required to Change States Given the Mass of Material

$$Q = m \times L_f \quad or \quad Q = m \times L_v$$

Where:
$$Q = \text{the energy required to change state (Joules)}$$
$$m = \text{mass (kg) [lb]}$$
$$L_f = \text{latent heat of fusion (J/g) [BTU/lb]}$$
$$L_v = \text{latent heat of vaporization (J/g) [BTU/lb]}$$

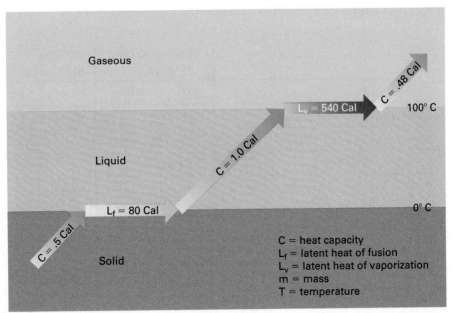

FIGURE 4.2 Latent heats of water: English system (BTUs).

Gaseous

L_v = 970 BTU   C = .48 BTU   212° F

Liquid

C = 1 BTU

L_f = 143 BTU   32° F

C = .5 BTU

Solid

C = heat capacity
L_f = latent heat of fusion
L_v = latent heat of vaporization
m = mass
T = temperature

Determining the energy absorbed by a mass during a change in state is relatively easy. Identify the starting temperature and the ending temperature. Then determine the total heat required to change the mass from the starting to the ending temperature. We use water for the purposes of this discussion; however, by learning the heat capacities and latent heats for other materials, you can determine the exact energy changes within that mass.

Using the following format can simplify the process (see Figure 4.2, Figure 4.3, Table 4.3, and Table 4.4). Calculate latent heats only when mass moves through the temperature where the phase change occurs. These examples are helpful in determining the fire extinguishing capabilities of water.

FIGURE 4.3 Latent heats of water: SI system (calories).

Gaseous

L_v = 540 Cal   C = .48 Cal   100° C

Liquid

C = 1.0 Cal

L_f = 80 Cal   0° C

C = .5 Cal

Solid

C = heat capacity
L_f = latent heat of fusion
L_v = latent heat of vaporization
m = mass
T = temperature

TABLE 4.3 — English Measurement: Tabulation Form

	TEMPERATURE CHANGE	ENERGY REQUIRED
Temperature rise in gaseous state	_____ − 212 = _____	× .48 = _____ BTU/lb
Latent heat of vaporization	970 BTU	
Temperature rise in liquid state	212 − _____ = _____	× 1 = _____ BTU/lb
Latent heat of fusion	143 BTU	
Temperature rise from starting point to freezing point	32 − _____ = _____	× .5 = _____ BTU/lb
Total energy		
× Mass		
Total energy required/absorbed		

EXAMPLE 4.4

Determine the amount of energy absorbed by increasing the temperature of 75 gallons of water from 70° to 215°F. The initial temperature is above freezing; thus, no energy is absorbed in the solid phase nor is energy required for the phase change from a solid to a liquid.

The energy required to raise the temperature of a liquid from 70°F to 212°F is one BTU per degree for each pound: 212 − 70 indicates a 142 degree differential; thus, 142 BTUs/lb are needed. For each pound, 970 BTUs/lb are required to change phase from a liquid to a vapor (LHV). In the vapor phase, water's heat capacity is .48 BTU/lb; thus, to raise the temperature 3°F requires 1.44 BTU/lb. Add these values together to derive 1,113.44 BTUs/lb, then multiply by the mass of liquid (75 gallons times 8.34 pounds per gallon). Results indicate that **696,456.72 BTUs** are required to raise the 75 gallons from 70°F to 212°F.

	TEMPERATURE CHANGE	ENERGY REQUIRED
Temperature rise in gaseous state	215 − 212 = 3	× .48 = _1.44_ BTU/lb
Latent heat of vaporization	970 BTU	_970_ BTU
Temperature rise in liquid state	212 − 70 = 142	× 1 = _142_ BTU/lb
Latent heat of fusion	143 BTU	
Temperature rise from starting point to freezing point	32 − __ = __	× .5 = _0_ BTU/lb
Total energy		1,113.44 BTU/lb
(75 gal * 8.34 lb/gal = 625.5 gallons) × mass		625.5 lb
Total energy required/absorbed		**700,000 BTU**

TABLE 4.4 | SI Measurement: Tabulation Form

	TEMPERATURE CHANGE	ENERGY REQUIRED
Temperature rise in gaseous state	_____ − 100 = _____	× .48 = _____ cal/g
Latent heat of vaporization	540 cal	
Temperature rise in liquid state	100 − _____ = _____	× 1 = _____ cal/g
Latent heat of fusion	80 cal	
Temperature rise from starting point to freezing point	0 − _____ = _____	× .5 = _____ cal/g
Total energy		
× Mass		
Total energy required/absorbed		

EXAMPLE 4.5

Convert the answer in Example 4.4 to SI units.

$$70°F = 21°C \ (70 − 32 = 38/1.8 = 21)$$
$$215°F = 101.7°C \ (215 − 32 = 183/1.8 = 101.7)$$

Seventy-five gallons is 285 liters (1 gallon = 3.8 liters). One liter of water has 1 kilogram of mass; thus, the total mass is 285,000 grams.

	TEMPERATURE CHANGE	ENERGY REQUIRED
Temperature rise in gaseous state	101.7 − 100 = 1.7	× .48 − .816 cal/g
Latent heat of vaporization	540 cal	540 cal
Temperature rise in liquid state	100 − 21 = 79	×1 = 79 cal/g
Latent heat of fusion	80 cal	
Temperature rise from starting point to freezing point	0 − __ = __	× .5 = 0 cal/g
Total energy		619.16 cal/g
× Mass		285,000 g
Total energy required/absorbed		176,460,600 cal

Comparing Examples 4.4 and 4.5 should reveal that they correlate. One BTU equals 252 calories: the proof we need to see that the examples are correct, 176,460,600 cal/252 cal/BTU = 700,240 BTU. Our English system version resulted in 696,457 BTU, a difference of 3,783 BTU due to rounding errors (.54 percent difference).

EXAMPLE 4.6

What is the energy required to raise the temperature of sixty pounds of water, with an initial temperature of 25°F and a final temperature of 213°F, at which point it is converted to steam?

The first step is to change the temperature from 25°F to the freezing/melting point of ice (32°F). We will label this as Q_1 and use the change in temperature formula:

$$Q_1 = mc\Delta T = (60 \text{ lbs})(0.5 \text{ BTU})(32 - 25°F) = \underline{\textbf{210 BTU}}$$

The second step is to calculate the phase change. Remember that there is no change in temperature at this point. The water will go from 32°F ice to 32°F liquid. We call this Q_2.

$$Q_2 = mLH_f = (60 \text{ lbs})(143 \text{ BTU}) = \underline{\textbf{8,580 BTU}}$$

The third step is to calculate the temperature change from 32°F to 212°F, with no phase change. This will be Q_3.

$$Q_3 = mc\Delta T = (60 \text{ lbs})(1 \text{ BTU})(212 - 32°F) = \underline{\textbf{10,800 BTU}}$$

The fourth step is to calculate the phase change from liquid to vapor. Again, with no change of temperature, we will go from 212°F (liquid state) to 212°F (vapor state). This point will be Q_4.

$$Q_4 = mLH_v = (60 \text{ lbs})(970 \text{ BTU}) = \underline{\textbf{58,200 BTU}}$$

The fifth step is to calculate the final change in temperature, from 212°F to 213°F, which we will note as Q_5.

$$Q_5 = mc\Delta T = (60 \text{ lbs})(0.48 \text{ BTU})(213 - 212°F) = \underline{\textbf{30 BTU}}$$

The final step is to add all of the Q values, for a total Q.

$$
\begin{aligned}
Q_{tot} &= Q_1 + Q_2 + Q_3 + Q_4 + Q_5 \\
&= 210 \text{ BTU} + 8,580 \text{ BTU} + 10,800 \text{ BTU} + 58,200 \text{ BTU} + 30 \text{ BTU} \\
&= \underline{\textbf{77,820 BTU}}
\end{aligned}
$$

Volume Expansion

Expansion of volume accompanies a phase change. Water changing from the liquid to the gaseous state expands, occupying 1,700 times more volume than the original liquid. Expansion, when water expands from liquid to gas, can be beneficial in fire attack because the expansion displaces air containing oxygen. This expansion can also produce hazardous conditions, especially by exposing firefighters inside compartments to steam conditions, which is ambient air above 100°C.

FOG FIRE STREAMS

Significant discussion continues in the fire service about which is better: fog or solid fire streams. The answer may depend on the intended results. Fog streams may prove beneficial when firefighters are attacking from exterior positions, yet these streams may produce undesirable results from firefighters' interior operations.

EXPANSION OF MATTER (SOLIDS AND LIQUIDS)

Changing temperature in matter produces a change in volume. An increase in temperature produces greater volume, an **expansion of matter**; conversely, a reduction in temperature results in less volume. Volume change depends on the chemical involved, which is expressed as a coefficient of change applicable to that material. To determine the volume of liquid or solid that results when the temperature of mass is changed (expressed as V_2) from the original volume, add the product derived by multiplying the original volume (V_1) by the change coefficient (α), by the change in temperature (Δt). See Equation 4-10.

expansion of matter
- Change in volume resulting from a change in temperature. Increasing the temperature in a unit of matter results in increased volume or pressure, while lowering temperature in that matter results in reduced volume or pressure.

Equation 4-10: Volume Expansion

$$V_2 = V_1 + (V_1 * \alpha * \Delta T)$$

One exception to this rule is water, which is most dense (greatest mass/volume) at 3.9°C. As water heats or cools from that temperature, it becomes less dense. This anomaly is why ice floats, rather than sinks, on warmer liquid water.

EXAMPLE 4.7

Five (5) gallons (19 liters) of gasoline ($\alpha = .0008$) is filled when the outside temperature is 20°F; what volume will be present if the gasoline is heated to 90°F?

$$V_2 = V_1 + (V_1 * \alpha * \Delta T)$$
$$V_2 = 5 \text{ gal} + (5 * .0008 * 70°F)$$
$$V_2 = 5 \text{ gal} + (5 * .0008 * 70°F)$$
$$V_2 = 5 \text{ gal} + .28$$
$$V_2 = 5.28 \text{ gallons}$$

In a container that provides vapor space (a place for expansion), this 5 to 6 percent increase in volume should pose no problem. However, if no vapor space exists, increased temperature may produce container damage or rupture (see Table 4.5).

GENERAL PROPERTIES OF THE GASEOUS STATE*

Of the three states of matter, only the gaseous state is capable of being described in comparatively simple terms. This description relates the volume of a gas to its temperature and pressure. Next, we will examine the laws of nature that apply to the gaseous state of matter: Boyle's law, Charles's law, and the combined gas law.

*Adapted from Eugene Meyer, *Chemistry of Hazardous Materials*, Fifth Edition. New York: Prentice Hall, 2010.

	COEFFICIENT OF VOLUME EXPANSION	
LIQUID	$\alpha\ (°F^{-1})$	$\alpha\ (°C^{-1})$
Acetic acid	0.00059	0.00107
Acetone	0.00085	0.00153
Benzene	0.00071	0.00128
Carbon tetrachloride	0.00069	0.00124
Diethyl ether	0.00098	0.00176
Ethanol	0.00062	0.00112
Gasoline	0.00080[a]	0.00144
Glycerine	0.00028	0.00051
Methanol	0.00072	0.00130
Pentane	0.00093	0.00168
Toluene	0.00063	0.00113
Water	0.00012	0.00022

TABLE 4.5 Coefficient of Volume Expansion for Selected Liquids

[a]The coefficient of volume expansion for gasoline varies from 0.00055 to 0.00090°F^{-1}. The coefficient listed in the table is a representative value.

Source: Meyer, *Chemistry of Hazardous Materials*, Fifth Edition, Brady, Table 2.9, page 74.

Boyle's Law

Robert Boyle, an Irish physicist, demonstrated experimentally that the volume of a confined gas varies inversely with its absolute pressure when the temperature remains fixed. The constant-temperature experiments performed by Boyle demonstrated that the mathematical product of the volume and absolute pressure of a gas is always constant. This observation is commonly called Boyle's law. It is expressed mathematically as follows:

Equation 4-11: Boyle's Law

$$V_1 \times P_1 = V_2 \times P_2$$

Here, V_1 and P_1 are the initial volume and initial absolute pressure, respectively; V_2 and P_2 are the final volume and final absolute pressure, respectively.

Boyle's law is used to determine the volume that a gas occupies when it remains at a fixed temperature but undergoes a change in pressure. Suppose the pressure of 40 cubic feet (40 ft^3) of an arbitrary gas is changed from 14.7 psi to 450 psi at a fixed temperature. What is the new volume assumed by the gas? Common sense tells us that the new volume must be less than 40 cubic feet because the application of any pressure squeezes the gas into a smaller volume. Using Boyle's law, we can readily compute the new volume as follows:

$$V_2 = (40\ ft^3 \times 14.7\ psi)/450\ psi = 1.3\ ft^3$$

Charles's Law

Jacques Charles and Joseph Gay-Lussac, two French scientists, independently demonstrated experimentally that the volume of a confined gas increases proportionately in relation to the increase in its absolute temperature when the pressure remains fixed. This statement is now known as Charles's law. It is expressed mathematically as follows:

Equation 4-12: Charles's Law

$$V_1 \times T_2 = V_2 \times T_1$$

Here, V_1 and T_1 are the initial volume and initial absolute temperature, respectively; V_2 and T_2 are the final volume and final absolute temperature, respectively.

Suppose an arbitrary gas at atmospheric pressure occupies a volume of 300 mL at 0°C. What volume does it occupy at 100°C and the same pressure? Common sense tells us that a heated gas always occupies a larger volume. We must first convert the temperatures in degrees Celsius into absolute temperatures on the Kelvin scale. In Chapter 2, we learned that the conversion to absolute temperature is accomplished by adding 273.15°C to the Celsius temperature readings. On the Kelvin scale, the temperature readings are as follows:

$$0°C = 273.15°K$$
$$100°C = 373.15°K$$

Now, we can use Charles's law to compute the new volume.

$$V_2 = 300 \text{ mL} \times 373.15°K/273.15°K = 410 \text{ mL}$$

Combined Gas Law

Boyle's and Charles's laws can also be combined into the following mathematical expression:

Equation 4-13: Combined Gas Law

$$V_1 \times P_1 \times T_2 = V_2 \times P_2 \times T_1$$

This is called the combined gas law. It is used to calculate the volume of a gas at a new temperature and pressure.

The volume of a gas can be forced to remain fixed, as when a gas is confined within a steel cylinder or other storage vessel. Under this circumstance, $V_1 = V_2$, and the combined gas law assumes the following form:

$$P_1 \times T_2 = P_2 \times T_1$$

The heating of a gas that is initially enclosed within a sealed metal drum at 1,000 psi_a at 70°F (21°C) is shown in Figure 4.4. When the gas is heated, its pressure increases accordingly. When the gas is heated to 570°F, its pressure nearly doubles; when it is heated to 1070°F, it nearly triples. These temperatures are akin to those routinely encountered during building fires. To relieve the strain on their walls, the second and third drums in the figure are likely to rupture.

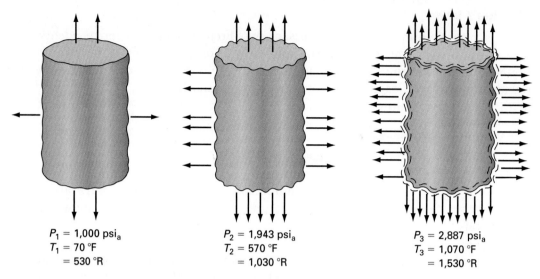

$$P_1 = 1,000 \text{ psi}_a \qquad P_2 = 1,943 \text{ psi}_a \qquad P_3 = 2,887 \text{ psi}_a$$
$$T_1 = 70 \text{ °F} \qquad T_2 = 570 \text{ °F} \qquad T_3 = 1,070 \text{ °F}$$
$$= 530 \text{ °R} \qquad = 1,030 \text{ °R} \qquad = 1,530 \text{ °R}$$

FIGURE 4.4 The effect of applied temperature on a gas confined to a constant-volume container, such as a sealed metal drum or barrel. An increase in the temperature of the gas by 500°F causes the internal pressure to nearly double, which often causes the container to rupture.
Source: Meyer, *Hazardous Materials,* Fifth Edition, 2009, Figure 2.13, page 79.

EXAMPLE 4.8

A welder purchases a gas cylinder of flammable hydrogen gas and chains it to a wall within a workshop. After periodic usage, the gauge pressure reads 235 psi_g when the temperature is 65°F. What is the gauge pressure reading when the temperature of the cylinder contents becomes 350°F during a fire?

Solution. Because a steel cylinder is a constant-volume container, $P_1 \times T_2 = P_2 \times T_1$. These symbols refer to the *absolute* initial and final pressures and temperatures, respectively.

$$P_1/T_1 = P_2/T_2$$
$$P_1 \times T_2 = P_2 \times T_1$$
$$P_1 = 235 \text{ psi}_g + 14.7 \text{ psi} = 250 \text{ psi}_a$$
$$T_1 = 65°F + 459.67°F = 525°R$$
$$T_2 = 350°F + 459.67°F = 810°R$$
$$P_2 = 250 \text{ psi}_a \times 810°F = 386 \text{ psi}_a \ 525°F$$

Then, the gauge pressure is determined by subtracting 14.7 psi from the absolute pressure, which yields 371 psi_g.

$$P_2 = 386 \text{ psi}_a - 14.7 \text{ psi} = 371 \text{ psi}_g$$

The absorption of heat causes the internal pressure of the cylinder contents to increase from 235 psi_g to 371 psi_g.

Source: Adapted from Meyers, E. (2010) *Chemistry of Hazardous Materials.* New York: Brady/Prentice Hall.

Pressure and Its Measurement

Pressure is expressed in the amount of force applied to a given area, that is, pounds per square inch (psi) or pascal (Newtons per meter squared). The most common force encountered is that of mass comprising the Earth's atmosphere, which at sea level, at 68°F, results in one atmosphere (1 atm), which is commonly expressed as 14.7 psi. Other units of measure for the same are:

14.7 psi
29.9 in. Hg
760 mm Hg (torr)
101,325 pascal
1.01 bar

Because atmospheric pressure is exerted equally in all directions and circumstances, humans commonly experience no perception of pressure. Changes in atmospheric conditions are detectable by the human senses. For example, humans can detect the change in pressure on their eardrums and sinuses.

When measuring pressure, we must distinguish between gauge pressure and absolute pressure. Gauge pressure (psig) indicates the differential between normal atmospheric pressure and that observed from the force applied. Gauge pressure is usually observed within a closed system, that is, a pressure tank or where fluids discharge. When the vessel is open and equalized with the atmosphere, the gauge pressure is 0. As the system is closed or as mass discharges from an orifice, pressure measurements rise based on the amount of force being exerted.

To determine absolute pressure (psia), addition of the atmospheric pressure (usually 14.7 psi) to the gauge pressure is required. When pressure within a system is lower than atmospheric conditions, the condition is often described as a vacuum. When the vacuum is opened, areas of greater pressure equalize with the lower pressure region through the infusion of mass.

EXAMPLE 4.9

A tank is constructed to hold liquid. A gauge is affixed 100 feet below the level reading "full." Before liquid is introduced, the gauge pressure indicates 0 psi.

When the tank is filled with water, the mass accumulated above the gauge area is 43.4 pounds for every square inch of surface area in the cross section. Thus, a gauge indicates 43.4 psi. If the same tank contained an equal volume of toluene (specific gravity of .86), the pressure gauge would indicate 37.3 psi.

The absolute pressure of the water is indicated by adding 14.7 psi (atmospheric pressure) to the 43.4 psi gauge reading. The result is 58.1 psia.

VAPOR PRESSURE

All liquid substances, and many solids such as naphthalene, possess sufficient molecular motion that molecules escape from their surfaces in the form of a vapor. When a liquid is left in an open container, its molecules leave the surface and the

TABLE 4.6 — Vapor Pressures for Common Liquids*

SUBSTANCE	VAPOR PRESSURE (mm Hg)	VAPOR PRESSURE (Pa)	VAPOR PRESSURE (bar)
Water	18.0	2,400	0.024
Ethanol	67.5	9,000	0.09
Methanol	93.8	12,505	0.13
Gasoline	403.4	53,780	0.54

*Values are shown for standard temperature and pressure

liquid evaporates. When the liquid is confined in a partially filled container, the molecules continue to escape from the surface. Because they cannot escape from the closed container, however, the vapor space within the container becomes saturated and some molecules return to the liquid. Within a short time, an equilibrium is set up between the number of molecules escaping from the surface and those returning to the liquid state.

As the molecules leave the surface of the liquid, pressure is created in the vapor space of the closed container. This is the **vapor pressure** of the liquid (see Table 4.6). In compliance with the kinetic theory of gases, the number of molecules and thus the magnitude of the vapor pressure increase with a rise in the temperature of the liquid. Liquids with low boiling points possess comparatively high vapor pressures, which means they evaporate easily. Liquids with high boiling points, on the other hand, possess relatively low vapor pressures and they evaporate more slowly.

Typically a gas or vapor is the most hazardous physical state for any material. It is the vapor of a flammable liquid that burns, and it is the vapor that can cause adverse health effects when people inhale them. Clearly, the vapor pressure of a flammable or toxic substance has a direct impact upon its potential fire and health hazards.

A flammable or toxic liquid that has a relatively low vapor pressure evolves little vapor at the given temperature. Consequently, the likelihood that the substance can ignite or be inhaled is low. However, when a liquid possesses a relatively high vapor pressure, a substantially greater volume of vapor evolves at the given temperature. This increases the potential risk of flammability and toxicity caused by inhalation comparative to lower vapor pressure substances. As a general rule, a vapor pressure of 10 mm Hg is often used to differentiate "high" from "low" when identifying the potential hazards of a substance.

Altering Phase Change Temperatures

SOLUTIONS AND COMPOUNDS

Molecular change that alters the compound may dramatically change heat capacity and phase change points, which is easily understood (see Table 4.7). However, change in solution and pressure can also make dramatic changes in characteristics of matter. For example, by adding small quantities of sodium chloride (NaCl), or common table salt, to water, the water's phase change temperatures change. The freezing point lowers while the boiling point rises. This is why salt is often added

TABLE 4.7	Heat Capacities of Some Common Liquids				
MATTER	COMPOSITION	HEAT CAPACITY—LIQUID (kJ/kg-K)	LATENT HEAT OF VAPORIZATION (kJ/kg)	HEAT CAPACITY—VAPOR (kJ/kg-K)	HEAT CAPACITY—SOLID (kJ/kg-K)
Acetone	C_3H_6O	2.12	501	1.29	—
n-butane	C_4H_{10}	2.30	26	1.68	—
Carbon disulfide	CS_2	1.00	315	1.68	—
n-heptane	C_7H_{16}	2.20	316	1.66	—
n-hexane	C_6H_{14}	2.24	335	1.66	—
Methanol	CH_4O	2.37	1101	1.37	—
Methyl ethyl ketone	C_4H_8O	2.30	434	1.43	—
n-octane	C_8H_{18}	2.20	301	1.65	—
n-pentane	C_5H_{12}	2.33	357	1.67	—
Propane	C_3H_8	2.23	343	1.67	—
Styrene	C_8H_8	1.76	356	1.17	—
Toluene	C_7H_8	1.67	360	1.12	—
Water	H_2O	4.18	2260	2.01	—
Polyurethane foams	—	—	—	—	~1.4
Wood, oak	—	—	—	—	2.0
Gold	Au	—	—	—	0.13

Source: Data derived from *NFPA Fire Protection Handbook*, Twentieth Edition, Table 6.17.1

during cooking, more to raise the temperature within cooking water and thus facilitate cooking.

PRESSURE CHANGE

Change in pressure also affects phase change temperatures. Because ambient atmospheric pressure is greater at sea level (1 atm), water boils at 212°F; however, the boiling temperature could be lowered to 202°F at one mile higher if all other factors remained the same. Conversely, when pressure increases, the boiling point rises. This can be illustrated by automotive radiators. Water (with additives to alter phase change points) exists at increased pressure. Warnings advise against opening the system when it is hot because, when pressure is released while the water temperature is above the boiling point in ambient pressure, a sudden phase change will occur, resulting in burns for those in close proximity to the steam expansion.

Summary

Though fire is a chemical reaction, its exothermic nature results in many changes that are explained by physics. Expansion that results in buoyancy, which then initiates fluid movement, is easily quantified when you understand the guiding principles. This chapter examined the principles of matter and physics related to fire development.

Review Questions

1. **True or false:** When calculating the heat capacity of a given mass, latent heats should be accounted for in each unit of mass (pound or gram).
2. **True or false:** Measurement of heat capacity applies only to water (H_2O).
3. **True or false:** The ratio of volume to mass describes weight.
4. When matter exists in a form in which it has definite shape without a container, it is said to be a:
   a. Gas
   b. Liquid
   c. Solid
5. The surface of an iron burns the skin at approximately 65°C, a temperature at which most individuals experience pain. This temperature is called the threshold for pain. Will an average person sense pain when touching a steel radiator whose surface temperature is 175°F?
6. The approximate temperature range of an incandescent metal surface is 800 to 1,000°C when the surface is a bright, cherry red color. What is the minimum temperature of this incandescent metal surface when expressed in degrees Fahrenheit?
7. The four reaction zones of color and temperature of a candle flame are illustrated in Figure 4.5. Incandescent soot particles burning in the yellow zone, sometimes called the luminous zone, provide most of the light from a burning candle.
   a. Identify the coolest and hottest zones of the flame.

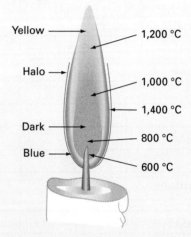

FIGURE 4.5 The zones of color and temperature of a candle flame.
Source: Meyer, *Hazardous Materials,* Fifth Edition, 2009, Figure 2.15, page 84.

   b. What is the temperature in degrees Fahrenheit of the blue zone?
   c. What is the temperature in degrees Rankine of the yellow zone?
8. A tanker holding gasoline is 4½ feet in diameter and 30 feet long. How many gallons does it contain when it is 80 percent full?
9. Airbags used to support a wrecked vehicle are deployed and filled with 1,200 liters of air to support the load when the temperature is 75°F. During the evening hours, temperatures drop to 31°F. Quantify what will happen to the volume within the bags.
10. Assume a compartment measuring 10 m × 15 m × 2.6 m has no openings and experiences

a fire. The temperature inside the compartment rises 30°C. If the initial pressure was 14.7 psi (1 atm), what maximum pressure is reached within the structure?

a. $\dfrac{P_1}{T_1} = \dfrac{P_2}{T_2}$ ($T_1 = 300°\text{K}$, $T_2 = 330°\text{K}$)

b. $\dfrac{14.7 \text{ psi}}{300 \text{ K}} = \dfrac{P_2}{330 \text{ K}}$

c. $\dfrac{14.7 \text{ psi} * .330 \text{ K}}{300 \text{ K}} = 16.2 \text{ psi}$

PEARSON
myfirekit™

For additional review and practice tests, visit **www.bradybooks.com** and click on MyBradyKit to access book-specific resources for this text!

Register your access code from the front of your book by going to **www.bradybooks.com** and selecting the mykit links.

References

Cengel, Y., and Boles, M. (2006). *Thermodynamics: An Engineering Approach,* Fourth Edition. New York: McGraw Hill.

Meyer, E. (2010). *Chemistry of Hazardous Materials,* Fifth Edition. New York: Prentice Hall.

NFPA (2009). *NFPA Fire Protection Handbook,* Twentieth Edition. Quincy, MA: NFPA.

NFPA 921 (2008). *The Guide for Fire and Explosion Investigations.* Quincy, MA: NFPA.

SFPE (2009). *The SFPE Handbook of Fire Protection Engineering,* Fourth Edition. Bethesda, MD: SFPE.

KEY TERMS

OBJECTIVES

After reading this chapter, you should be able to:

- Define fire in scientific terms.
- Describe the similarities and differences in the fire triangle and fire tetrahedron models used to describe fire.
- Describe and explain the similarities and differences in premixed and diffusion flames.
- Describe and explain laminar and turbulent flames.
- List the five classes of fire and describe what differentiates each class.
- Define the terms *flame plume*, *fire plume*, *flame jet*, and *flame spread*.
- Describe and explain the term *heat of combustion* as it relates to fuel.
- Describe and explain the term *combustion efficiency* as it relates to fuels.

The use and danger of fire are recorded in some of the earliest historical writings. Use of fire for light, heat, and cooking are well documented throughout the oldest religious writings, as are the consequences for improper use. Though fire has been with us for all recorded time, how to define the term is not as clear.

Merriam-Webster's dictionary (1998) defines the noun *fire* as:

(1): the phenomenon of combustion manifested in light, flame, and heat (2): one of the four elements of the alchemists. Etymology: Middle English, from Old English *fȳr;* akin to Old High German *fiur* fire, Greek *pyr.*

In fact, the Greeks considered fire to be so important that they listed it as one of the major elements in the universe, along with earth, air, and water.

Dr. James Quintiere defines fire as "an uncontrolled chemical reaction producing light and sufficient energy" (Quintiere, 1998). Dr. Vyto Babrauskus's *Ignition Handbook* uses the American Society for Testing Materials (ASTM) (ASTM E176) definition of fire, "destructive burning as manifested by any of the following: light, flame, heat, or smoke" (Babrauskas, 2003). "Rapid oxidation with the evolution of heat and light; uncontrolled combustion" is used by Dr. John DeHaan to describe fire (DeHaan, 2007). Rather than using the term *fire* in defining the process, Dr. Raymond Friedman refers to combustion as "an exothermic (heat producing) chemical reaction between some substance and oxygen" (Friedman, 1998).

The definition of fire promulgated by the National Fire Protection Association's (NFPA's) National Fire Code© component document NFPA 921, the *Guide for Fire and Explosion Investigations,* will be used throughout this text. **Fire** is a "rapid oxidation process, which is a chemical reaction, resulting in the evolution of heat and light in varying intensities" (NFPA 921, 2008). The reason this definition is chosen over the others centers on the concept that many other definitions indicate that fire is an uncontrolled reaction, which is not precise. Controlled fire is used daily for providing heating, cooking, and lighting in domestic settings, and it is used extensively in industrial settings to produce goods and wares commonly needed by society. When uncontrolled, fire quickly destroys lives, natural settings, buildings, vehicles and/or commodities used by humans.

fire
- A rapid oxidation process, which is a chemical reaction, resulting in the evolution of light and heat in varying intensities.

The Fire Triangle and Fire Tetrahedron

A common way to describe fire is to consider the three components required for combustion to occur. The **fire triangle** depicts the components required for fire as *heat, fuel,* and *air* (oxygen) (see Figure 5.1). The model indicates that a fuel must be present in a gaseous form to react chemically with oxygen in the presence of

fire triangle
- A model that describes three components necessary for ignition: fuel, heat, and oxygen.

FIGURE 5.1 The fire triangle.

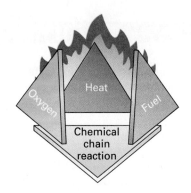

FIGURE 5.2 The fire tetrahedron.

sufficient heat to initiate the oxidation reaction. The model indicates that removal of one component extinguishes the flame or ceases the reaction.

The fire triangle was the fundamental training point for the fire service and safety professions until the 1960s, when Walter Haessler introduced a fourth component. The **fire tetrahedron** incorporates this fourth component for combustion to occur (see Figure 5.2). The concept that a continuous chemical chain reaction must occur between the fuel and oxygen in the presence of heat is depicted by this new model. Extinguishment results when any component is removed, which is similar to the fire triangle, with the exception that, if the chemical chain reaction is interrupted, the fire will also be extinguished. Within these models, heat is a form of energy that is capable of initiating and supporting chemical changes and changes of state (see Figure 5.3).

Fuel is an atom or molecule in the gaseous state, within its flammable limits, that is readily able to react chemically with oxygen. Oxygen (O_2) for the majority of combustion reactions is commonly provided in the form of atmospheric air (atmospheric air is comprised of approximately 21 percent O_2). Sometimes, however, oxygen is provided chemically within the fuel or as a separate oxidizing agent within the reaction, especially within condensed phase fuels. The chemical chain reaction is the result

fire tetrahedron

■ A model that describes four components necessary for continued combustion: fuel, heat, oxygen, and uninterrupted chemical chain reaction.

FIGURE 5.3 Surface heating and the evolution of pyrolysis gases.
Source: Adapted from Bengtsson, L., *Enclosure Fires,* 2001, Raddnings Verket.

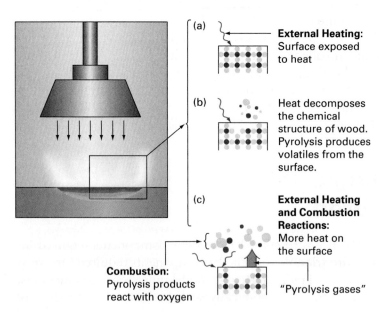

(a) **External Heating:** Surface exposed to heat

(b) Heat decomposes the chemical structure of wood. Pyrolysis produces volatiles from the surface.

(c) **External Heating and Combustion Reactions:** More heat on the surface

Combustion: Pyrolysis products react with oxygen

"Pyrolysis gases"

of the combustion of the fuel and the effectiveness of that combustion, which in turn releases energy in the form of heat. This heat is used to further break down the solid or liquid fuel to produce fuel in a gaseous state for combustion to continue.

Classification of Fires

To better understand the combustion reaction of particular fuels and situations, fires are classified into five categories, or classes (**classification of fire**), using letters to differentiate between the types. Four categories (Classes A, B, D, and K) of fires are denoted by the fuel involved. The remaining category, Class C, gains its designation based on hazards associated with attempting to extinguish the fire.

CLASS A

Fires occurring with ordinary combustible fuels (or more specifically, solid fuels) are categorized as Class A fires. The primary distinguishing factor for Class A fires is a solid material that must undergo pyrolysis, which results in char or ash formation. These fires are extinguished primarily by cooling the fuel to cease pyrolysis. To readily identify fire extinguishers rated for Class A fires, a large A is situated in a triangle that is green in color.

CLASS B

Burning liquids and gases comprise Class B fires. Class B fires burn on the surface of fuel packages or within an area where the proper gaseous mixture exists. Soot may result when unburned carbon remains following incomplete combustion, but the remaining fuel is not charred. Extinguishment is accomplished by removing air, cooling the fuel, preventing fuel from reaching the combustion zone, or interrupting the chemical chain reaction. Hazards are identified with a large B emblazoned in a red square.

CLASS C

Class C fires involve any other type of fuel (Class A, B, D, or K) but they have energized electrical equipment in the area of burning. Hazards associated with Class C fires are electrical shock to those in the area and reignition of extinguished fuels when electrical energy reheats the fuel package. Disconnecting the electrical supply eliminates this hazard. A blue circle with a large C distinguishes these fire hazards and the extinguishing agents that put them out.

CLASS D

Class D fires involve combustible metals. Titanium, magnesium, zirconium, and aluminum are examples of metals that burn. Burning metals evolve intense heat; thus, the use of conventional extinguishing media is not generally recommended. Dry powder compatible with the particular metal is applied at low velocity to extinguish smaller fires. Larger fires may require large amounts of dry sand or dirt to effect extinguishment. Water applied at a low rate can result in explosive reactions with metal fires; however, large amounts (copious quantities) of water may extinguish some metals such as magnesium. Conversely, sodium and lithium are water-reactive; thus, application of any water results in exothermic reactions, often very violent reactions. Combustible metals are designated with a large D in the center of a yellow star.

CLASS K

Class K fires are relatively new to fire protection. Evolution of cooking methods at restaurants, food-processing plants, and other places where large quantities of food are fried in deep fryers with vegetable oils rather than animal fats resulted in fires with unanticipated characteristics. Fire protection systems and portable fire extinguishers used in kitchen situations were typically the same as those used with other flammable or combustible liquids. However, experience showed that those systems were not sufficient to suppress the flames and then prevent reignition. Materials that produce saponification (that is, change the vegetable oil to soap) are used in kitchen systems and fire extinguishers.

Oxidation Reaction

endothermic
- A chemical reaction that occurs with the absorption of heat.

exothermic
- A chemical reaction that releases heat energy.

Fire is a chemical reaction known as an oxidation reaction. In general, chemical reactions can be either **endothermic** or **exothermic**. Endothermic reactions absorb energy, while exothermic reactions give off heat. When organic compounds react with oxygen and undergo complete combustion, water, carbon dioxide, light, and heat are the products given off during the reaction. Thus, combustion is considered an exothermic reaction (see Figure 5.4).

We can use equations to simplify the understanding of reactions. In the equations, the reactants are shown on the left of the arrow and the products are shown on the right. Balanced equations indicate equal amounts of a chemical on either side of the reaction.

EXAMPLE 5.1

Simple Oxidation of Ethane

$$C_2H_6 + 3.5O_2 \rightarrow 2CO_2 + 3H_2O$$

or

$$2C_2H_6 + 7O_2 \rightarrow 4CO_2 + 6H_2O$$

Though we use the simple reactant-to-product model to represent combustion reactions, oxidation actually involves a far greater number of elementary or intermediate steps. These steps represent the molecular level of the combustion reaction.

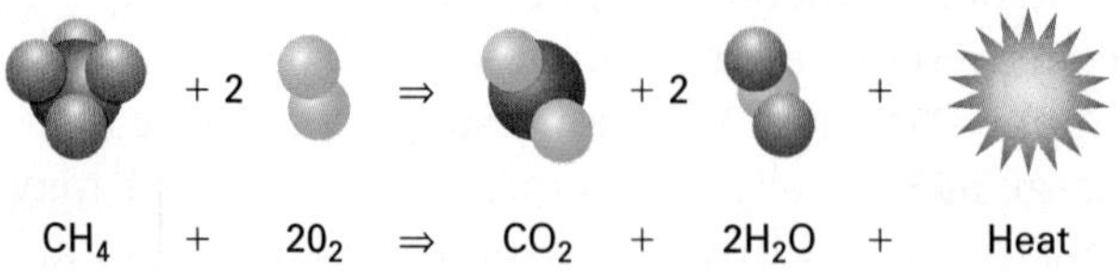

FIGURE 5.4
Complete combustion (nitrogen excluded).
Source: Adapted from Bengtsson, L., *Enclosure Fires,* 2001, Raddnings Verket.

$$CH_4 \quad + \quad 2O_2 \quad \Rightarrow \quad CO_2 \quad + \quad 2H_2O \quad + \quad \text{Heat}$$

Step	Reaction		
1	$O_2 + H$	$\rightarrow$	$OH + O$
1b	$OH + O$	$\rightarrow$	$O_2 + H$
2	$O + H_2$	$\rightarrow$	$H + OH$
2b	$H + OH$	$\rightarrow$	$O + H_2$
3	$OH + H_2$	$\rightarrow$	$H + H_2O$
3b	$H + H_2O$	$\rightarrow$	$OH + H_2$
4	$OH + OH$	$\rightarrow$	$H_2O + O$
4b	$H_2O + O$	$\rightarrow$	$OH + OH$
5	$H + O_2 + M$	$\rightarrow$	$HO_2 + M$
6	$HO_2 + H$	$\rightarrow$	$OH + OH$
7	$HO_2 + H$	$\rightarrow$	$H_2 + O_2$
8	$HO_2 + H$	$\rightarrow$	$H_2O + O$
9	$HO_2 + OH$	$\rightarrow$	$H_2O + O_2$
10	$CO + OH$	$\rightarrow$	$CO_2 + H$
10b	$CO_2 + H$	$\rightarrow$	$CO + OH$
11	$CH_4 + H$	$\rightarrow$	$H_2 + CH_3$
11b	$H_2 + CH_3$	$\rightarrow$	$CH_4 + H$
12	$CH_4 + OH$	$\rightarrow$	$H_2O + CH_3$
13	$CH_3 + O$	$\rightarrow$	$CH_2O + H$
14	$CH_3 + OH$	$\rightarrow$	$CH_2O + H + H$
15	$CH_3 + OH$	$\rightarrow$	$CH_2O + H_2$
16	$CH_3 + H$	$\rightarrow$	CH_4
17	$CH_2O + H$	$\rightarrow$	$CHO + H_2$
18	$CH_2O + OH$	$\rightarrow$	$CHO + H_2O$
19	$CHO + H$	$\rightarrow$	$CO + H_2$
20	$CHO + OH$	$\rightarrow$	$CO + H_2O$
21	$CHO + O_2$	$\rightarrow$	$CO + HO_2$
22	$CHO + M$	$\rightarrow$	$CO + H + M$
23	$CH_3 + H$	$\rightarrow$	$CH_2 + H_2$
24	$CH_2 + O_2$	$\rightarrow$	$CO_2 + H + H$
25	$CH_2 + O_2$	$\rightarrow$	$CO + OH + H$
26	$CH_2 + H$	$\rightarrow$	$CH + H_2$
26b	$CH + H_2$	$\rightarrow$	$CH_2 + H$
27	$CH + O_2$	$\rightarrow$	$CHO + O$
28	$CH_3 + OH$	$\rightarrow$	$CH_2 + H_2O$
29	$CH_2 + OH$	$\rightarrow$	$CH_2O + H$
30	$CH_2 + OH$	$\rightarrow$	$CH + H_2O$
31	$CH + OH$	$\rightarrow$	$CHO + H$

FIGURE 5.5 Thirty-one intermediate steps in the oxidation of methane.
Source: Data derived from Linan and Williams, *Fundamental Aspects of Combustion.*

Figure 5.5 represents the elementary or intermediate steps required to oxidize methane completely. When you are learning about combustion, we recommend exploring the elementary steps. For understanding basic fire science, however, simple oxidation reactions are sufficient to represent concepts.

Simple chemical reaction equations do not account for other chemicals present in the reaction area. The most common example is nitrogen. When ambient air brings oxygen to the reaction, approximately 21 percent of the volume is oxygen. Nitrogen from the air typically does not chemically react; however, heat is lost to the nitrogen. Impurities in the air and/or fuel may also become reactants that alter the combustion process (see Figure 5.6).

FIGURE 5.6 Fire plume.
Source: Photo by authors.

Flames

Flames are the visible, luminous body where the oxidation reaction is occurring. It is also known as the combustion zone. Gases within the combustion zone increase in temperature, resulting in an expanded volume. The expanded volume causes the gases to become less dense than the surrounding air, wherein buoyant forces cause the fluids to rise. Flame characteristics result from the fuel's ability to mix with adequate oxygen to produce complete combustion. Oxygen is introduced in the combustion reaction in one of two methods: *diffusion* or *premixed*. The most common flames that the fire safety professional encounters result from diffusion of oxygen into the fuel gas region to support combustion. As with the Bunsen burner flame shown in Figure 5.7(b), air is introduced to the fuel at the combustion site when the air inlet at the base of the tube is shut off. The resulting combustion occurs where the fuel and air mixture are within the fuel's flammable range. The interior of this reaction is the source of the fuel for the reaction, and it will be fuel-rich. Thus, it requires air or oxygen to be diffused into the reaction zone for combustion to occur. This is known as a **diffusion flame**. The location where the combustion reaction can take place by the diffusion of oxygen into the reaction is known as the flame sheet (see Figure 5.8). The interior portion of a diffusion flame is fuel-rich and not undergoing flaming combustion; outside the reaction zone, it is fuel-lean and no combustion can occur. Reactions are still occurring in the center of the flame, however, and they are not visible as a flame.

diffusion flame
▪ Flame resulting from fire where oxygen mixes with the fuel at the combustion zone.

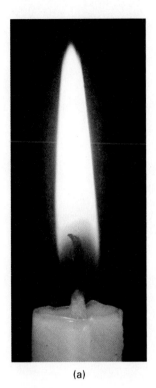 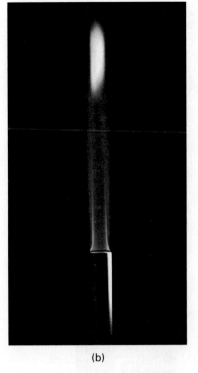

(a)   (b)

FIGURE 5.7 Diffusion versus premixed flame: (a) candle flame—diffusion, flame; (b) Bunsen burner—premixed flame.
Source: Photo by authors.

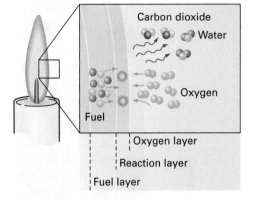

FIGURE 5.8 Candle flame: flame sheet illustrating fuel-rich and fuel-lean regions of a flame.
Source: Adapted from Bengtsson, L., *Enclosure Fires*, 2001, Raddnings Verket.

Premixed indicates that fuel gases and oxygen (air) are mixed before they reach the combustion zone. When air is introduced before the combustion point and the mixture is allowed to form prior to ignition, the result is **premixed flames**. Premixed flames are flames where the fuel and air are mixed prior to the ignition point. Air is brought into the fuel stream, often through a venturi effect, where turbulence in a conduit allows for mixture of the fuel and air. This turbulent mixture produces more efficient burning. Premixed flames can also occur if a gas is allowed to leak into a volume of air prior to reaching an ignition source, such as that found during residential and industrial explosions. Premixing results in greater combustion efficiency. Solid- and liquid-phase fuels can involve only diffusion flames; gaseous fuels may involve either diffusion or premixed flames.

Figure 5.9 shows propane gas utilized in a Bunsen burner. When the air inlet at the base of the Bunsen burner tube is closed, the flame is a diffusion flame, which shows a yellowish-orange flame (Figure 5.9a). Remember that diffusion flames are caused when the air is being diffused into the reaction zone during the combustion reaction. When the air inlet at the base of the burner tube is open, the fuel and air can mix prior to the combustion reaction; thus, the flame is a premixed flame, which shows a blue flame (Figure 5.9b).

Combustion zones (flames) depicted by constant shape and low velocity (flames that have little side-to-side movement) are considered laminar diffusion flames. A candle flame is an example of a laminar flame (see Figure 5.10a). **Laminar flames**

premixed flames
■ A flame for which the fuel and oxidizer are mixed prior to combustion, as in a laboratory Bunsen burner or a gas cooking range. Propagation of the flame is governed by the interaction among flow rate, transport processes, and chemical reaction.

laminar flames
■ Streamlines of the flames, usually small ones, are smooth and fluctuations are negligibly small.

FIGURE 5.9 Different flame types of Bunsen burner and flame color: (a) diffusion flame (air inlet closed), (b) premixed flame (air inlet open).
Source: Photos by authors.

(a)   (b)

FIGURE 5.10 Laminar versus turbulent flames: (a) laminar flame, (b) turbulent flames.
Source: Photos by authors.

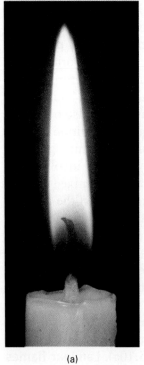

(a)   (b)

are characterized by a clearly defined combustion zone. Lines separating combustion areas from noninvolved gases are distinct. Premixed and smaller diffusion flames are examples of laminar flames. Laminar flames are steady because the fuel and air mixture are present in a configuration that provides a relatively constant combustion zone.

When the flow of fuel is at a faster rate than the air can be mixed into the reaction, the mixing process occurs in whirls, which form turbulent diffusion flames. Turbulence occurs when the fuel gases released are less constant, which results in whirls in the zone where fuel and air mixtures support combustion. Oxidation occurring within a generalized area but not in a steady geometric pattern exemplifies **turbulent flames** (see Figure 5.10b). Friction between heated gases moving upward and cooler surrounding air results in turbulence. Moving flames and transient flaming may occur as the fuel/air mixture, combined with proper heat, is present for varying periods. Combustion occurs where the fuel/air mixture is even and sufficient heat is present to initiate the reaction. Turbulence in larger fires forms swirling currents labeled *eddies*. Eddies entrain fresh air above the primary combustion zone (see Figure 5.11).

Flame jets result when fuel resting in a pressurized state is released. Examples are pressurized gases stored in cylinders and then directed through torches. These pressurized gases are often premixed flames. The factor determining whether a flame jet is laminar or turbulent depends greatly on the exit velocity of the fuel, as you can see in Figure 5.12.

turbulent flames
- Larger flames with higher velocity gas flow. These flames undergo many changes in direction rather than display straight or curving lines.

flame jets
- Horizontal flame movement due to convective heat transfer.

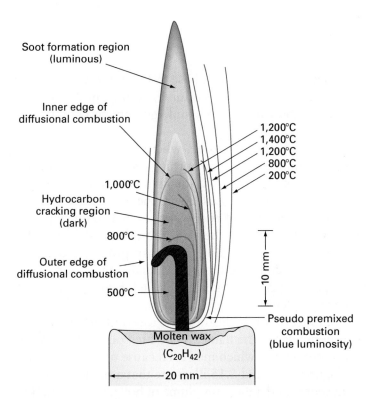

FIGURE 5.11 Temperature distribution in laminar candle flame showing a zone of a very high temperature.
Source: Fire, Rossotti, H. St. Anne's College, Oxford University, United Kingdom.

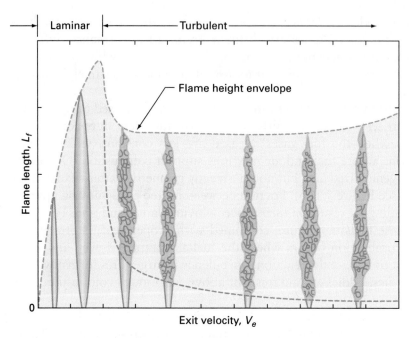

FIGURE 5.12 Flame lengths for gaseous jets issuing from a fixed diameter pipe. *Source:* Hottel, H., Hawthorne, W., "Diffusion in Laminar Flame Jets," in Third Symposium on Combustion and Flames and Explosion Phenomena. (Baltimore, MD: Williams & Wilkins, 1949), 254–256.

Fire Plume

fire plume
■ The column of hot gases, flames, and smoke rising above a fire.

flame plume
■ A body or stream of gaseous material involved in the combustion process and emitting radiant energy at specific wavelengths determined by the combustion chemistry of the fuel.

The **fire plume** designates the area from the flame's base to the point where ambient atmospheric conditions (temperature/pressure/volume) recur. The fire plume includes the flame plume and the buoyant smoke and incomplete combustion gases (see Figure 5.13).

The **flame plume** identifies the combustion zone, where flaming combustion is occurring, and represents the luminous body of burning gases. Heat from the exothermic chemical reaction causes electrons within involved atoms to move more quickly. This increase in movement, when coupled with the specific free radicals produced in the reaction, results in emission of energy waves in a frequency visible to humans. Temperatures and the types of chemical radicals produced in the reaction determine the color observed. Thus, flame color can provide indicators of combustion efficiency; however, it does not serve as a reliable indicator of the fuels involved. Changes in the percentage of oxygen available for combustion and fuel composition are also factors in determining flame color.

The height of the flame is a function of the heat release rate and the amount of air entrained into the sides or perimeter of the flame. As pointed out earlier, flames are affected greatly by the air entrainment, which may even cause turbulent mixing and the formation of eddies. Figures 5.14 and 5.15 illustrate that a flame has a persistent flame area, intermittent flame, and a buoyant plume of hot gases and smoke. The heat release rate, the diameter of the burning fuel, and the mixing of the entrained air into the reaction zone drive the pulsing structure along its perimeter and at the top of the flame plume. The height, temperature, and distance of the flame

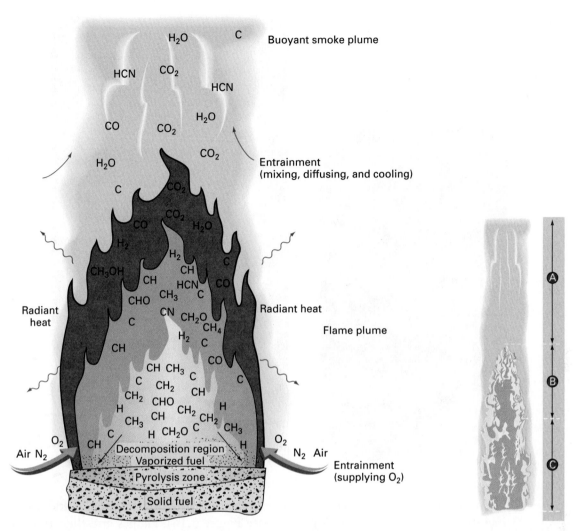

FIGURE 5.13 Typical flaming combustion of a solid fuel showing pyrolysis where volatized fuel decomposes to simpler species prior to combustion.
Source: Data derived from DeHaan (2007), *Kirk's Fire Investigation,* Figure 3.1, page 25.

FIGURE 5.14 Different sections in a fire plume: A, gas flow plume; B, fluctuating flame; C, continuous flame.
Source: Adapted from Bengtsson, L., *Enclosure Fires,* 2001, Raddnings Verket.

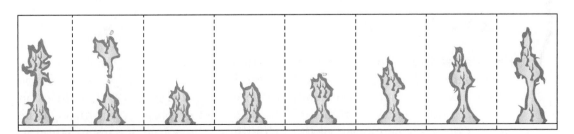

FIGURE 5.15 Intermittency of a buoyant diffusion flame burning on a 0.3 m porous burner. The sequence represents 1.3 s of cine film, showing 3 Hz oscillation.
Source: Data derived from McCaffrey, B. Purely Buoyant Diffusion Flames. NBSIR 79–1910, Gaithersburg, MD: National Bureau of Standards, Center for Fire Research, October 1979, page 32.

FIGURE 5.16 (a) Methanol and (b) heptane fires. *Source:* Photos by authors.

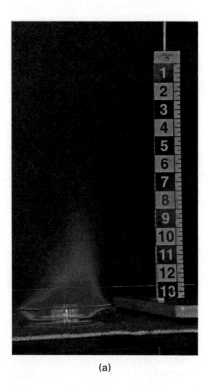

(a)

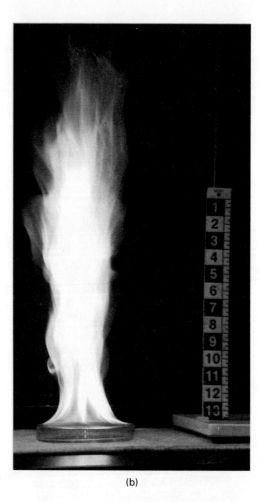

(b)

smoke

■ The airborne solid and liquid particulates and gases evolved when a material undergoes pyrolysis or combustion, together with the quantity of air that is entrained or otherwise mixed into the mass.

plume from other fuel items are also driving factors for radiant ignition of additional fuel items that promote a fire's growth (see Figure 5.16). Flames and their radiant heat transfer mechanism will be covered in more detail in later chapters.

Smoke designates the by-products of combustion, including fire gases (carbon dioxide, water vapor, carbon monoxide, and other products of a combustion reaction) and particulates such as carbon and unburned particulates that are transported by convected currents. The unburned pyrolysis products (pyrolyzates) are also contained within smoke, especially in ventilation-controlled fires.

Flame Spread

flame spread

■ The movement of flaming combustion from one place along a fuel item to another.

Flame spread refers to the speed with which the flame front (oxidation zone) moves across a fuel's surface. It is expressed in a linear measure per unit of time (m/s or ft/s). Flame speed is generally greater in premixed flames than in diffusion flames. Explosions result from rapid flame spread or propagation through the gaseous fuels. Flame movement through diffused gases (fuel gas/air mixture) in near stochiometric mixtures results in rapid temperature increases and thus the overpressurization of containing structures.

Fuels that contain oxygen in their chemical structure can experience flame movement through the fuel without atmospheric air. These types of fuels thus pose a danger for rapid fire development. When flame fronts are sufficiently rapid and/or the escape of fire gases is restricted, explosions may result. More discussion of flame spread will be introduced in Chapter 11.

HEAT OF COMBUSTION

The heat of combustion varies with fuels and is reported in kilojoules per gram. **Heat of combustion** indicates the amount of energy per unit of mass contained within a fuel, or how much energy a fuel can release when oxidized (see Table 5.1).

heat of combustion
- The quantity of heat evolved by the complete combustion of one mole (gram molecule) of a substance.

TABLE 5.1	Heat of Combustion	
MATERIAL	**HEAT OF COMBUSTION** $\Delta H_{c,eff}$ **(kJ/kg)**	
Cryogenics		
Liquid H_2	12,000	
LNG (mostly CH_4)	50,000	
LPG (mostly C_3H_8)	46,000	
Alcohols		
Methanol (CH_3OH)	20,000	
Ethanol (C_2H_5OH)	26,800	
Simple Organic Fuels		
Butane (C_4H_{10})	45,700	
Benzene (C_6H_6)	40,100	
Hexane (C_6H_{14})	44,700	
Heptane (C_7H_{16})	44,600	
Xylene (C_8H_{10})	40,800	
Acetone (C_3H_6O)	25,800	
Dioxane ($C_4H_8O_2$)	26,200	
Diethyl ether ($C_4H_{10}O$)	34,200	
Petroleum Products		
Benzine	44,700	
Gasoline	43,700	
Kerosene	43,200	
JP-4	43,500	
JP-5	43,000	
Transformer oil, hydrocarbon	46,400	
Fuel oil, heavy	39,700	
Crude oil	42,500–42,700	
Solids		
Polymethylmethacrylate ($C_5H_8O_2)_n$	24,900	
Polypropylene ($C_3H_6)_n$	43,200	
Polystyrene ($C_8H_8)_n$	39,700	
Miscellaneous		
561® silicon transformer fluid	28,100	

Source: Data Derived from SFPE, Babrauskas 2008.

Theoretical heats of combustion, otherwise known as the chemical heat of combustion, are easily determined for solids by a controlled experiment in an oxygen bomb calorimeter. This apparatus allows a material to undergo nearly complete combustion by providing oxygen at higher levels and measuring mass loss and energy released.

The effective heat of combustion is the actual amount of energy that is released when fuel oxidizes in air. Theoretical heats of combustion are reduced by the amount of energy lost to the process of converting liquid and solid fuels to the gaseous state (pyrolysis), which is then lost to ambient atmospheres. The effective heat of combustion factors in the energy required to convert a fuel into a gaseous form that will combust in liquids and solids and that cannot mix efficiently with the ambient air.

Combustion Efficiency

In fires, complete combustion is never attained. In other words, the efficiency of combustion in a fire is never 100 percent. This is mainly due to the loss of heat to the ambient atmosphere and the inefficient diffusion of oxygen in the chemical reaction. The efficiency of a fuel burning in the air is driven by the amount of oxygen available and the fuel. Table 10.2 lists different fuels and their chemical heats of combustion (theoretical) compared to their effective heats of combustion. Combustion efficiency is usually represented by the lowercase Greek letter chi: χ. This value ranges from 0 to 1. The number 1 represents perfect or 100 percent efficient combustion, while 0 represents completely inefficient combustion; however, both limits are impractical. Therefore, most fuels are in the middle of this range. This concept will be discussed in further detail in Chapter 9.

Summary

We defined fire as a rapid oxidation process, which is a chemical reaction, that results in the evolution of heat and light in varying intensities. We believe this definition is the best basis for understanding combustion. Fire can be uncontrolled, as in an emergency situation, but it can also be controlled so that it is useful rather than destructive. Students of fire behavior must understand the factors described in this chapter to be able to predict the conditions that can result from combustion in a particular area. The nature of fire is such that it often happens in unintended areas, so students (firefighters) must be able to quickly identify factors that facilitate accurate assumptions on which to base their actions to counter the fire's effects.

Review Questions

1. Define fire in scientific terms.
2. Describe the differences and similarities between laminar and turbulent flames.
3. Describe the differences and similarities between the fire plume and the flame plume.
4. Specifically, analyze the temperature differences between the heated combustion gases and the luminescent flaming combustion.
5. Review the combustion efficiency for various fuels. Is there a relationship to those fuels that burn more completely versus those that do not?
6. What type of flame typically results from Bunsen burners used in laboratory experiments?
7. Soot primarily results from which chemical?
8. Fire is:
   a. an endothermic reaction
   b. an exothermic reaction
   c. both an endothermic and exothermic reaction
   d. neither an endothermic nor an exothermic reaction
9. The fire triangle has been used for years to describe the combustion process. The three elements of the fire triangle include:
   a. fuel, heat, and the chemical chain reaction
   b. fuel, oxygen, and the chemical chain reaction

   c. fuel, heat, and oxygen
   d. heat, oxygen, and the chemical reaction
10. A premixed flame occurs when:
    a. the fuel and oxidizer are mixed at the point of combustion
    b. the fuel and oxidizer are mixed prior to the point of combustion and are not dependent on atmospheric oxygen for combustion
    c. when the flame is luminous
    d. none of the above
11. The amount of heat energy that is released during complete combustion is the material's:
    a. heat of combustion
    b. latent heat of vaporization
    c. specific heat
    d. all of the above
12. A fire that involves combustible metals would be classified as a
    a. Class A fire
    b. Class B fire
    c. Class C fire
    d. Class D fire
13. When energized electrical equipment is involved in a fire, the fire would be classified as a
    a. Class A fire
    b. Class B fire
    c. Class C fire
    d. Class D fire

14. If during the oxidation process, a material releases heat energy, the reaction is considered
    a. endothermic
    b. exothermic
    c. heat of combustion
    d. none of the above

15. If the flow of a fluid, liquid or gas is straight, the flow is said to be
    a. turbulent
    b. eddy
    c. lateral
    d. laminar

PEARSON
myfirekit™

For additional review and practice tests, visit www.bradybooks.com and click on MyBradyKit to access book-specific resources for this text!

Register your access code from the front of your book by going to www.bradybooks.com and selecting the mykit links.

References

ASTM E176. *Standard Terminology of Fire Standards*. West Conshohocken, PA: American Standards of Testing and Materials.

Babrauskas, V. (2003). *Ignition Handbook*. Issaquah, WA: Fire Science Publishers.

Babrauskas, V. (2008). "Heat Release Rate." *The Handbook of Fire Protection Engineering*, Fourth Edition. Bethesda, MD: SFPE.

DeHaan, J. (2007). *Kirk's Fire Investigation*, Sixth Edition. Upper Saddle River, NJ: Pearson/Prentice Hall.

Friedman, R. (1998). *Principles of Fire Protection Chemistry*, Third Edition. Quincy, MA: NFPA.

Kuvshinoff, B. (1977). *Fire Sciences Dictionary*. Wiley-Interscience Publications: New York.

Merriam Webster's Dictionary (1998). Retrieved on July 16, 2009, from: http://www.merriam-webster.com/dictionary/fire.

NFPA 921 (2008). *Guide for Fire and Explosion Investigations*. Quincy, MA: National Fire Protection Association.

Purdue. Retrieved on July 16, 2009, from http://www.chem.purdue.edu/gchelp/atoms/states.htm.

Quintiere, J. (1998). *Principles of Fire Behavior*. Albany, NY: Delmar.

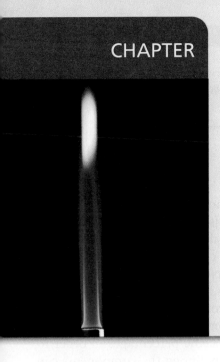

CHAPTER 6

Gaseous Combustion

KEY TERMS

autoignition temperature, *p. 89*

deflagration, *p. 94*

detonation, *p. 94*

flame spread, *p. 93*

flammable limits (lower flammable limit [LFL] and upper flammable limit [UFL]), *p. 85*

gas, *p. 84*

minimum ignition energy (MIE), *p. 89*

stoichiometric mixture, *p. 85*

vapor density, *p. 91*

OBJECTIVES

After reading this chapter, you should be able to:

- Describe and explain the conditions that separate matter existing in the gaseous state from matter existing in other states.
- Describe and explain the flammable limits of a gaseous fuel, including the upper flammable limit, the lower flammable limit, and stoichiometric mixture.
- Diagram a simple oxidation reaction.
- Describe and explain three atmospheric conditions that affect gaseous combustion.
- Explain the difference between autoignition temperature and minimum ignition energy for gaseous fuels.
- Explain the term *vapor density* and how it affects fire hazards associated with gaseous fuels.
- Describe and explain two types of explosions associated with gaseous fuels and the pressures that are anticipated from each.
- Explain the difference between a detonation explosion and a deflagration explosion.

Gaseous Combustion

gas
■ The physical state of a substance that has no shape or volume of its own and expands to take the shape and volume of the container or enclosure it occupies.

Matter that is in a state where atoms collide but are not in a fixed arrangement, and where the matter assumes the shape of a container, is in the gaseous state (Meyer, 2010). **Gases** are the simplest form of matter and are therefore the easiest fuel to discuss. Gases are elemental to the understanding of how combustion occurs in all states of matter and thus will be discussed first. Fuels can typically enter the combustion reaction only when they are in a gaseous state; liquids and solids must gasify before combustion is possible, and these two states will be covered in subsequent chapters. Combustion of gaseous fuels may involve diffusion or premixed flames (see Table 6.1).

Mixture of fuel gases and oxygen must be in specific percentages in order to support a combustion reaction, and the relationship among the components can be compared to cooking food. Proper amounts of each ingredient in a recipe, when heated to the proper temperature, can result in wonderful eating experiences.

TABLE 6.1	Common Gases at Standard Temperature and Pressure					
GAS	**FLAMMABLE LIMITS (% BY VOLUME IN AIR)**		**SPECIFIC GRAVITY (AIR = 1)**	**IGNITION TEMPERATURE (°C)**	**HEAT OF COMBUSTION (MJ/kg)**	**MIE (mJ)**	
	LOWER	**UPPER**					
Acetylene	2.5	100	0.91	305	48.2	0.017 at 8.5%	C_2H_2
Anhydrous ammonia	15–16	28	0.6	651	18.6	680	NH_3
Carbon monoxide	12.5	74	0.97	609	10.1		CO
Methane	4.7	15	0.56	640	50	0.21 at 8.5%	CH_4
Hydrogen	4	75	0.07	500	120	.016 at 28%	H_2
Ethane	3	12.4		515	47.5	.24 at 6.5%	C_2H_6
Propane	2.1	9.6	1.52	500	46.3	.25 at 5.2%	C_3H_8
Butane	1.8	8.5	2	408	45.7	.25 at 4.7%	C_4H_{10}

Sources: Fire Protection Handbook, Twentieth Edition, (2008), Quincy, MA: National Fire Protection Association. Babrauskas, Vytenis, *Ignition Handbook* (2003), Issaquah, WA: Fire Science Publishers. NFPA 77, Recommended Practice on Static Electricity (2007), Quincy, MA: National Fire Protection Association, PraxairMSDS http://www.praxair.com/praxair.nsf/0/BD40C55346B4930F85256E5F007E8C2A/$file/Methane+-+Canada+-+2007.pdf

FIGURE 6.1
Combustion within
flammable range.

However, improper percentages of specific components yield undesirable products. The exact proper mixture (mass of each reactant) of chemicals for a chemical reaction, where all reactants chemically change to yield a new product or products, is the **stoichiometric mixture** (Babrauskas, 2003). In other words, the stoichiometric quantity of oxidizer (air) is just that amount needed to burn a quantity of fuel completely. In stoichiometry, all mass from the reactants chemically change, becoming part of the product. Balanced chemical equations represent stoichiometric mixtures for chemical reactions.

In nature, stoichiometric mixtures are rare; instead, some of the reactants remain unaltered after the reaction. Combustion reactions occur most commonly with oxygen in the air, and they can occur within a range reported as a volume percentage of gas to air (see Figure 6.1). The lowest percentage of fuel in air that will undergo combustion is called the **lower flammable limit (LFL)**. Conversely, the highest percentage of fuel in air where combustion can occur is the **upper flammable limit (UFL)**. Mixtures with an insufficient percentage of fuel gas are too *lean* to burn; mixtures with insufficient air are too *rich* to burn (see Figures 6.2 and 6.3).

> NOTE: The terms *flammable range, lower flammable limit (LFL),* and *upper flammable limit (UFL)* are interchangeable with *explosive range, lower explosive limit (LEL),* and *upper explosive limit (UEL).*

The lower flammable limit is typically determined experimentally for pure gases. However, a fuel gas is rarely composed of a pure gas. Most fuel gases are mixtures with various percentages of different gases. The lower flammable limit of a mixture is calculated based on Le Chatelier's law:

$$L_m = \frac{100}{\sum_i \dfrac{P_i}{L_i}}$$

stoichiometric mixture
■ Chemical conditions where the proportion of reactants is such that there is no surplus of any reactant after the chemical reaction is complete.

flammable limits (lower flammable limit [LFL] and upper flammable limit [UFL])
■ The range of gas to air mixtures where combustion can occur. These mixtures exist between the upper flammable limit and lower flammable limit of a fuel.

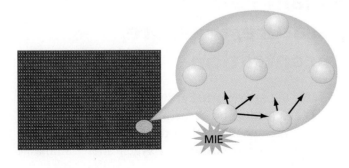

FIGURE 6.2 MIE without combustion—too lean to burn.

FIGURE 6.3 MIE without combustion—too rich to burn.

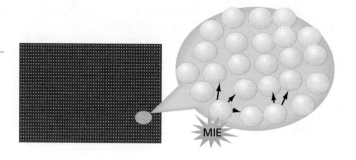

MIE

Where:

L_m = the lower flammable limit of the mixture of hydrocarbons in air

Σ = the sum of the components

P_i = the percentage composition of component i (i = type of gas)

L_i = lower flammable limit for i

This law requires that the percentage composition of the mixture and the LFL for each gas are known.

EXAMPLE 6.1

Calculate the lower flammable limit of a mixture of hydrocarbon gases containing 50 percent propane, 30 percent methane, and 20 percent ethane.

Known: Composition of gases, LFL for each gas (Table 6.2): propane = 2.1, methane = 5.0, ethane = 3.0

Find: LFL of mixture.

Analysis:

$$L_m = \frac{100}{\sum_i \dfrac{P_i}{L_i}}$$

$$L_m = \frac{100}{\dfrac{50}{2.1} + \dfrac{30}{5.0} + \dfrac{20}{3.0}} = 2.7\%$$

Dependence of Flammability Limits on Temperature, Pressure, and Oxygen Concentration

The flammable ranges listed in Table 6.2 are reported at standard temperature and pressure (STP), and the reaction is assumed to be occurring in air. If the pressure, temperature, or the percentage of oxygen changes, then the flammable range will also change. Changes in the percentage of atmospheric oxygen alter flammable

	MINIMUM IGNITION TEMP, F/C	FLAMMABILITY LIMITS, % FUEL GAS BY VOLUME		MAXIMUM FLAME VELOCITY, fps AND m/s	
FUEL		Lower	Upper	In Air	In O_2
Acetylene, C_2H_2	581/305	2.5	81.0	8.75/2.67	—
Blast furnace gas	—	35.0	73.5	—	—
Butane, commercial	896/480	1.86	8.41	2.85/0.87	—
Butane, n-C_4H_{10}	761/405	1.86	8.41	1.3/0.40	—
Carbon monoxide, CO	1128/609	12.5	74.2	1.7/0.52	—
Carbureted water gas	—	6.4	37.7	2.15/0.66	—
Coke oven gas	—	4.4	34.0	2.30/0.70	—
Ethane, C_2H_6	882/472	3.0	12.5	1.56/0.48	—
Gasoline	536/280	1.4	7.6	—	—
Hydrogen, H_2	1062/572	4.0	74.2	9.3/2.83	—
Hydrogen sulfide, H_2S	558/292	4.3	45.5	—	—
Mapp gas, C_3H_4‡	850/455	3.4	10.8	—	15.4/4.69
Methane, CH_4	1170/632	5.0	15.0	1.48/0.45	14.76/4.50
Methanol, CH_3OH‡	725/385	6.7	36.0	—	1.6/0.49
Natural gas	—	4.3	15.0	1.00/0.30	15.2/4.63
Producer gas	—	17.0	73.7	0.85/0.26	—
Propane, C_3H_8	871/466	2.1	10.1	1.52/0.46	12.2/3.72
Propane, commercial	932/500	2.37	9.50	2.78/0.85	—
Propylene, C_3H_6	—	—	—	—	—
Town gas (Br. coal)^d	700/370	4.8	31.0	—	—

TABLE 6.2 Flammability Data (Minimum Ignition Temperature, Flammable Limits, Maximum Flame Velocity) for Common Gases

Source: Data derived from Drysdale (1998), Quintiere (1998), Zalosh (2002).

ranges and combustion temperatures. Increased atmospheric oxygen concentration (percentage) results in an expanded flammable range; conversely, decreased amounts of oxygen narrow the flammable range to the stoichiometric mixture.

Changes in atmospheric pressure also have an impact on the flammable range. Lower atmospheric pressures do not have a substantial impact on the flammable range, unless great changes occur. For example, kerosene inside fuel tanks at sea level does not have a vapor pressure higher than the atmospheric pressure; therefore, it lies in the nonflammable range (too lean). However, when an airplane carrying this liquid takes off and reaches a substantial elevation, the atmospheric pressure drops substantially, but the vapor pressure remains the

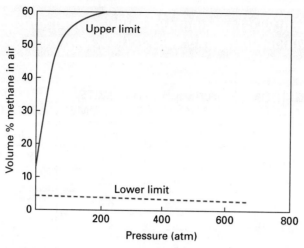

FIGURE 6.4 Variation of flammability limits with pressure: methane in air at super-atmospheric pressures.
Source: Data derived from Zabetakis, 1965.

same and can produce a flammable mixture. Of course, this danger remains until the associated drop in temperature takes place in the tank and reduces the vapor pressure back to a nonflammable range. Very high pressures, on the other hand, have a very significant impact on the upper flammable limit for fuels (see Figure 6.4). In both high pressures and low pressures, the lower limit of the fuel is relatively unaffected.

Temperature changes also have an impact on the flammable ranges of fuels (see Figure 6.5). Higher temperatures increase the flammability limits. The upper

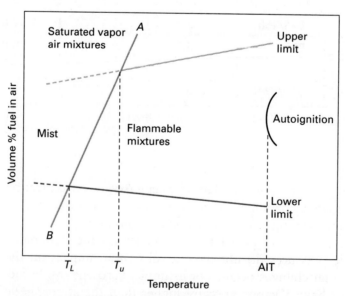

FIGURE 6.5 Effect of temperature on the limits of flammability of a flammable gas/air mixture at a constant initial pressure.
Source: Data derived from Zabetakis, 1965.

flammable limit is substantially increased, while the lower flammable limit is minimally affected. However, the total range is increased with increasing temperatures. This concept greatly affects flaming combustion during post-flashover or fully involved compartment fires. Specifically, flaming combustion inside a fully involved compartment requires less oxygen to burn the same fuel than was required when temperatures were lower at the beginning of the enclosure fire.

COMBUSTION OF METHANE

The theoretical combustion of hydrocarbon gases (carbon and hydrogen reactants) yields carbon dioxide and water as products. This relationship is commonly depicted in stoichiometric proportions that would prove extremely difficult to attain; however, it facilitates understanding of the underlying reaction. For methane (CH_4), this relationship is depicted as follows:

$$CH_4 + 2O_2 \rightarrow CO_2 + 2H_2O$$

Because combustion rarely occurs in an atmosphere containing pure oxygen, other chemicals involved or affected by the reaction are considered. Ambient air contains about 21 percent oxygen and about 79 percent nitrogen. Nitrogen has a relatively high thermal capacity (32.7 J/mol · K); it absorbs a substantial amount of the energy from the combustion reaction and limits flame propagation. In fact, many explosion prevention techniques are to introduce a substantial amount of nitrogen into the atmosphere to slow or stop the heat transfer through the fuel, which results in the cessation of flame propagation. The stoichiometric mixture for the combustion of methane in air is approximately 9.5 percent methane to 91.5 percent air.

$$CH_4 + 2(O_2 + 3.76N_2) \rightarrow CO_2 + 2H_2O + 7.52N_2$$

Though more thorough in explaining the combustion of methane in air, the preceding discussion is a simplification of the chain of events in the oxidation of methane. Figure 6.6 indicates thirty-two intermediate steps that actually occur during the reaction. In fact, the simple combustion process of methane includes approximately forty reactions involving over fifty different species (molecules, atoms, or radicals). Free radicals (i.e., OH, CHO·) that are produced in the reactions are less understood than the overall chemical reaction. Free radicals are simply those molecular fragments with unsatisfied chemical valences (created during the combustion reaction), which rapidly combine with other molecules to satisfy their chemical need. These free radicals are part of the reason for the speed of the reaction. These free radicals are also the portion of the combustion reaction that flame retardant additives are intended to combine with to slow the progression of the flame and continued combustion. Flame retardant additives work by breaking the chemical chain reaction.

Energy released from combustion remains relatively similar, regardless of oxygen concentration; however, temperatures resulting from combustion vary with concentration. Increased oxygen concentration results in less energy lost to other atmospheric gases (i.e., nitrogen). Thus, greater temperature is achieved from combustion of equal quantities of fuels. Conversely, when a lower concentration of oxygen is present, more energy is lost heating inert gases, which results in a lower temperature.

For each gram of methane oxidized, 50 kilojoules of energy are released (heat of combustion).

(continued)

a	CH_4 + M	=	$\cdot CH_3$ + H$\cdot$ + M				
b	CH_4 + $\cdot OH$	=	$\cdot CH_3$ + H_2O				
c	CH_4 + H$\cdot$	=	$\cdot CH_3$ + H_2				
d	CH_4 + $\cdot O\cdot$	=	$\cdot CH_3$ + $\cdot OH$				
e	O_2 + H$\cdot$	=	$\cdot O\cdot$ + $\cdot OH$				
f	$\cdot CH_3$ + O_2	=	CH_2O + $\cdot OH$				
g	CH_2O + $\cdot O\cdot$	=	$\cdot CHO$ + $\cdot OH$				
h	CH_2O + $\cdot OH$	=	$\cdot CHO$ + H_2O				
i	CH_2O + H$\cdot$	=	$\cdot CHO$ + H_2				
j	H_2 + $\cdot O\cdot$	=	H$\cdot$ + $\cdot OH$				
k	H_2 + $\cdot OH$	=	H$\cdot$ + H_2O				
l	$\cdot CHO$ + $\cdot O\cdot$	=	CO + OH				
m	$\cdot CHO$ + $\cdot OH$	=	CO + H_2O				
n	$\cdot CHO$ + H$\cdot$	=	CO + H_2				
o	CO + $\cdot OH$	=	CO_2 + H$\cdot$				
p	H + $\cdot OH$ + M	=	H_2O + M				
q	H + H$\cdot$ + M	=	H_2 + M				
r	H + O_2 + M	=	HO_2 + M				

This reaction scheme is by no means complete. Many radical-radical reactions, including those of the HO_2 radical, have been omitted.

M is any third body participating in radical recombination reactions (rows p–r) and dissociation reactions such as row a.

FIGURE 6.6 Intermediate reactions that occur in the combustion of methane.
Source: Bowman, C. T. (1975). Non-equilibrium radical concentration in shock-initiated methane oxidation. 15th Symposium (International) on Combustion, pp. 869–882. The Combustion Institute, Pittsburgh.

IGNITION ENERGY

Most ignition sequences begin when sufficient heat energy contacts a fuel gas in the presence of oxygen, where concentrations are sufficient to support combustion. The energy required to initiate flame is defined and/or measured in multiple ways, including piloted ignition temperature, **autoignition temperature,** and **minimum ignition energy (MIE).**

Minimum Ignition Energy

The most reactive mixture for all gases is the slightly fuel-rich level stoichiometry and therefore requires the lowest energy for ignition. Most fuel gases commonly encountered have a minimum ignition energy of 0.25 to 0.3 milijoules. This energy, although minute, must be present in an area where the fuel/air mixture is within the flammable range. This is also known as piloted ignition.

Autoignition (Autogenous) Temperature

Autoignition temperature is the temperature to which a mixture of fuel gas and air, which will burn, must be raised to initiate combustion without any external heat source (i.e., a spark, a pilot flame) (Babrauskas, 2003). In other words, the autoignition temperature relates to the atmospheric temperature that provides

autoignition temperature
■ The lowest temperature at which a combustible material ignites in air without a spark or flame.

minimum ignition energy (MIE)
■ The amount of heat required to initiate piloted combustion within a mixture of flammable gas and air.

enough heat transfer to the fuel, which in turn allows the combustion reaction to occur without other energy sources.

Differences and Similarities

When the MIE is introduced to a small portion of a combustible fuel/air mixture, combustion is initiated. The flame temperature then becomes the pilot that serves to ignite the remaining volumes of the mixture. The ambient temperature of the entire mixture is inconsequential for piloted ignition, but it is critical for autoignition.

VAPOR DENSITY (SPECIFIC GRAVITY OF VAPORS AND GASES)

The characteristic of matter in the gaseous state, regardless of its combustion characteristics, that indicates the density when compared to air at the same temperature and pressure is **vapor density** (also known as the specific gravity of vapors). Air at 20°C and 1 atmosphere is designated as having a vapor density of 1. Gases that have less density than air (<1) are lighter and thus rise in ambient air. A common example of a gas with a lower vapor density than air is natural gas. The primary constituent of natural gas is methane, which has a vapor density of 0.55. Conversely, gases that sink in ambient air or displace the air because they are denser are assigned a vapor density more than 1. One can calculate the vapor density by comparing the density (grams per cubic meter) of gases and air at an equal temperature and pressure.

vapor density
■ The ratio of the average molecular weight of a given volume of gas or vapor to the average molecular weight of an equal volume of air at the same temperature and pressure. Also known as *specific gravity of vapors and gases.*

Vapor density involves comparing the molecular weight of a substance with the molecular weight of air, which is 29. Simple division of the molecular weight (MW) of the chosen substance by that of air derives the ratio that indicates buoyancy related to air at 14.7 psi (101 kPa) at 68°F (20°C).

$$\text{Vapor density} = MW/29$$

EXAMPLE 6.2

Methane (CH_4) is composed of carbon, with a molecular weight of 12, and hydrogen, with a molecular weight of 1. Because four atoms of hydrogen are present, the weight of each must be factored.

Carbon	$12 \times 1 = 12$
Hydrogen	$1 \times 4 = 4$
Methane	$= 16$

Vapor density of methane = 16 (MW of methane)/29 (MW of air) = .55. Thus, methane is more buoyant than air and will rise.

Propane (C_3H_8) is composed of carbon, with a molecular weight of 12, and hydrogen, with a molecular weight of 1. All atoms must be accounted for when you are determining molecular weight.

Carbon	$12 \times 3 = 36$
Hydrogen	$1 \times 8 = 8$
Propane	$= 44$ MW

Vapor density of propane = 44 (MW of methane)/29 (MW of air) = 1.5. Thus, propane is less buoyant than air and will sink in air. Both gases are commonly used for heating and cooking, but they have vastly differing characteristics when released into the atmosphere.

Burning characteristics change greatly with a change in the chemical structure (the molecular construction) of similar gases. An example comes from C_2 derivatives, some of which are listed below:

Ethane (C_2H_6)

Ethylene (C_2H_4)

Acetylene (C_2H_2)

Acetylene is very flammable (see Table 6.3). It is so easily ignitable that approved cylinders are filled with porous material (pumice) and liquid (methyl ethyl ketone), within which the acetylene gas is dissolved to prevent fire entering into and then burning in the tank. Acetylene has a flammable range of 1 to 99 percent (some

TABLE 6.3	Vapor Densities of Some Common Gases or Vapors
SUBSTANCE	**VAPOR DENSITY (AIR = 1)**
Acetylene	0.899
Ammonia	0.589
Carbon dioxide	1.52
Carbon monoxide	0.969
Chlorine	2.49
Fluorine	1.7
Hydrogen	0.069
Hydrogen chloride	1.26
Hydrogen cyanide	0.938
Hydrogen sulfide	1.18
Methane	0.553
Nitrogen	0.969
Oxygen	1.11
Ozone	1.66
Propane	1.52
Sulfur dioxide	2.22

Source: Meyer, *Chemistry of Hazardous Materials*, Fifth Edition, Brady, Table 2.5, page 57.

texts even put it at 100 percent). Though easily ignitable, acetylene does not burn cleanly in ambient air. The molecular structure (triple bonds) disassociates easily and often at a point where insufficient air is present to result in fire gases; thus, excessive carbon is released. It should be noted that when used in an oxy-acetylene cutting torch, acetylene is premixed with oxygen and thus a clean, high-temperature flame results.

Flame Propagation

When a fuel exists within its **flammable limits (lower flammable limit and upper flammable limit)** and a competent ignition source (above the MIE for the fuel) is introduced, ignition of the fuel occurs. The flame will spread or propagate outward in all directions from this ignition source through the fuel, in a process called **flame spread,** unless an object (i.e., a wall or a ceiling) obstructs it. Remember that fluids follow the path of least resistance (see Figure 6.7).

flame spread
■ The movement of flaming combustion from one place along a fuel item to another. Also known as *flame propagation.*

Let's assume that this flammable mixture is outside in the open atmosphere. As the flame propagates through the flammable mixture, heating of the gases involved in ignition and heating of surrounding gases causes a flame front to expand beyond the area where the proper mixture was originally located. The increase in temperature associated with the combustion reaction causes an increase in the volume of the gas. The flame front or combustion reaction continues through the flammable mixture until the fuel is exhausted. The propagation of a flame through the fuel/air mixture without the production of damaging pressure is commonly known as a *flash fire.*

Next, assume the same mixture is inside a confining structure (maybe an industrial warehouse or a residential house). The mixture is within its flammable limits and an ignition source has been introduced. The flame will propagate through the fuel similar to what occurred through the mixture in the open; in this case, however, the confining structure limits the expansion of the gases. The increase in temperature caused by the combustion reaction and flame front propagation causes an increase in the volume of the gas, which in turn causes an increase in pressure on the confining structure. If the pressure increase is fast enough and it is of adequate force, then the confining structure may fail.

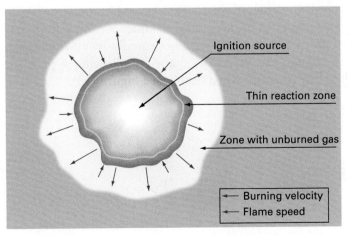

FIGURE 6.7
Illustration of the ignition of a flammable mixture and flame propagating through the fuel.

Flame propagation speed depends on the propensity of a fuel to enter into the oxidation reaction and on the heat transfer through the mixture. Table 6.2 indicates the flame propagation rates of various gases in their stoichiometric range. Drysdale (1998) reports that the maximum burning velocity and flame propagation occurs for a slightly fuel-rich mixture. Therefore, the greatest increase in pressure and the potential for greatest damage for any given fuel is when it is slightly fuel rich, which is known as the *optimum mixture* for an explosion.

Gas Explosions

Fire safety personnel encounter two common types of explosions: mechanical and chemical. Mechanical explosions (or boiling liquid expanding vapor explosions [BLEVEs]) will be covered in Chapter 8. Chemical explosions are based on the generation of high-pressure gas from an exothermic chemical reaction. The most common chemical explosion is the combustion explosion, where the reaction is driven by the combustion of a hydrocarbon fuel mixed with either air or some other oxidizer. A combustion reaction can also involve a flame propagation through dusts, which will be discussed in Chapter 9.

The combustion reactions are classified as either **detonation** or **deflagration**, depending on the velocity of the flame front through the fuel/oxidizer mixture. Deflagrations are the propagation of a combustion zone at a velocity that is less than the speed of sound in the unreacted medium (NFPA 68, 2007). Detonation is the propagation of a combustion zone at a velocity that is greater than the speed of sound in the unreacted medium (NFPA 68, 2007). The major difference between the two is burning velocity and the associated increase in pressure. Deflagrations more commonly occur with hydrocarbon fuel and air mixtures that produce a slower burning velocity and lower pressure rise. Deflagrations are controlled by the rate of heat release in the flame and the propagation of the flame through the fuel mixture. Detonations typically occur with condensed phase fuels or hydrocarbon fuel and oxygen mixtures that produce faster burning velocities and higher pressure rises. (A detailed discussion of condensed phase fuels or explosives is beyond the scope of this book; see Thurman [2006] for more information.) Detonations are independent of the rates of heat-generating chemical reactions and are primarily driven by the shock wave or shock front that forms ahead of the flame. The shock wave causes an abrupt change in temperature and pressure due to a flow instability caused by speeds in excess of the speed of sound (Quintiere, 1998).

The shape of the pressure wave from an explosion under ideal conditions is spherical, moving outward from the point of ignition equally in all directions. Obstructions or vents may change this spherical shape by allowing a path for lesser resistance, which alters the pressure wave shape (see Figure 6.8).

detonation
- Instantaneous combustion or conversion of a solid, liquid, or gas into larger quantities of expanding gas, accompanied by heat, shock, and noise. Flame propagation in detonations is more than 3,300 feet per second.

deflagration
- Rapid burning, faster than open-air burning of the material but slower than detonation.

FIGURE 6.8 Ideal blast front spherical propagation.
Source: Adapted from Ron Hopkins.

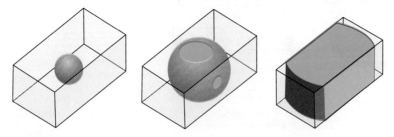

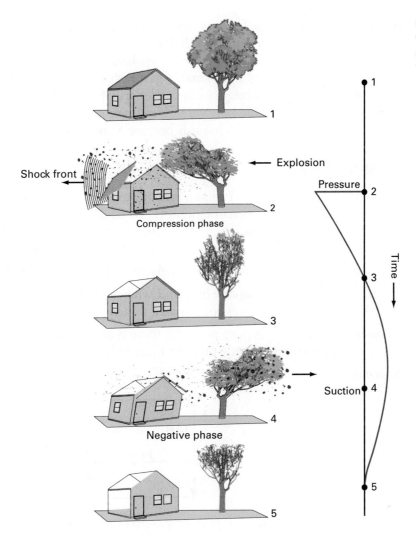

FIGURE 6.9 Blast wave generated by an explosion.

The blast wave generated by an explosion consists of a shock front or a positive pressure wave that rises almost instantaneously. It is followed by an expansion wave in which the pressure returns to its ambient value (Zalosh, 2003) (see Figure 6.9).

The entire volume of the confining structure does not have to be filled with gas within its flammable limit for a combustion explosion to occur. Lemoff (2008) writes that most combustion explosions of conventional structures typically occur with less than 25 percent of the enclosure occupied by the flammable mixture.

Warning About Flammable Gases

Though not listed as a flammable gas, some substances can ignite. Flammable gases are defined as "[a]ny material that is a gas at 68°F (20°C) or less and 17.7 psi (101 kPa) of pressure and is ignitable at 14.7 psi (101 kPa) when in a mixture of 13 percent or less by volume of air; or has a flammable range at 14.7 psi (101kPa) of at least 12 percent regardless of the lower flammable limit (U.S. Department of Transportation, 2006). This indicates the ease of ignition for most chemicals; however, it does not preclude substances such as ammonia (anhydrous ammonia) from igniting. With an ignition temperature of 1,204°F (651.1°C) and the flammable range from 16 to 25 percent (a range of 9 percent) (Fertilizerworks.com, 2001), anhydrous ammonia is not classified as a flammable gas; rather, it is classified as a combustible gas, but it can burn with relative ease under specific conditions.

Use caution when evaluating gases. Determine the ignition temperature and flammable ranges. Do not dismiss hazards because a chemical is not listed as flammable.

EXAMPLE 6.3

A room (16 ft by 13 ft by 13 ft) has a propane gas leak. Assume no air exchanges and that the room is closed (no ventilation). (1) At what point can this room have an explosion? How much gas is required for this event to occur? (2) If the leak were at a rate of 0.05 ft³/s, how long would it take for the room to undergo an explosion?

Known: Volume of the room; gas involved; LFL and UFL (Table 6.2); leak rate.

Find: (1) The LFL of propane gas, and what volume of gas is required to reach the LFL; (2) the time it takes for the leak to achieve LFL.

Assumptions: The gas is pure propane, there is no leakage of gas out of the room, and 25 percent of the volume must be filled before the explosion can occur.

Schematic:

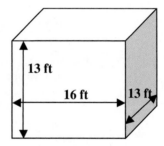

Analysis: Propane: LFL–UFL = 2.1−9.5 percent; room volume is 13 ft × 16 ft × 13 ft = 2,704 ft³; 25 percent of room volume is (2,704 ft³ * 0.25) = 676 ft³.

The gas required to reach LFL for the compartment: 25 percent volume * 2.1 percent = 676 ft³ * 0.021 = **14.2 ft³** gas.

The gas required to reach UFL for the compartment: 25 percent volume * 9.5 percent = 676 ft³ * 0.095 = **64.22 ft³** gas.

(1) Between **14.2 ft³** and **64.22 ft³** of propane gas is required for an explosion to occur.

(2) Time it takes to reach LFL: 14.2 ft³/0.05 ft³/s = 284 s/60 s = **4.7 min**; UFL = 1,284 s = **21.4 min**.

Summary

Hydrocarbons in the gaseous state can readily ignite and burn when the correct percentage is present with oxygen. This percentage varies according to ambient oxygen content. An increased oxygen percentage in the atmosphere facilities burning. When too much fuel or too much air is blended with the fuel, burning cannot occur. Ignition is initiated when the fuel/air mixture is heated to the autoignition temperature or when piloted ignition results from a portion of the vapor cloud experiencing minimum ignition energy.

When ignition of gaseous fuels occurs in confined areas, explosions can result. Explosions occur when the flame front moves through the fuel/air mixture at a rate of speed sufficient to result in a rapid pressure rise, thus overpressurizing the confining structure.

Review Questions

1. Define and describe matter in the gaseous state.
2. The maximum amount of fuel gas that burns in a mixture of air is called the:
   a. Stoichiometric mixture
   b. Lower flammable limit
   c. Upper flammable limit
   d. Autoignition temperature
3. When only trace amounts of fuel gas exist in the atmosphere and will not ignite, the mixture is:
   a. A Stoichiometric mixture
   b. Below the lower flammable limit
   c. Above the upper flammable limit
   d. At the autoignition temperature
4. Diagram a simple oxidation reaction. Balance the following oxidation reaction:

$$C_2H_6 + O_2 \rightarrow CO_2 + H_2O$$

5. **True or false:** The oxidation reaction is a very simple process that is accurately represented by balancing carbon, hydrogen, and oxygen molecules that enter the reaction with the carbon dioxide and water molecules that result.

6. Significantly lowering atmospheric temperature will likely change the burning characteristics of fuel gas by:
   a. Reducing the amount of vapors that evolve
   b. Reducing the autoignition temperature
   c. Increasing the amount of vapors that evolve
   d. Raising the lower flammable limit
7. Autoigintion temperature is the temperature at which:
   a. Gaseous fuels evolve sufficient vapors to ignite
   b. Gaseous fuel vapors ignite with a heat source
   c. Gaseous fuels reach their upper flammable limit
   d. Gaseous fuels ignite without an external heat source
8. Minimum ignition energy values are expressed:
   a. In watts
   b. In BTUs
   c. In joules
   d. As a percentage
9. If unaffected by wind currents, a fuel gas with a vapor density greater than 1:
   a. Remains at the source of vapor production
   b. Flows downward from the source

c. Flows upward from the source.

d. Is of no concern when it comes to vapor movement.

10. Ignition of a fuel/air concentration within a containing structure can result in:

a. A BLEVE

b. A detonation that destroys all contents within that structure

c. A deflagration that breaches the structure from overpressure

d. Will not damage the structure

PEARSON

myfirekit™

For additional review and practice tests, visit **www.bradybooks.com** and click on MyBradyKit to access book-specific resources for this text!

Register your access code from the front of your book by going to **www.bradybooks.com** and selecting the mykit links.

References

Babrauskas, V. (2003). *Ignition Handbook*. Issaquah, WA: Fire Science Publishers.

Drysdale, D. (1998). *An Introduction to Fire Dynamics*. Chichester, West Sussex, England: Wiley.

Fertilizerworks.com (2001). Retrieved February 21, 2009, from http://www.fertilizerworks.com.

Hopkins, R. (n.d.). *Explosion Investigation Presentation*. Sarasota, FL: National Association of Fire Investigations.

Lemoff, T. (2008). Gases (Section 6, Chapter 10) in *Fire Protection Handbook*. Quincy, MA: NFPA.

Meyer, E. (2010). *Chemistry of Hazardous Materials,* Fifth Edition. New York: Brady Fire, Pearson.

NFPA 68 (2007). *Standard on Explosion Protection by Deflagration Venting*. Quincy, MA: National Fire Protection Association.

NFPA 921 (2008). *Guide for Fire and Explosion Investigations*. Quincy, MA: NFPA.

Quintiere, J. (1998). *Principles of Fire Behavior*. Albany, NY: Delmar Learning.

Thurman, J. T. (2006). *Practical Bomb Scene Investigation*. Boca Raton, FL: Taylor and Francis Group.

U.S. Department of Transportation (2006). *Part 173-Shippers—General Requirements for Shipments and Packages*. Retrieved February 21, 2009, from www.fmcsa.dot/gov/rules-regulations/administration/fmcsr/fmcsrruletext.asp?chunkKeu=9001633480047559.

Zabetakis, M. (1965). *Flammability Characteristics of Combustible Gases and Vapors*. US Bureau of Mines, Bulletin 627.

Zalosh, R. (2003). *Industrial Fire Protection Engineering*. Chichester, West Sussex, England: Wiley.

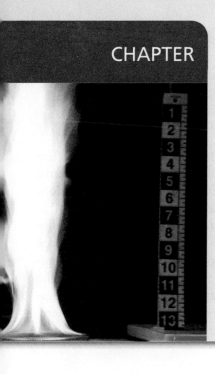

KEY TERMS

OBJECTIVES

After reading this chapter, you should be able to:

- Explain changes in matter that must occur for ignitable liquids to enter into an oxidation reaction.
- List and describe factors that differentiate flammable and combustible liquids.
- Explain the similarities and differences between an ignitable liquid's flash point and its fire point.
- Describe and explain the terms *vapor pressure* and *boiling point of ignitable liquids*.
- Describe and explain the ignition hazards associated with aerosols.
- Calculate the heat release rate of a pool fire for a given ignitable liquid.

Liquids that produce vapors that can undergo combustion with relative ease are considered ignitable liquids. It is important to remember that liquids do not burn; rather, the vapors produced by the liquids are actually what undergo combustion. Because the vapors ignite, the flammable limits and many of the same principles discussed in Chapter 6 are imperative to consider when dealing with the ignitability and flame spread of liquids. As the liquid increases in temperature, vapors are evolved from the surface of the liquid. Once these vapors are within their flammable limits, ignition can occur. Combustion of these vapors causes heating of the liquid mass, which causes the continued release of vapors that can enter the combustion reaction. Heating can come from an external source, ambient atmosphere, or radiant heat feedback from the flame to the fuel surface (see Figure 7.1).

Classification of Liquids

combustible liquid
▪ Any liquid that has a closed cup flash point at or above 100°F (37.8°C), as determined by the test procedures and apparatus.

flammable liquid
▪ Any liquid that has a closed cup flash point below 100°F (37.8°C).

To classify hazards associated with ignitable liquids, the U.S. Department of Transportation provides a system for identifying flammable or **combustible liquids**. **Flammable liquids** have a flash point of 100°F (38°C) or below, while liquids with flash points between 101°F and 200°F are classified as combustible liquids. The significance of proper flash point testing cannot be overstated. The flash point of ignitable liquids is often used as the main consideration when determining the relative danger of the liquid. Two standard flammability classification systems are in use today: *NFPA 30: Flammable and Combustible Liquids Code* (National Fire Protection Association) and the Federal Hazardous Substances Act enacted in 1990. Both systems provide uniform classifications for the labeling of ignitable liquids according to their flash point. The National Fire Codes, as promulgated by the National Fire Protection Association (NFPA), distinguishes classes of ignitable liquids as flammable or combustible, with a cutoff flash point of 100°F (37.8°C). Under National Fire Code, *NFPA 30: Flammable and Combustible Liquids*

FIGURE 7.1 Flame over a fuel, with vapors emitting from the surface and radiant energy applied to the fuel's surface.

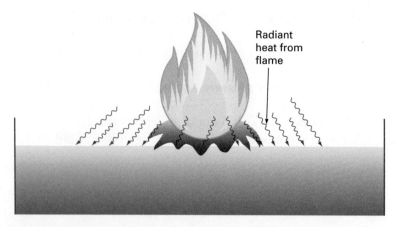

Radiant heat from flame

Code, there are six classifications for ignitable liquids as determined by their Tagliabue (TAG) closed cup flash points.

Flammable Liquids

> *Class IA*—liquids with flash points below 73° and boiling points below 100°F (22.8° and 37.8°C, respectively).
> *Class IB*—liquids with flash points below 73° and boiling points above 100°F (22.8° and 37.8°C, respectively).
> *Class IC*—liquids with flash points at or above 73° and below 100°F (22.8° and 37.8°C, respectively)

Combustible Liquids

> *Class II*—liquids with flash points at or above 100° and below 140°F (37.8° and −60.5°C, respectively).
> *Class IIIA*—liquids with flash points at or above 140° and below 200°F (60.5° and 93.3°C, respectively).
> *Class IIIB*—liquids with flash points at or above 200°F (93.3°C).

The Federal Hazardous Substances Act enacted in 1990 is part of the same federal law that created the Consumer Products Safety Commission. It distinguishes classes of ignitable liquids as extremely flammable, flammable, or combustible. The Hazardous Substances Act currently lists three classifications of flammability, as determined by their Setaflash Closed Cup tester.

> *Extremely flammable*—liquids with flash points at or below 20°F (−6°C).
> *Flammable*—liquids with flash points above 20°F (−6°C), and below 100°F (38°C).
> *Combustible*—liquids with flash points at or above 100°F (38°C), up to and including 150°F (65°C).

Liquids with flash points above 150 degrees Fahrenheit are not classified as hazardous substances under the Act.

Numerous state and federal regulations and industry standards deal with the appropriate labels and warnings on flammable or combustible liquid products, based on these classification systems. Therefore, it can be deduced that the classification, labeling, and warning of a material's relative flammability danger is established solely by its flash point. A more detailed discussion of the hazards and safety concerns for mislabeled materials can be found later in this chapter.

An additional hazard classification comes from the National Electric Code (NFPA 70). Some chemicals that are within the criteria specified for Class 1, Division 1 possess ignition characteristics that pose abnormal dangers for ignition. An example is carbon disulfide (CS^2), which has an autoignition temperature of 212°F.

Liquid Ignitability

FLASH POINT AND FIRE POINT

Liquids volatize (change to vapor) when sufficient energy is present to cause conversion. A common example is water left uncovered; the water evaporates into the atmosphere over time. The rate of conversion is the main consideration when

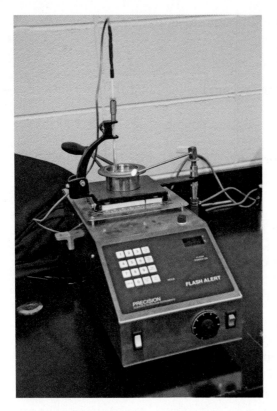

FIGURE 7.2 Open cup flash point test apparatus.
Source: Photo by authors.

FIGURE 7.3 Closed cup flash point test apparatus.
Source: Photo by authors.

latent heat of vaporization
■ The amount of heat required to change a liquid to a vapor.

differentiating the major classes of liquids. The energy required for evaporation is known as the **latent heat of vaporization.**

The flash point of a liquid is the lowest temperature of a liquid at which the liquid gives off vapors at a sufficient rate to support a momentary flame across its surface (source). Flash points are determined by specific laboratory test protocols that produce a momentary flash of flame across the surface of the liquid (see Figures 7.2 and 7.3). There are several different types of flash point tests. Each test method may produce slightly different flash points for the same liquid. When reporting the flash point, it is important to specify which test was used.

The test apparatus fall into two categories (determined by the construction of the test chamber): open cup or closed cup. In the open cup tests, the vapors are totally exposed to the atmosphere (Figure 7.2). In the closed cup tests, the test vapors are confined within the test apparatus, and the test igniter flame is introduced into the vapors near the surface of the liquid through a small mechanical door (Figure 7.3).

Closed cup tests involve placing a small quantity of liquid into a cup that rests in a water bath to facilitate even heating. The cup is then covered with a device that allows introduction of a small pilot flame periodically to a shutter opening on top of the cover. A heating unit is adjusted to raise the liquid's temperature slowly. The operator periodically records liquid temperature, then rotates a knob that opens the shutter as the pilot flame approaches. The temperature recorded immediately before a flash fire in the cup is noted as the flash point. Careful observation of the closed cup test reveals increased luminance as the pilot flame reaches the shutter opening just before ignition. This luminance is caused by the ignition of fuel gases that have not yet reached concentration to flash.

The open cup test apparatus functions much the same as the closed cup tester; however, no cover is present above the liquid surface. Thus, you must be careful to prevent dilution of the vapors emitted as the liquid heats. As with the closed cup test, the operator records liquid temperature periodically, then introduces a pilot flame across the surface. The **flash point** is recorded as the temperature that occurs immediately before a flash fire is witnessed. Open cup flash points are typically 3° to 9°C (6° to 16°F) higher than the closed cup flash points (Zalosh, 2003). If the open cup test continues with higher liquid temperatures, the sample will reach a point when it will continue to burn after the pilot flame is removed, which is known as the material's fire point. The **fire point** is the temperature to which the substance must be raised to produce sufficient vapors to sustain burning after the initial flash. Fire points are typically 5° to 31°C (9° to 56°F) higher than the corresponding closed cup flash points. Five main test apparatus designs are in general use for flash point testing: TAG (short for Tagliabue, the last name of the person who developed the test apparatus) closed cup, TAG open cup, Cleveland open cup, Pensky-Martens closed cup, and Setaflash (rapid tester).

Although the individual flash point tests differ by the apparatus and test protocol, the same basic method is used in all tests to determine the flash point (see Table 7.1). The temperature of the liquid specimen is gradually increased in a controlled manner, and a small gas ignition flame is introduced into the vapor space just above the surface of the liquid. When a momentary flash of flame is transmitted within the vapors across the surface of the liquid specimen, the temperature is recorded. The lowest temperature of the liquid at which this flash occurs is the flash point. This recorded temperature is then adjusted to the standard sea level atmospheric pressure of 760 mmHg (1 atm).

Test method selection is based on considerations such as the maximum temperature capabilities of the various pieces of equipment, the particular properties of the liquid (i.e., high viscosity), or the specifications of the code with which the liquid must comply. Although both closed and open cup tests yield valid data, differences in values are common. Closed cup testing generally results in lower flash point temperatures than does open cup testing. The reason is that vapors are constrained in the vessel as heating occurs and thus do not disperse in air currents as easily as those from open cup tests.

Flash point tests and fire point tests are good relative measures of a liquid's ignitability; however, a single test such as a flash point should not be the only source for determining the definitive flammability danger of a material. Manufacturers and/or suppliers of such materials should be held accountable for reviewing the process and handling conditions of the material and designate any additional testing required. For instance, if a material generates no distinguishable flash point, the manufacturers and/or suppliers should review other physical characteristics of the material to aid in correctly distinguishing the relative flammability danger of the material.

Improper labeling or warning of a material's relative flammability danger based on inaccurate flash points or improper flash point testing has been the main issue in many product liability lawsuits. The responsibility for a fire or explosion

flash point
■ The temperature at which the liquid and solid states of a substance coexist at one atmosphere or 101.3 kPa.

fire point
■ The lowest temperature at which a liquid ignites and achieves sustained burning when exposed to a test flame.

TABLE 7.1 — Flammable/Combustible Liquid Ignitability Parameters

LIQUID	CLOSED CUP FLASH POINT (F)[a]	OPEN CUP FLASH POINT (F)	FIRE POINT (F)	BOILING POINT (F)[a]	AUTOIGNITION TEMPERATURE (F)[a]
Acetone	0	15	—	133	869
Acrylonitrile	23	32	—	171	898
Benzene	12	—	—	176	1040
n-Butanol	84	110	110[c]	243	689
Carbon disulfide	−22	—	—	115	194
Cyclohexane	−4	—	—	179	473
n-Decane	115	131	142[c]	345	410
Ethanol	55	71	—	173	685
Ethyl ether	−49	—	—	95	320
Ethylene glycol	232	240	—	387	752
Fuel oil—no. 2	255[b]	—	264[b]	—	500[b]
Fuel oil—no. 6	295[b]	—	351[b]	—	629[b]
Gasoline[e]	−46	—	—	Varies	536
Glycerol	320	350	—	554	698
Heptane-n	25	30	—	198	536
Hexane-n	−7	−14	—	156	437
Isopropanol	53	60	—	181	750
Jet A fuel[d]	100–150	—	—	335–570	>435
JP-4 jet fuel[d]	−10 → +30	—	—	140–455	>445
Kerosene	>100	—	—	304–574	410
Methanol	54	60	—	148	725
Methyl ethyl ketone	28	34	—	175	960
Methyl salicylate	214	—	—	432	850
Mineral spirits	104	—	—	300	473
Motor oil	420[b]	—	435[b]	—	690[b]
Naptha, V.M. & P	28	—	—	211–320	450
Phenol (carb. acid)	175	185	—	358	817
Styrene	90	100	—	293	914
Toluene	40	45	—	231	1026
Turpentine	95	—	—	300	488
Vinyl acetate	18	30	—	163	801
Xylene-o	63	75	—	291	867

[a]Closed cup flashpoint, boiling point, and ignition temperature data are from NFPA 325M-1977 unless otherwise noted
[b]Data from Modak, EPRI NP-1731, 1981
[c]Data from Glassman and Dryer (1980)
[d]Jet A and Jp-4 data from AGARD-AR-132-Volume 2
[e]Data from NFPA 497-1997

incident may well rest with the manufacturer or supplier of such a liquid if the ultimate user had not been sufficiently warned of the product's danger. Responsibility could also be attributed to the current literature and test standards that fail to adequately address the effects of the outgassing phenomenon.

IGNITION CONCEPTS

Ignition temperature is the heat energy required for autoignition of flammable vapors released by the liquid. Remember that autoignition is the ignition of a material in the absence of an ignition source. The autoignition temperature of liquids is well above the boiling point temperature of the liquid; therefore, the temperature recorded is actually the temperature required for spontaneous ignition of the vapor/air mixture. Determination of the ignition temperature can be determined by raising the temperature of the liquid in a controlled environment until ignition occurs. This test is conducted inside an autoignition temperature apparatus, known as the Setchkin test apparatus, which consists of a furnace, test flask, and instrumentation (see Figure 7.4). In this test, no pilot flame is introduced; rather, ignition occurs when temperatures throughout the vapor mixture are sufficient to initiate an exothermic reaction great enough to initiate combustion. The temperature at which flaming combustion begins is recorded as the autoignition temperature.

Vapor pressure describes the pressure exerted by vapors leaving the liquid's surface. Vapor pressure is recorded for a specific temperature and pressure. When the pressure of vapors being emitted from a liquid's surface exceeds the atmospheric pressure, the substance's **boiling point** is recorded in terms of temperature. Water

vapor pressure
■ The pressure exerted within a confining vessel by the vapor of a substance at equilibrium with its liquid. A measure of the substance's propensity to evaporate.

boiling point
■ The temperature at which the vapor pressure of a substance equals the average atmospheric pressure.

FIGURE 7.4 Setchkin (autoignition) test apparatus.
Source: Photo by authors.

TABLE 7.2	Properties of Some Common Liquids		
LIQUID	**SPECIFIC GRAVITY [68°F (20°C)]**	**MISCIBLE/IMMISCIBLE WITH WATER**	**FLAMMABLE/ NONFLAMMABLE**
Acetone	0.79	Miscible	Flammable
Carbon disulfide	1.26	Immiscible	Flammable
Chlorobenzene	1.11	Immiscible	Flammable
Cyclopentanone	0.95	Immiscible	Flammable
2-Ethylhexanol	0.83	Immiscible	Flammable
Heptane	0.68	Immiscible	Flammable
Hydrochloric acid	1.18	Miscible	Nonflammable
Methyl acetate	0.97	Miscible	Flammable
Methyl ethyl ketone	0.81	Miscible	Flammable
Sulfuric acid	1.84	Miscible	Nonflammable
Trichlorofluoromethane	1.49	Immiscible	Nonflammable

Source: Meyer, *Chemistry of Hazardous Materials*, Fifth Edition, Brady, Table 2.4, page 59.

vapor density
■ The ratio of the average molecular weight of a given volume of gas or vapor to the average molecular weight of an equal volume of air at the same temperature and pressure. Also known as *specific gravity of gas*.

specific gravity
■ The ratio of the average molecular weight of a given volume of liquid or solid to the average molecular weight of an equal volume of water at the same temperature and pressure.

solution
■ A homogeneous mixture of two or more substances.

has a boiling point of 212°F (100°C) in normal atmospheric conditions (1 atm at 20°C). An increase in pressure results in higher temperatures to facilitate boiling. Conversely, at lower atmospheric pressure (i.e., in a high-altitude location like Denver, Colorado), the boiling point (temperature) is lower.

The terms **vapor density** and *flammable range* were described in Chapter 6; however, they apply to vapors produced by ignitable liquids. **Specific gravity** is a comparison of liquids to water. Water has a density of 1 kg/L (8.43 lbs/gal). Substances that have greater density (more mass per unit of volume) tend to sink in water; those with less density (lower mass per unit of volume) float on water (see Table 7.2).

Miscibility and *solubility* are terms used interchangeably to describe the tendency of materials to become **solutions** or to retain their initial characteristics. Chemicals that are miscible or soluble in another substance become part of the mixture but are not chemically combined; an example is salt in water. Ocean water contains dissolved salt, but the salt (solute) can be separated from that water (solvent) without chemical change of either substance. Another common example is the paint used on the walls of a home. After the application of paint on the wall, the solvent volatizes (evaporates), leaving the solute.

The miscibility of liquids depends mostly on their electrical polarity, and they are called polar or nonpolar solvents. Water and alcohols are examples of polar solvents. Petroleum hydrocarbons are examples of nonpolar solvents. Simple experiments reveal that water and oil do not mix: Water is polar and oil is nonpolar. No matter what you do, the oil/water mixture separates when mechanical agitation ceases (see Figure 7.5). Yet water mixes with alcohols, each dissolving and diluting the other. The International Association of Fire Chiefs has published information on their website (www.iafc.org) for training firefighters regarding issues related to gasoline blended with ethanol.

FIGURE 7.5 Water and oil.
Source: Photo by authors.

Combustible Liquids

High flash point–high boiling point liquids are less likely to ignite under typical atmospheric conditions. However, when the ambient temperature is elevated, those liquids typically classified as combustible are prone to act similarly to flammable liquids. Examples include diesel fuel (fuel oil number 2), with a flash point of 120°F dispersed on asphalt pavement during hot summer days. External heating can raise the fuel above its flash point, and thus sufficient vapors would be present to sustain flame should the minimum ignition energy be introduced.

MIXTURES

Characteristics associated with ignitable liquids are relatively simple to understand when a single element or compound is involved; mixtures, however, are more difficult to understand. Flammability characteristics of the most volatile element or compound in the mixture may persist. These characteristics become the danger indicator for the entire volume of mixture.

AEROSOLS

Finely divided particles of liquid dispersed in air are known as **aerosols**. Unlike the requirement for heating a significant mass of liquid to the flash point, smaller particles lose less energy to surrounding mass and thus convert to vapor with greater ease. Once ignition of a few aerosol droplets occurs, flame propagation can ensue.

aerosol
- A fine mist or spray containing minute particles.

Aerosols are not always from intended sources, such as those commonly found in commercial products like pressurized cans and spray nozzles. Often, high-pressure fluid transfer systems experience small ruptures, which provide an opportunity for small particles to be released in aerosol form.

THIN FILM

Combustible liquids that are applied in a very thin film over heated surfaces may ignite at lower temperatures than those commonly listed for the liquid. Ignition temperature tends to become lower because of the heat loss mass present when the thickness of the product is greater.

WICKING

Wicks are commonly constructed from organic materials that allow the liquid from a bulk mass (i.e., a pool) to rise through thin tubes of the organic structure, commonly known as capillary action. As the combustible liquid begins to rise from the pool of liquid through the wick, its mass lessens and the exposed surface area of the liquid increases. Combustible liquids that are brought up through wicks require a lower ignition energy than when they are part of pools. Wicks may be purposely designed as part of a lamp, or they may be formed through the wicking action present in vegetation when liquids are released into the environment. Whether the wicking action is intended or not, thin layers of the liquid reach an area where they change to vapor without heating additional liquid mass; thus, some fuels that would not normally ignite in pools will burn readily from the wick. An example is found in kerosene lamps and heaters. At typical ambient temperatures, the fuels do not vaporize sufficiently to burn, but fuels brought into a wick burn easily.

POOL FIRES

The heat release rate of fires involving burning liquids is derived by calculating the potential energy of the fuel and the surface area exposed to air. When both values are known, deriving the heat release rate is relatively simple. Ignitable liquids that exhibit a specific shape are considered pool fires. Pools may have physical restrictions that establish boundaries, but they also include those situations when liquids are spilled without restriction. When liquid flow is not an issue, the pool is well defined; however, liquids that are still flowing may increase the surface area of the pool.

The burning rate is first determined by identifying the burning characteristics for specific fuels and the pool area.

$$\dot{m} = A_p \dot{m}''$$

Where:

A_p = pool surface (m^2 or ft^2)
$\dot{m}''$ = the burning flux of the pool (kg/m^2-s or lb/ft^2-s)

The heat release rate (Q) is then determined by multiplying the fuel's heat of combustion (ΔH_C) by the burning rate (m).

$$\dot{Q} = \Delta H_C \dot{m}$$

Where: $\dot{Q}$ = heat release rate (kJ/s which is reported as kW)
ΔH_C = heat of combustion (kJ/kg)
$\dot{m}$ = mass burning rate (kg/s)

APPLICATION

The burning rate can be helpful when identifying a strategy and tactics for combating liquid fuel fires. The application of water can raise the liquid level in tanks because water is more dense and not miscible with petroleum fires. Thus, the application of water, even foam, may cause burning fuel to overflow the confines of its vessel. Fuel removal may be the best option. An assessment of the burning rate can assist fire professionals in determining the anticipated duration, and calculating the HRR can lead to determining the heat flux that will be imposed on surrounding surfaces. When fire damage to surrounding items is less important, the best option may be fuel depletion.

BOILOVER, FROTHING, AND SLOPOVER

In large tanks that are storing unrefined petroleum or heavy petroleum products (i.e., combustible liquids), a fire on the surface can heat water and other contaminants in and through the vessel. Water or foam is used to fight fires, and water sinks below the fuel surface. Heated fuel causes water to boil, which results in frothing. The bubbles of the ignitable liquid burn easily. Expansion at upper regions of the fuel can also expand sufficiently to cause fuel to eject from the tank. The worst-case scenario, **boilover**, results when heat transfers downward through the fuel to the water in the tank's bottom. Temperatures can be well above water's boiling point, and sudden heating of the water causes rapid expansion of the liquid to vapor (steam). The steam forces fuel out of the tank. Boilover is a violent occurrence, resembling an eruption from the vessel. One such incident occurred in Tacoa, Venezuela, on December 19, 1982, where boilover of a fuel storage tank sent burning liquid downhill toward an occupied area, killing more than 150 people and destroying nearby structures.

boilover
■ The phenomenon associated with the production of steam during a crude petroleum fire; the steam drives the burning petroleum up and over the walls of its confining tank.

Summary

This chapter reviewed characteristics associated with liquids that burn. Understanding these characteristics is important for those involved in preventing and handling fires involving liquids that burn. This understanding will help you anticipate conditions that can result in danger to employees and firefighters.

Review Questions

1. Flammable liquids have a flash point that is:
   a. Less than 100°C
   b. More than 100°C
   c. Less than 100°F
   d. More than 100°F
2. Combustible liquids in aerosol burn more easily than the same liquid in pools because:
   a. The droplets have a lower flash point
   b. The droplets have a higher surface-to-mass ratio
   c. The droplets have a lower surface-to-mass ratio
   d. There is no difference in burning
3. The fire point of an ignitable liquid is:
   a. Higher than the flash point
   b. Lower than the flash point
   c. The same as the flashpoint
   d. Applicable only to flammable liquids
4. **True or false:** The flash point is the only indicator of the dangers associated with a flammable liquid.
5. The specific gravity of combustible and flammable liquids is generally:
   a. 1
   b. Less than 1
   c. Greater than 1
   d. They have no specific gravity.
6. Vapor density (specific gravity of vapors) in combustible and flammable liquids is
   a. 1
   b. Less than 1
   c. Greater than 1
   d. Vapor density does not apply to these liquids.
7. The boiling point of a liquid refers to:
   a. 212°F
   b. The temperature where vapor pressure is equal to the atmospheric pressure.
   c. Nothing related to ignitable liquids
8. **True or false:** The ignition temperature and flash point are the same for all liquids.
9. The flash point temperature recorded by an open cup tester generally is:
   a. Exactly the same as a closed cup
   b. Higher than a closed cup
   c. Lower than a closed cup
10. **True or false:** The autoignition temperature of vapors from ignitable liquids indicates a spark initiated the combustion.

References

NFPA 30. (2008). *Flammable and Combustible Liquids Code*. Quincy, MA: NFPA.

NFPA 70. (2008). *National Electric Code*. Quincy, MA: NFPA.

Zalosh, R. (2003). *Industrial Fire Protection Engineering*. New York, Wiley.

8
Solid Combustion

KEY TERMS

cellulosic fuels, *p. 117*

char, *p. 115*

dehydration, *p. 117*

fire retardant, *p. 127*

heat of gasification (L_g or L_v), *p. 114*

melting, *p. 116*

moisture content, *p. 117*

polymer, *p. 113*

pyrolysis, *p. 113*

smoldering combustion, *p. 116*

surface area–to–mass ratio, *p. 122*

thermal inertia, *p. 123*

OBJECTIVES

After reading this chapter, you should be able to:

- Explain thermal decomposition in solids.
- Explain and describe pyrolysis.
- Distinguish between and categorize different types of solid fuels and their fuel characteristics.
- Identify and explain the variables affecting solid combustion.

PEARSON
myfirekit™

For additional review and practice tests, visit **www.bradybooks.com** and click on MyBradyKit to access book-specific resources for this text!

Solid is matter that has a definite shape and volume. The solid material is essentially a collection of molecular bonds in a distinct structure. When heat is applied to a solid material, the temperature begins to increase at a rate dependent on the material properties and the heat applied. As heat is applied and the temperature begins to increase, the energy may be great enough to cause the molecular bonds to break down into lower molecular weight. The breakdown or decomposition of the molecular bonds by thermal energy is better known as thermal decomposition. Depending on their composition, solids are expected to undergo one or more of the following changes when exposed to a sufficient heat source: melt, dehydrate, char, and/or smolder.

The majority of solids undergo a chemical decomposition, better known as **pyrolysis**, which is the chemical decomposition of the solid fuel by the application of a heat source (see Figure 8.1). As the molecular bonds begin to break down, smaller molecules (lower molecular weights) are released in the gaseous form; they will ignite if they are released from the solid at a sufficient rate that the flammable limit is reached (see Figure 8.2). Most solids undergo this process of pyrolysis when they are exposed to a sufficient heat source; however, some may melt first and vaporize like a liquid. For example, cellulosic materials (i.e., wood, paper) and thermoset **polymers** undergo pyrolysis and leave behind a carbonaceous residue (char) in the presence of a sufficient heat source, but thermoplastics (soda bottle containers) melt when exposed to heat and vaporize, similar to liquids.

pyrolysis
- The chemical decomposition of a compound into one or more other substances by heat alone; pyrolysis often precedes combustion.

polymer
- High molecular weight substances produced by the linkage and cross-linkage of its multiple subunits (monomers).

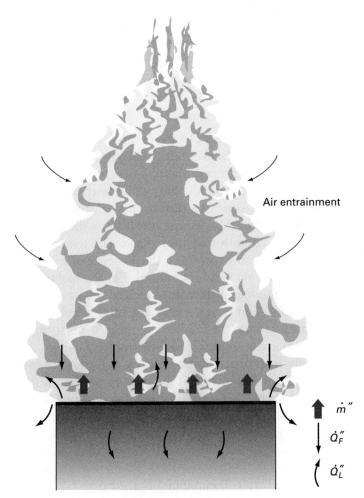

Air entrainment

$\dot{m}''$

$\dot{Q}_F''$

$\dot{Q}_L''$

FIGURE 8.1
Schematic representation of a burning surface, showing the heat and mass transfer processes: $\dot{m}$ = mass loss rate from the surface; $\dot{Q}_F$ = heat flux from the flame to the surface; $\dot{Q}_L$ = heat losses (expressed as a heat flux from the surface).
Source: Data derived from Drysdale, D. (1998).

A few rare solids (for example, methenamine) can transition directly from a solid to a gas, known as sublimation, when heated.

Solids can also on occasion burn directly, without vaporization/pyrolysis, by smoldering or glowing combustion involving direct oxidation at the surface of the material.

Pyrolysis

Most solids must gasify before they can ignite and burn. The process of changing state from a solid to a gaseous form by means of chemically decomposing molecular bonds is known as pyrolysis. Pyrolysis is an endothermic process. Solids typically consist of a complex matrix of molecular bonds that, when heated, begin to break down, releasing smaller chains of hydrocarbons (lower molecular weights) in a gaseous form. Once the rate and amount of this gaseous fuel leaving the fuel surface exceeds the lower flammable limit, ignition may take place. The temperature and pyrolysis rate needed for ignition to occur depends on the presence or lack of a competent ignition source (piloted ignition versus autoignition). For the fuel to sustain combustion, the heat returning to the fuel surface must exceed the heat that is lost into the fuel mass via conduction and that which is reflected off the fuel source.

The molecules being lost (or mass lost from the solid) are in the gaseous form, so a certain amount of energy had to enter the solid to cause the molecular bonds to break down. This energy is known as the **heat of gasification (L_g or L_v)**. The heat of gasification is used to describe the amount of energy that is required to produce a unit mass of flammable vapor from a combustible that is initially at ambient temperatures (Drysdale, 2008). This value is obtained from experimental means and is given in units of kJ/g, representing the amount of energy required to produce a unit mass of volatiles. Liquids have lower heats of gasification than solids because it require less energy to break the molecular bonds of the liquids than those of the solids. Consequently, lower heats of gasification also pose a greater flammability hazard (see Table 8.1).

heat of gasification (L_g or L_v)
- The quantity of heat required to cause a mass unit of solid or liquid to convert to the gaseous state; expressed in J/g or BTU/lb. Also known as the heat of vaporization.

	L_v (kJ/g)	$\dot{Q}_F''$ (kW/m²)	$\dot{Q}_L''$ (kW/m²)	$\dot{m}_{ideal}''$ (g/m² · s)
COMBUSTIBLES				
FR phenolic foam (rigid)	3.74	25.1	98.7	11[a]
FR polyisocyanurate foam (rigid, with glass fibers)	3.67	33.1	28.4	9[a]
Polyoxymethylene (solid)	2.43	38.5	13.8	16
Polyethylene (solid)	2.32	32.6	26.3	14
Polycarbonate (solid)	2.07	51.9	74.1	25
Polypropylene (solid)	2.03	28.0	18.8	14
Wood (Douglas fir)	1.82	23.8	23.8	13[a]
Polystyrene (solid)	1.76	61.5	50.2	35
FR polyester (glass-fiber reinforced)	1.75	29.3	21.3	17
Phenolic (solid)	1.64	21.8	16.3	13
Polymethylmethacrylate (solid)	1.62	38.5	21.3	24
FR polyisocyanurate foam (rigid)	1.52	50.2	58.5	33
Polyurethane foam (rigid)	1.52	68.1	57.7	45
Polyester (glass-fiber reinforced)	1.39	24.7	16.3	18
FR polystyrene foam (rigid)	1.36	34.3	23.4	25
Polyurethane foam (flexible)	1.22	51.2	24.3	32
Methyl alcohol (liquid)	1.20	38.1	22.2	32
FR polyurethane foam (rigid)	1.19	31.4	21.3	26
Ethyl alcohol (liquid)	0.97	38.9	24.7	40
FR plywood	0.95	9.6	18.4	10[a]
Styrene (liquid)	0.64	72.8	43.5	114
Methylmethacrylate (liquid)	0.52	20.9	25.5	76
Benzene (liquid)	0.49	72.8	42.2	149
Heptane (liquid)	0.48	44.3	30.5	93

Source: Data determined by Tewarson and Pion (1976).
[a]Charring materials. $\dot{m}_{ideal}''$ taken as the peak burning rate.

Char

char
■ Carbonaceous material resulting from pyrolysis, often appearing to be blackened blisters, on the surface of cellulose and other solid organic fuels.

Char is a carbonaceous material that has been burned or pyrolyzed and has a blackened appearance (NFPA 921, 2008). Solid organic compounds (for example, cellulosic materials, wood products, thermoset plastics) form a layer of char as the fuel is pyrolyzed. As these materials are heated, both fuel and water molecules are

FIGURE 8.3 Examples of smoldering combustion.

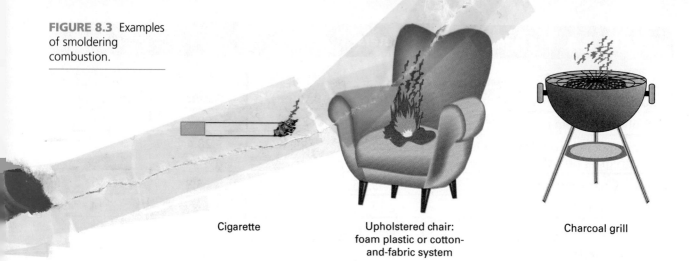

Cigarette

Upholstered chair: foam plastic or cotton-and-fabric system

Charcoal grill

liberated, leaving behind mostly a solid carbon residue. The char layer serves as an insulating layer between the virgin material (not yet pyrolyzed) and the heat source. Therefore, when an item forms a char layer, it increases the energy required to break down the existing fuel lying behind the now insulating layer of char. Burning rates slow down because of this behavior. For this reason, people use fire pokers in fireplaces to knock off the char layer on burning logs and thus expose the virgin wood beneath, which in turn allows for better combustion of the fuel.

SMOLDERING COMBUSTION

smoldering combustion
■ Combustion without flame, usually with incandescence and smoke. Also known as *nonflaming combustion*.

Smoldering is defined as combustion without flame, usually with incandescence and smoke (NFPA 921, 2008) (see Figure 8.3). Drysdale (1998) states that the "principal requirement for smoldering is that the material must form a rigid char when heated" (p. 287). It has been shown that **smoldering combustion** typically consists of three distinct zones. Zone 1 is the pyrolysis zone, where the temperature rise is steep and volatiles are released to the atmosphere. Zone 2 is the charred zone, where glowing combustion is occurring with no volatiles being released. Temperatures in this zone are at their highest, reaching between 600° and 750°C. Finally, zone 3 consists of the area that already underwent smoldering combustion; temperatures are low; and the fuel is exhausted, with only residual ash remaining (see Figure 8.4).

Melting

melting
■ Changing the physical state of matter from a solid to a liquid.

Melting is a physical change of the substance. It does not change the chemical makeup of the materials, just the physical form. Melting causes a solid to form a liquid pool of the same material, just in a different form. When exposed to heat, thermoplastic solids soften and melt prior to giving off vapors and igniting. Even though thermoplastics are solids, once heated, the material melts into a liquid pool and behaves from that point as a liquid fuel. This means that enough heat must be imposed on what is now a liquid to vaporize enough vapors for ignition and sustained combustion.

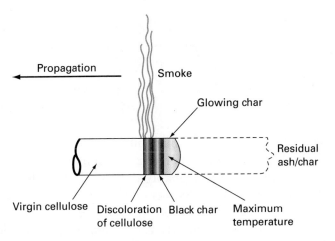

FIGURE 8.4
Representation of steady smoldering along a horizontal cellulose rod.
Source: Data derived from Drysdale, D. (1999).

Zone 1 Pyrolysis zone in which there is a steep temperature rise and an outflow of visible airborne products from the parent material.

Zone 2 A charred zone where the temperature reaches a maximum, the evolution of the visible products stops, and glowing occurs. Heat is released in Zone 2, where surface oxidation of the char occurs. Zone 2 is where the temperature reaches a maximum, typically in the region of from 600° to 750° for smoldering.

Zone 3 A zone of very porous residual char and/or ash that is no longer glowing and whose temperature is falling slowly.

Dehydration

Many solids have water inherent in their structures (absorbed or adsorbed); some solids have water molecules chemically bound into their molecular structures. As heat is exposed to one of these solids, the water molecules increase in temperature and may vaporize out of the solid in a process called **dehydration**. This property of water is what makes some solids good fire barriers. For instance, gypsum wallboard (i.e., drywall) consists of chemically bound water that, when exposed to heat, absorbs much of the energy by breaking the molecular bonds and vaporizing this water (i.e., dehydrating). This allows the material to prevent heat transferring quickly through the material itself and affecting the structural members behind the wall lining.

The burning characteristics of many solid fuels depend greatly on the amount of moisture contained within their structure, also known as the **moisture content**. The dryer the solid, or the less moisture inside the solid, the better the solid will burn. This property is very evident in the burning characteristics, including the flame spread and heat release rate, of **cellulosic fuels**.

dehydration
- Removal of water; in the fire prevention context, the removal of water from solid fuel through heating the fuel.

moisture content
- Amount of moisture intrinsic to an object, expressed in percentage of total mass for that object.

cellulosic fuels
- Hydrocarbon fuels that contain cellulose molecules, which are chains of molecules with the chemical formula of $C_6H_{10}O_5$.

Characteristics of Cellulosic Fuels

Cellulosic materials, including wood, paper, and cotton, are composed of cellulose molecules, which consist of molecules of glucose ($C_6H_{12}O_6$) that are chemically bonded together in long-chain polymers (see Figure 8.5). The percentage of cellulose in these materials depends on the type of material. For example, cotton and paper consists of high percentages of cellulose (90 percent), while wood contains between 40 and 50 percent cellulose.

FIGURE 8.5
Cellulose: a polymer of glucose units produced by plants.
Source: Data derived from Cholin (2008).

Wood is a major component in our built environment. It is used as a structural element in our buildings and in the construction of much of the furniture within our structures. Therefore, wood is probably the fuel that the fire safety professional will encounter most frequently. As mentioned above, wood consists of 40 to 50 percent cellulose; the remaining wood structure contains approximately 25 percent hemicellulose and approximately 25 percent lignin (Drysdale, 1998). Wood has many complex compounds combined within its structure, but because most woods are constructed primarily of these three molecules, a generalization can be performed for ignition and spread issues.

As wood is heated, the molecules begin to increase in temperature and the molecules begin to move faster. When the wood's temperature is increased high enough, both fuel and water molecules are liberated, leaving behind mostly a solid carbon residue. The char layer serves as an insulating layer between the virgin material (not yet pyrolyzed) and the heat source. Wood discolors and chars at temperatures between 200° and 250°C (392° and 482°F). Wood begins to break down rapidly at temperatures above 300°C (572°F) (Drysdale, 1998) (see Figure 8.6).

As mentioned previously, the moisture content within cellulosic fuels has a major impact on burning. In Chapter 9, we will discuss the fact that, when the moisture content is lowered by 10 percent in pine Christmas trees, there is a significantly different heat release rate: Both its peak and the rate required to reach the peak are affected. The tree with the lower moisture content peaks at 3.2 megawatts (MW) within 60 seconds, while the tree with the higher moisture content reached only 2 MW and took 90 seconds to reach its peak. The moisture within a solid can also affect how easily the radiant feedback from the flame or from an external heat source can influence the fuel receiving this heat. Typically, the moisture absorbs much of the heat that is applied to the solid. When the moisture is not present, then all of this heat can be absorbed directly by the fuel, which can instantly begin the pyrolysis process. In other words, when a solid dehydrates during heating, the pyrolysis process is slowed. It is reported that the wood structural members typically have moisture content below 10 percent, while "green"

FIGURE 8.6 Cross-section through a slab of burning, or pyrolyzing, wood.
Source: Data derived from Drysdale, D. (1999).

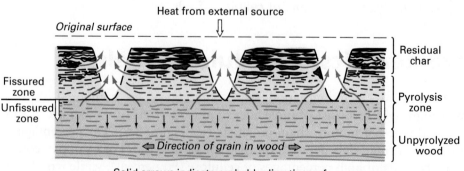

Solid arrows indicate probable directions of movement of volatile products.

wood may have up to 50 percent moisture content (Cholin, 2008). The following equation is a way to calculate the moisture content (MC) (Cholin, 2008):

$$MC = \left(\frac{\text{original weight} - \text{dry weight}}{\text{dry weight}} \right) \cdot 100\%$$

Characteristics of Upholstered Furniture

Upholstered furniture is of particular interest because of its high involvement and influence on a developing fire. Upholstered furniture, including sofas and over-stuffed chairs, is generally discussed in conjunction with mattresses, and both will be discussed here. The basic construction of upholstered furniture consists of a wood frame, interior padding supported by metal springs, and an exterior covering. This basic construction has not changed substantially over the past 70 years, but the materials used in their construction have changed drastically. Prior to 1970, the majority of furniture items were constructed with cotton padding and some form of fiber exterior covering. Since the 1960s, the application of polyurethane foams as an interior padding has steadily increased in the construction of upholstered furniture. In fact, the majority of furniture items today consist of some sort of polyurethane foam lining. Dan Madryzkowski from the National Institute for Science and Technology (NIST) once commented during a presentation that most furniture items today are blocks of polyurethane foam with seats carved into them and with a little bit of wood added to keep them from falling over. He was joking, but his description is frighteningly correct from a fire safety professional's perspective. The fire characteristics of polyurethane foam and its influence on fire growth rates are discussed briefly in this chapter and then in greater depth in Chapter 9.

Frames for furniture items today are still constructed of wood, metal, and some plastics. The choice of the frame construction material depends mostly on the manufacturer and the furniture item. Many different types of exterior coverings, including cotton, leather, or mixed-fiber fabrics, are used. The interior padding or cushion of the furniture item is the driving factor behind the burning characteristics (heat release rates, combustion by-products) of the fuel, while the exterior covering plays a major role in the ignition propensity of the item. Many flammability and combustibility tests have been developed to analyze upholstered furniture's fire characteristics, and these tests are discussed at length in the National Fire Protection Association (NFPA) *Fire Protection Handbook* (2008).

The important consideration for a fire safety professional when it comes to upholstered furniture is that furniture items consist of several different fuels. When these fuels are exposed to heating, they react and burn differently. Most important, polyurethane foam is a tremendous fuel, and often results in a fast-growing fire. Polyurethane foam melts and drips when exposed to heat and behaves similar to a liquid fuel.

Characteristics of Polymer Fuels

Polymers can be broken into three categories: thermosetting plastics, thermoplastics, and elastomers. Thermosetting plastics are those plastics that have been hardened into a defined shape during manufacturing; when they are heated, they do not typically soften or melt. These plastics char when exposed to heat. Thermoset plastics include phenolics and epoxies. Thermoplastics are not cured or hardened during the manufacturing process and soften and melt when exposed to heat. Common examples include acrylics, nylons, and polystyrene. Elastomers are those items that have been created to match the properties of natural rubbers; they include butadiene and neoprene. As a fuel, elastomers are typically difficult to ignite and burn, and they produce a dense smoke output. Tire fires are a perfect example.

Most plastics burn rapidly and produce high heat release rates and smoke output. When exposed to heating, thermoplastics melt and flow. This characteristic greatly influences the burning characteristics of the fuel, depending on its orientation. If the flowing now-liquid fuel is flowing away from the heat source, then ignition and flame spread may be mitigated. However, if the flow is toward the heat source, then the potential for ignition, faster flame spread, and energy release is increased. When exposed to heat, thermoset plastics form a char layer, similar to cellulosic fuels. This char layer also forms an insulating layer that slows the rate of burning.

Plastics are often formed into foams or expanded plastics. Foams, or cellular plastics, are created with pockets of air throughout the fuel (see Figure 8.7). This type of orientation greatly influences the burning and ignition characteristics of the fuel. Ignition is easier due to the change of the surface area–to–mass ratio. The plastic in a solid form has its molecules tightly packed, thus allowing for an easier path for heat to be conducted away from the surface through the rest of the mass. This property serves to delay the temperature increase at the surface, thus delaying ignition. However, when plastics are expanded with a matrix of air pockets dispersed throughout the fuel, this thermal property is voided. Instead, when a heat source is applied to the surface, the energy conducted into the material cannot be dispersed as easily and begins to increase the temperature at the surface much more easily, resulting in faster ignition. The density of this fuel/air matrix influences the ignition capability of the fuel and the ease of flame spread throughout the material.

FIGURE 8.7 Air/fuel matrix in cellular plastics.
Source: Photo by authors.

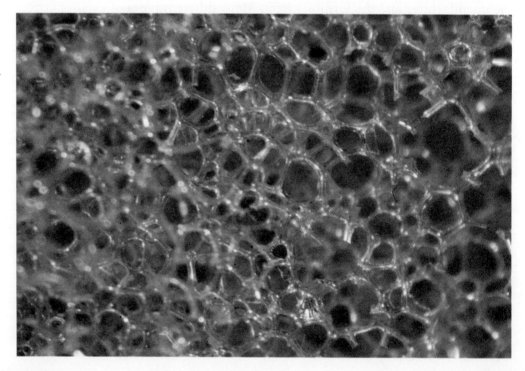

Characteristics of Combustible Metals

Nearly all metals burn under the right conditions. The most common metals considered combustible include calcium, hafnium, lithium, magnesium, niobium, plutonium, potassium, sodium, tantalum, thorium, titanium, uranium, zinc, and zirconium (Christman, 2008). Typically, a metal in a bulk mass is difficult to ignite, so the most common means to ignite and burn these metals is when they are in the form of dust, fine particles, or thin layers. Most metal dusts are extremely dangerous and if suspended, they can result in massive explosions. The temperatures associated with the burning of metals are typically higher than those temperatures found when burning normal combustibles.

The alkali metals, which include lithium, sodium, potassium, rubidium, cesium, and francium (see Figure 8.8), are highly reactive with water. The reactivity increases as you descend the periodic series; in other words, francium is the most reactive. As these metals are introduced to water or even the moisture in the air

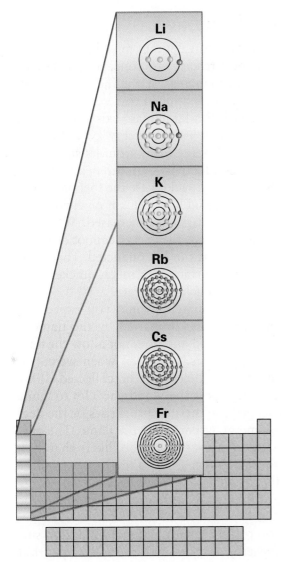

FIGURE 8.8 Alkali metal series from the periodic table.

(for the more reactive metals), an exothermic reaction occurs. Water's molecular structure is broken down into its elements (hydrogen and oxygen), which increases the reaction rate. Typically, the oxygen combines with the metal, and the hydrogen can burn. Small amounts (1 to 2 grams) of cesium and rubidium, when introduced to water, can result in explosive reactions, which in turn result in damage consistent with seated explosions. Other metals, most notably magnesium, also react violently when water is introduced. It is important to recognize these characteristics when choosing an extinguishing agent.

Variables Affecting Solid Combustion

The configuration, orientation, surface area–to–mass relationship, density, and thermal properties influence the ease of ignition and burning characteristics of solid combustibles. Greater detail regarding ignition characteristics will be discussed in Chapter 11, so the concepts will only be introduced here.

SURFACE AREA–TO–MASS RELATIONSHIP

surface area–to–mass ratio
■ The ratio of surface area to the total mass of an object. Thin objects have more surface per unit of mass than do thick objects comprised of like matter.

The **surface area–to–mass ratio** is an important concept that must be considered when analyzing ignition, flame spread, and heat release rate concepts. As the surface area increases and the mass is lessened (ratio increases), the fuel particles become smaller or more finely divided. When heat is applied to a solid or even a liquid, the heat is conducted into that material. If a large mass exists behind the surface, the material has a greater ability to dissipate the energy away from its surface. The ability of this material to conduct energy away from its surface relates to its thermal properties (i.e., thermal conductivity, specific heat) and the density of that material (i.e., how closely or loosely packed are the fuel particles), which is discussed later in the chapter. If the material has been cut or divided out of the large mass, then its surface area–to–mass ratio has increased. When the same heat as before is now applied to this lesser mass, the energy cannot dissipate as much due to the lack of mass into which the heat can be conducted. Therefore, the energy can increase the temperature at the surface at a much quicker rate, which may result in ignition of the material at a much faster rate.

For example, take a wood log and apply a match. Would you expect the wood to ignite? Most likely not because the energy imparted by the flame is being dissipated through the large mass of the wood and does not allow the temperature to increase to the point to cause pyrolysis. However, when we cut or sand this wood log, wood shavings and fine particles may be collected. If we apply the exact same type of match, do you expect the wood particles to ignite now? Bear in mind that we have not changed the thermal properties or the density of the wood, nor the temperature or energy imparted by the flame. The only thing that has changed is the surface area–to–mass ratio. More likely than not, the wood shavings will ignite. Now, there is less mass to allow for heat to be conducted away and dissipated from this localized location. Because the energy cannot be dissipated, the temperature of the wood can increase and enable the pyrolysis process.

The same principle applies when you try to ignite a material on a bulk surface area versus an edge. For example, when you try to ignite a piece of paper, it is much easier to ignite the paper on its edge rather than in the center. The material properties of the paper do not change from the center of the paper to its edge. The only characteristic that changes is the ability of the energy to be dissipated easily through the material. The center of the paper allows heat to be conducted in every direction away from the heat source, preventing a fast temperature rise. The edge of this paper has only certain directions that the heat can be conducted into the remaining mass, which delays the energy dissipation from the paper's surface.

Dusts

The effect of the surface area–to–mass relationship is most evident when looking at dusts and their explosion tendencies. Practically any combustible (including metals), when finely divided and suspended in the air as a dust cloud, can ignite and lead to an explosion. The concentration of the fuel has to be within its flammable limits for an explosion to occur. The concentration is most commonly listed from explosion test data as a required particle size within a specific volume. A list of explosion test data for common materials is found in Table 8.2. A nontechnical way of determining if the dust cloud is within its flammable limits is if it is visible to the naked eye (Hopkins, n.d.). At this point, the dust cloud is more than likely within its flammable limits.

The flame propagates in a dust explosion very similar to the reaction witnessed in a diffuse fuel (i.e., propane and air) explosion. Ignition occurs, and a flame begins to propagate through the fuel by the flame front heating the fuel. This raises the fuel to its ignition temperature and spreads through the fuel. The same fuel characteristics that affect burning solids (i.e., moisture content and particle size) influence the ignition and burning velocity of a dust explosion.

ORIENTATION

The orientation of the fuel has a major impact on how the fuel will behave once it is ignited. A fuel oriented vertically has its virgin fuel heated by the flowing heated gases via direct contact including both convection and conduction heat transfer, as well as a much greater radiant heat view factor associated with the parallel fuel and flame source. A fuel oriented horizontally has only the radiant feedback from the flame itself, which is approximately 30 percent. The position of the fuel anywhere in between the horizontal plane and the vertical plane, known as an inclined plane, also influences the flame spread capabilities. Orientation will be discussed in greater detail later in Chapter 11 because it relates strongly to the fire spreading across the material.

THERMAL INERTIA

Thermal properties that affect the rate of heat transfer through a material play a significant role in the ignition and spread of fire on a solid fuel. The most important of these properties include the density (ρ), specific heat (c_p), and the thermal conductivity (k). These three thermophysical properties for fire dynamics are typically more important when the product of the three properties are calculated and substituted into the fire formulas, which is termed the **thermal inertia**.

thermal inertia
- The properties of a material that characterize its rate of surface temperature rise when exposed to heat; related to the product of the material's thermal conductivity (k), its density (ρ), and its heat capacity (c).

TABLE 8.2 Explosion Test Data for Common Solids

DUST TYPE	PARTICLE SIZE (μm)	1 m³ OR C_{min} (g/m³)	20 LITER P_{max} [Bar (g)]	VESSEL K_{St} (BAR m/sec)	DUST IGNITION G.G. (°C)	CLOUD TEMPERATURE BAM (°C)	VDI MIE (mJ)	GLOW TEMPERATURE (°C)	FLAMMABILITY <250:M CLASS
Pharmaceuticals/ Cosmetics/Pesticides									
Acetyl salicylic acid		15	7.9	217		550			2(5)
Ascorbic acid, L+	14	60	6.6	48	490			Melts	2(2)
Ascorble acid	39	60	9	111	460			Melts	2(2)
Caffeine		30	8.2	165		>550		Melts	2(5)
Digitalis leaves	46	250	8.5	8.5	73				
Methionine	<10	30	8.7	128	390		100	Melts	5
Intermediate Chemical Products									
Adipinic acid	<10	60	8	97	580			Melts	2(5)
Anthracene	235	15	8.7	231	600			>450	
Anthrachinone	<10		10.6	364					
Calcium acetate	92	500	5.2	9	730			>460	2
Casein	24	30	8.5	115	560			>450	
Methyl cellulose	22		10	157	400		12	380	3
Methyl cellulose	37	30	10.1	209	410		29	450	5
Ethyl cellulose	40		8.1	162	330			275	5
Cyanoacrylicacid methyl ester	260	30	10.1	269	500			>450	5
Fumaric acid	215	–100							5
Cellulose ion exchange resin	<10	60	10	91	410			>450	5

Source: Data derived from Schwab, 2008.

DUST TYPE	PARTICLE SIZE (µm)	1 m³ OR C_{min} (g/m³)	20 LITER P_{max} [Bar (g)]	VESSEL K_{St} (BAR m/sec)	DUST IGNITION G.G. (°C)	CLOUD TEMPERATURE BAM (°C)	VDI MIE (mJ)	GLOW TEMPERATURE (°C)	FLAMMABILITY <250:M CLASS
Hexamethylene tetramine	27	30	10.5	286					
Melamine	<10	1000	0.5	1	>850			>450	2
Barium/lead stearate		15	8.1	180					2(2)
Paraformaldehyde	23	60	9.9	178	460			>480	5
Lead stearate	12	30	9.2	152	630			Melts	5
Calcium stearate	145	30	9.2	155	550		12	>450	
Sodium stearate	22	30	8.8	123	870			Melts	2
Zinc stearate	13	30			520		5	Melts	
Metal dusts									
Aluminum powder	<10	80	11.2	515	560			430	
Aluminum/iron	21	250	9.4	230	760			>450	2
Aluminum/nickel	<10		11.4	300					
Aluminum powder	29	30	12.4	415	710			>450	4
Bronze powder	18	750	4.1	31	390			260	4
Iron carbonyl	<10	125	6.1	111	310			300	3
Magnesium	28	30	17.5	508					
Manganese (electrolyte)	16	30	6.3	157	−330			285	
Zinc dust from collector	<10	250	6.7	125	570			440	3

(continued)

TABLE 8.2   Explosion Test Data for Common Solids (continued)

DUST TYPE	PARTICLE SIZE (μm)	1 m³ OR C_{min} (g/m³)	20 LITER P_{max} [Bar (g)]	VESSEL K_{St} (BAR m/sec)	DUST IGNITION G.G. (°C)	CLOUD TEMPERATURE BAM (°C)	VDI MIE (mJ)	GLOW TEMPERATURE (°C)	FLAMMABILITY <250:M CLASS
Other Inorganic Products									
Sulfur	20	30	6.8	151	280				5
Cotton/Wood/Peat									
Cotton	44	(100)	7.2	24	560			350	3
Cellulose	51	60	9.3	66	500		250	380	5
Wood dust	33				500		100	320	
Lignin dust	18	15	8.7	208	470			>450	5
Paper dust	<10		5.7	18	580			360	
Peat (15% moisture)	58	60	10.9	157	480			320	4
Food/Feed									
Dextrose (ground)	22								2
Dextrose (ground)	80	60	4.3	18	500			570	3

K_{St} = maximum rate of pressure rise in a nonvented vessel times the cube root of the volume [bar meter/s]; C_{min} = concentration of the dust determined in the 1.2-L Hartmann bomb; P_{max} = maximum pressure if the dust is ignited at the optimum concentration in a nonvented test vessel [bar]; G.G. = Godbert-Greenwald Furnace; BAM = BAM Furnace; VDI = VDI method of determining Minimum Ignition Energy experimental technique described by Berthold (1987); MIE = minimum ignition energy; Flammability class = dusts are classified according to their ability to propagate a combustion wave when deposited as a layer-Class 1: No self-sustained combustion, Class 2: Local combustion of a short duration, Class 3: Local sustained combustion, but no propagation, Class 4: Propagating smoldering combustion, Class 5: Propagating open flame, Class 6: Explosive combution

$$\text{thermal inertia} = k \cdot \rho \cdot c_p$$

Where:
k = thermal conductivity (W/m $\cdot$ K)
ρ = density (g/m^3)
c_p = specific heat (J/g $\cdot$ K)

The thermal inertia of a material characterizes its ability to conduct energy away from its surface and through its mass. More specifically, it provides a means of determining the temperature increase at the surface of a material, which is the principal issue when looking at ignition. The higher the thermal conductivity, the faster energy can transfer away from the material's surface and into the mass of the material. And how closely the molecules are packed into a material (i.e., the density) also influences its ability to transfer the energy through its mass and away from the exposed surface. Therefore, the higher the thermal inertia, the easier heat can be transferred through the material's mass away from the surface, which results in a lesser chance of ignition. For instance, metals are good conductors of thermal energy and can dissipate the energy through their mass faster, resulting in lower temperatures at the surface. Wood has a low thermal conductivity and density, and thus wood has a low thermal inertia. Any energy imposed on the face of a piece of wood is dissipated from its surface slowly, which allows the temperature to rise quickly. Greater pyrolysis results in ignition sooner than would occur in products with a higher thermal inertia. These properties are paramount to ignition and estimating when an object will ignite. They are covered in much more detail in Chapter 11.

Another important thermal property of solid fuels that is important in modeling ignition times and heat transfer through a solid is if the fuel can be regarded as thermally thin or thermally thick. The classification of a material as thermally thin corresponds to this object's thermal properties. Thermally thin does not necessarily relate to the thickness of the solid, but rather to its inability to absorb heat energy very well, and to how quickly it transfers heat from one side to its opposite side. Typically, these materials are also thin in thickness, but not always. Thermally thick solids relate to a solid that absorbs heat energy more readily and does not transfer it as quickly through the material to the opposite side.

FIRE RETARDANTS

Fire retardants are intended to reduce either the ignitability or combustibility of a substance. Most often, these materials are chemical additives that are either added during the production of materials (i.e., chemically bound) or applied after the material is in its final form (i.e., sprayed on). These fire retardants act in various ways, but they typically strive to either reduce the ignition potential of the item or slow the combustion process. Many different types of additives can be used as fire retardants; thus, these additives can slow the burning process or resist ignition via several mechanisms. Some additives can form a char layer or insulating layer when exposed to heat, which in turn reduces the heat that is reaching the fuel. Other chemical additives are released when heated, and they slow or extinguish the reaction by either diluting the flammable gases or by chemically inhibiting chain reactions (Friedman, 1998). Finally, other additives absorb the energy, which prevents heat going directly toward the pyrolysis process.

It is important to recognize that adding these retardants to combustible materials does not prevent the item from burning. It may reduce the ignition potential or slow the combustion reaction, but it does not prevent it completely.

fire retardant
■ A liquid, solid, or gas that tends to inhibit combustion when applied on, mixed in, or combined with combustible materials.

Summary

Fuels involved in fire are most commonly in the solid state. Solids range from dense, strong materials to those that are very lightweight and flexible. Yet most solids must become vapor before entering into the combustion reaction. A fuel's propensity to undergo pyrolysis is related to chemical composition, the surface area–to–mass ratio of the fuel package, orientation, heat capacity, thermal inertia, and the critical heat flux for that substance. This chapter has addressed these factors to increase your understanding of how they affect fire development.

Review Questions

1. The burning of a piece of wood is an example of a:
   a. chemical change
   b. physical change
   c. heat energy
   d. none of the above
2. A general characteristic of all ordinary combustible nonmetal solids is:
   a. they are easy to ignite
   b. they all have to be surface heated until they produce sufficient vapors to be ignited
   c. there is a relationship between surface area–to–mass ratio and the rate of combustion
   d. both b and c
3. The moisture content of wood:
   a. has a dramatic effect on the ignition characteristics and on the initiation energy required for combustion
   b. has only a minor effect on the ignition characteristics and the activation energy required
   c. has no effect on the burning characteristics of wood
4. Which of the following properties is of primary concern when dealing with combustible metal dust or any other dust in suspension?
   a. Flash point and fire point
   b. Ignition temperature and flammability range
   c. Lower explosive limit
   d. All of the above
5. If fuels such as wood, paper, rubber, or other materials leave an ash after combustion, these would be examples of which class of fuel?
   a. Class A
   b. Class B
   c. Class C
   d. Class D
6. For a given solid fuel, the surface area–to–mass ratio of the fuel package affects burning in the following way:
   a. There is no change, regardless of the surface area–to–mass ratio.
   b. Greater mass with less surface area burns more readily.
   c. Greater surface area with less mass burns more readily.
   d. None of the above.
7. The reason that fuel orientation is a major factor in fire development relates to the:
   a. flame's ability to heat fuel surfaces to pyrolysis.
   b. ability of air (oxygen) to reach the fuel surface.
   c. flame height that results from burning on the fuel.
   d. loss of mass to fuel consumption.
8. Which of the following statements is true about a fuel's thermal inertia?
   a. The thermal inertia of a fuel does not affect the ignition of fuels.
   b. Fuels with higher thermal inertia tend to ignite more easily.

c. Fuels with lower thermal inertia tend to ignite more easily.
d. Fuels with lower thermal inertia ignite but do not continue to burn.

9. Heat of gasification refers to the heat:
   a. generated from burning gases
   b. required to initiate burning in gases
   c. released from pyrolyzed gases of solid and liquid fuels
   d. required to change the state of solids and liquids so that they will burn

10. Fire retardants are:
    a. changes in the fuel at the molecular level
    b. additives that are chemically formulated into fuels
    c. additives applied to the surface of fuels
    d. all of the above

References

Cholin, J. (2008). "Wood and Wood-Based Products" (Chapter 4, Section 6). *NFPA Fire Protection Handbook*. Quincy, MA: NFPA.

Christman, T. (2008). "Metals." *Fire Protection Handbook*, Twentieth Edition. Quincy, MA: NFPA.

Drysdale, D. (1999). *An Introduction to Fire Dynamics*. New York: Wiley.

Drysdale, D. (2008). "Physics and Chemistry of Fire" (Section 1, Chapter 1). *Fire Protection Handbook*, Twentieth Edition. Quincy, MA: NFPA.

Friedman, R. (1998). *Principles of Fire Protection Chemistry and Physics*, Third Edition. Quincy, MA: NFPA.

NFPA 921 (2008). *The Guide for Fire and Explosion Investigations*. Quincy, MA: National Fire Protection Association.

Schwab, R. (2008). "Dusts" (Chapter 8, Section 6). *Fire Protection Handbook*, Twentieth Edition. Quincy, MA: NFPA.

Tewarson, A., and Pion, R. F. (1976). "Flammability of Plastics. I. Burning Intensity." *Combustion and Flame*, 26, 85–103.

burning flux (ṁ″), *p. 131*

combustion efficiency (χ), *p. 131*

fire growth, *p. 147*

heat of combustion (ΔH$_c$), *p. 131*

heat of gasification, *p. 133*

heat release rate (HRR), *p. 131*

pool fires, *p. 163*

After reading this chapter, you should be able to:

- Define the term *heat release rate*.
- Describe how a cone calorimeter test relates to the burning behavior of a solid fuel.
- Calculate heat release rates.
- Explain heat release rate tests and measurement techniques.
- Explain different methods of identifying fire growth curves.
- Calculate the heat release rates and burning duration of pool fires.
- Calculate flame heights and plume correlations.

PEARSON

For additional review and practice tests, visit **www.bradybooks.com** and click on MyBradyKit to access book-specific resources for this text!

The heat release rate (HRR) is the single most important variable in a fire (Babrauskas and Peacock, 1991). Generally, the heat release rate provides fire safety professionals a way to quantify and intelligently discuss the complexity of fires, including their size, potential impact, and correlation with other variables (Icove and De-Haan, 2009). This chapter will introduce you to the importance of understanding and recognizing heat release rates and their measurement.

In the early years of fire research, there were many attempts to quantify the hazard of a room or occupancy by looking at the total amount of combustible fuels located in a building or room, and then equating that amount to the relative danger that existed in the structure. This was known as a fuel load or fire load, which is defined as the "total quantity of combustible contents of a building space or fire area, including interior finish and trim, expressed in heat units or the equivalent weight in wood" (Kuvshinoff, 1977). This simply provided someone the ability to recognize the total amount of energy that could be emitted if all the fuel in that area were to burn. While this value may provide a general benchmark for a fire inside a compartment that burned to completion, it is greatly limited in providing fire safety professionals with useful information regarding the relative hazard of different fuels. It is far more important for a fire safety professional to know how fast the energy would be given off and thus how much time should be allotted for occupants to exit safely.

General Introduction

The more formal definition of **heat release rate (HRR)** is the "rate at which the heat energy is generated by burning" (NFPA 921, 2008). In other words, the heat release rate is the power (energy over time) of the fire, which has units of Joules per seconds (J/s) or watts (W). Typically, fire sizes are given in kilowatts (kW; 1,000 watts) or in megawatts (MW; 1 million watts). Discussing a fire in terms of kilowatts allows the fire safety professional the ability to describe and better quantify the size of the fire. Unlike density or specific heat, the heat release rate is not an inherent material property. In other words, the heat release rate may be very different for the same material under different conditions (i.e., ventilation, orientation). The combination of several variables must be taken into consideration to determine a fuel's heat release rate, including the **heat of combustion** (ΔH_c), mass loss flux or **burning flux** ($\dot{m}''$), the surface area (orientation, surface area–to–mass ratio), and the **combustion efficiency** (χ). Each of these elements has been touched on in previous chapters. However, this chapter will present a more detailed analysis of each variable to provide a better understanding of how they affect the heat release rate during a fire. We will also show that it is possible to mathematically calculate the energy that can be released during a fire throughout its growth when these variables are known.

Many times, these variables are not specifically known and must either be estimated or measured under ideal (laboratory) conditions, which inevitably leads to approximate results. When a single material (for example, heptane) is burning, it is relatively easy to calculate precisely the heat release rate. However, fire safety professionals are often dealing with items constructed of multiple materials (for example, a polyurethane couch). Each of the materials has different physical

heat release rate (HRR)
- The rate at which heat energy is generated by burning.

heat of combustion (ΔH_c)
- The quantity of heat evolved by the complete combustion of one mole (gram molecule) of a substance.

burning flux ($\dot{m}''$)
- The mass of fuel consumed per unit time in the fire.

combustion efficiency (χ)
- The ratio of the output of a combustible's kinetic energy to the total input (potential energy).

properties, which makes it difficult to calculate the heat release rate. Further complicating the calculations is the fact that various blends and percentages of materials are used by different manufacturers to create similar products, which prevents standard heat release rates to be compiled for all objects. At this point in the evolution of fire science, for fuel configurations that are constructed of multiple materials, it is more common and precise to burn that particular item or a small portion of that item and measure its heat release rate. Thus, fire safety professionals must understand heat release rate testing and how to interpret the results to be effective in their positions.

Heat release rates are typically discussed in terms of either a fuel item burning, a fuel package burning, or an enclosure fire. Heat release rates for fuel items and fuel packages are often reported in tables as peak heat release rates (see Table 9.1). It is important to recognize the peak heat release rates of different fuel items, but it may be more important to recognize how each fuel item influences an enclosure fire. Both heat release rate concepts will be discussed throughout this chapter.

Babrauskas and Peacock (1991) make a compelling argument that the heat release rate is the single most important variable in fire hazard. They say that the faster the energy is released, the faster the heat can be transferred back to the fuel's surface, thus allowing for the production of more fuel in a form capable of ignition, which in turn allows for the production of more heat. This heat feedback loop is a very important issue in enclosure fires and results in the production of more smoke and products of combustion. Finally, the faster this cycle occurs, the sooner the compartment may transition into full-room involvement. Increased speed of fire development prevents escape and increases the rate of injuries and deaths. Knowledge of heat release rates is the foundation to understanding the dangers associated with fuels and how those fuels affect people and property.

Methods for Determining Heat Release Rate

There are two general methods for determining the heat release rate of a fuel: (1) estimate the heat release rate by calculations or (2) test the fuel item and measure its heat release rate. This first subsection will focus on the calculation of heat release rates.

ESTIMATING HEAT RELEASE RATES: CALCULATIONS

The heat release rate can be calculated by the following equation:

$$\dot{Q} = \chi \cdot \dot{m}'' \cdot A_f \cdot \Delta H_c$$

Where:

$\dot{Q}$ = heat release rate (kW)

χ = factor to account for incomplete combustion (<1.0) (unitless)

$\dot{m}''$ = mass loss flux (g/m^{2}s)

A_f = fuel surface area (m^2)

ΔH_c = heat of combustion of the volatiles (kJ/g)

TABLE 9.1 | Peak Heat Release Rates

FUEL	HEAT RELEASE RATE ($\dot{Q}$)
A burning cigarette	5 W
A typical light bulb	60 W
A burning candle	80 W
A human being at normal exertion	100 W
A burning wastepaper basket	100 kW
A burning 1 m² pool of gasoline	2.5 MW
Burning wood pallets, stacked to a height of 3 m	7 MW
Burning polystyrene jars, in 2 m² cartons 4.9 m high	30–40 MW

FUEL	WEIGHT		PEAK HRR (kW)
	kg	lb	
Trash bags, 42 L (11 gal) with mixed plastic and paper trash	1.1–3.4	2½–7½	140–350
Cotton mattress	11.8–13.2	26–29	40–970
TV sets	31.3–32.7	69–72	120 to over 1500
Plastic trash bags/paper trash	1.2–14.1	2.6–31	120–350
PVC waiting room chair, metal frame	15.4	34	270
Cotton easy chair	17.7–31.8	39–70	290–370
Gasoline/kerosene in 0.185 m² (2 ft²) pool	19	—	400
Christmas trees, dry	6–20	13–44	3000–5000
Polyurethane mattress	3.2–14.1	7–31	810–2630
Polyurethane easy chair	12.2–27.7	27–61	1350–1990
Polyurethane sofa	51.3	113	3120
Wardrobe, wood construction	70–121	154–267	1900–6400

Note: 1 kW = 1,000 W and 1 MW = 1,000 kW
Source: Karlsson and Quintiere, 1999, © CRC Press, LLC. Used with permission.

The heat release rate depends on material properties, such as the heat of combustion and **heat of gasification**. The heat release rate also depends on combustion processes under the specific conditions, such as the combustion efficiency and heat feedback to the fuel surface (see Figure 9.1).

heat of gasification
- Energy required to produce a unit mass of fuel vapor from a solid or a liquid.

Heat of Combustion

The total amount of heat generated in a fire is a result of the chemical reactions in the combustion of a material and is known as the chemical energy. The heat of combustion is the amount of energy emitted per unit of fuel mass consumed (kJ/g) and

FIGURE 9.1

Schematic representation of a burning surface, showing the heat and mass transfer processes: $\dot{m}''$ = mass loss rate from the surface; $\dot{Q}_F''$ = heat flux from the flame to the surface; $\dot{Q}_L''$ = heat losses (expressed as a heat flux from the surface).

Source: Data derived from Drysdale, D. (1998).

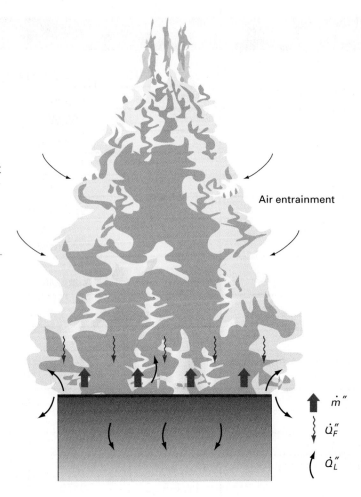

Air entrainment

$\dot{m}''$

$\dot{u}_F''$

$\dot{a}_L''$

varies by fuel. The theoretical heat of combustion (ΔH_T) is easily determined from a controlled experiment in an apparatus known as an oxygen bomb calorimeter (see Figure 9.2). This apparatus allows a material to undergo complete combustion by providing an atmosphere of elevated oxygen and measuring mass loss and energy released. This value assumes that all of the fuel is consumed and that no residue exists after the reaction. However, reactions always have some form of inefficient combustion and associated losses (i.e., energy lost to the surroundings), which prevents complete reaction of the fuel. Thus, the theoretical value is never attained. Therefore, the chemical heat of combustion (ΔH_{ch}), also known as the effective heat of combustion, should be used in fire problems. The subscript for the effective heat of combustion is often given as shown here: ΔH_{ch} or ΔH_c.

Remember that, during the combustion reaction, heat is given off as either convection or radiation. Thus, the chemical heat of combustion has two components: the radiative energy (ΔH_{rad}) and the convective energy (ΔH_{con}). As you can see in Table 9.2, the heats of combustion (ΔH_{ch}) for different fuels range from 10 kJ/g to 50 kJ/g. This range is quite extensive, which prevents an easy means to generalize energy released by the fuel. This table also shows the components of convective and radiative heat for well-ventilated fires for the different

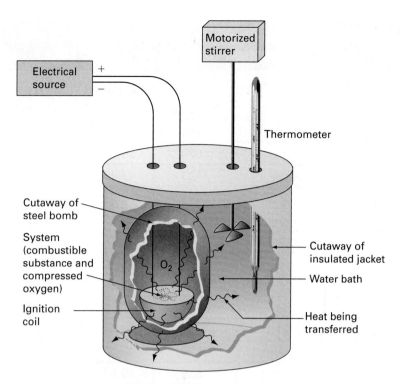

FIGURE 9.2 Typical oxygen bomb calorimeter.

Motorized stirrer

Electrical source

Thermometer

Cutaway of steel bomb

System (combustible substance and compressed oxygen)

O_2

Cutaway of insulated jacket

Water bath

Ignition coil

Heat being transferred

fuels. For a more complete list of heats of combustion, refer to the work performed by Tewarson (2008).

Switching from measuring the energy released when the fuel is consumed to measuring the energy released when a unit mass of air or oxygen is used in the combustion reaction provides a different perspective that proves to be very useful. In reviewing Table 9.2 again, you can see that the energy released for the consumption or reaction of a unit mass of air is a much smaller range of heats of combustion ($\Delta H_{c,\text{air}}$): between 2.91 and 4.10 kJ/g, which is commonly rounded to 3 kJ/g. If the energy released is measured for each gram of oxygen used in the reaction, you can see that the heats of combustion ($\Delta H_{c,\text{O}_2}$) is also a small range: between 11 kJ/g and 17 kJ/g, which is commonly averaged to 13.1 kJ/g $\pm$ 5 percent for most hydrocarbon fuels. In other words, when measuring the energy released when a unit mass of air or oxygen is consumed during a reaction, the range is relatively small. This indicates that when most fuels are undergoing combustion, a similar amount of energy is released per unit mass of oxygen or air used in the reaction. Specifically, it allows the fire professional to analyze the amount of energy emitted, regardless of the ventilation conditions, as opposed to attempting to identify the heat release rate by fuel consumption. This is the basis for oxygen consumption calorimetry, which will be discussed in more detail later in this chapter (American Society for Testing and Materials [ASTM] E1345).

Combustion Efficiency

In fires, complete combustion (100 percent efficiency) is never attained. This is mainly due to the loss of heat to the ambient atmosphere and the inefficient diffusion of oxygen into the chemical reaction because most fires are diffusion flames.

TABLE 9.2 | Heats of Combustion[a]: Chemical, Convective, and Radiation Heats of Combustion for Well-Ventilated Fires

		$-\Delta H_c$ (kJ/mol)	$-\Delta H_c$ (kJ/g)	$-\Delta H_{c,air}$ (kJ/g(air))	$-\Delta H_{c,ox}$ (kJ/g(O_2))
Carbon monoxide	CO	283	10.10	4.10	17.69
Methane	CH_4	800	50.00	2.91	12.54
Ethane	C_2H_6	1423	47.45	2.96	11.21
Ethene	C_2H_4	1411	50.35	3.42	14.74
Ethyne	C_2H_2	1253	48.20	3.65	15.73
Propane	C_3H_8	2044	46.45	2.97	12.80
n-Butane	$n\text{-}C_4H_{10}$	2650	45.69	2.97	12.80
n-Pentane	$n\text{-}C_5H_{12}$	3259	45.27	2.97	12.80
n-Octane	$n\text{-}C_8H_{18}$	5104	44.77	2.97	12.80
c-Hexane	$c\text{-}C_6H_{12}$	3680	43.81	2.97	12.80
Benzene	C_6H_6	3120	40.00	3.03	13.06
Methanol	CH_3OH	635	19.83	3.07	13.22
Ethanol	C_2H_5OH	1232	26.78	2.99	12.88
Acetone	$(CH_3)_2CO$	1786	30.79	3.25	14.00
D-Glucose	$C_6H_{12}O_6$	2772	15.4	3.08	13.27
Cellulose		—	16.09	3.15	13.59
Polyethylene		—	43.28	2.93	12.65
Polypropylene		—	43.31	2.94	12.66
Polystyrene		—	39.85	3.01	12.97
Polyvinylchloride		—	16.43	2.98	12.84
Polymethylmethacrylate		—	24.89	3.01	12.98
Polyacrylonitrile		—	30.80	3.16	13.61
Polyoxymethylene		—	15.46	3.36	14.50
Polyethyleneterephthalate		—	22.00	3.06	13.21
Polycarbonate		—	29.72	3.04	13.12
Nylon 6,6		—	29.58	2.94	12.67

[a]The initial states of the fuels correspond to their natural states at normal temperature and pressure (298°C and 1 atm pressure). All products are taken to be in their gaseous state—thus these are the net heat of combustion.

The efficiency of a fuel burning in the air is controlled by the amount of oxygen available and the conversion of fuel to the gaseous state (i.e., vaporization rate or pyrolysis rate). Table 9.2 lists different fuels and their theoretical heats of combustion compared to their chemical heats of combustion (or effective heat of combustion). If combustion were perfect, or 100 percent efficient, all of these values would be the same. But you can see in Table 9.2 that these values are never the same.

Combustion efficiency is usually represented by the Greek letter chi, Χ. This is a factor that accounts for incomplete combustion, and ranges from 0 to 1. The number 1 represents perfect or 100 percent efficient combustion, while zero represents completely inefficient combustion; however, both limits are impractical. Therefore, most fuels will be in the middle of this range. Dividing the effective heat of combustion (ΔH_{ch}) by the theoretical heat of combustion (ΔH_T) provides an estimate of the combustion efficiency under well-ventilated conditions (see Table 9.2). Drysdale (2008) states that this factor is found to be in the range of 0.3 to 0.4 for materials that have fire retardants added, and it is close to 0.9 for fuels with chemically bound oxygen.

Mass Burning Flux

One method of determining the heat release rate is to calculate or measure the mass loss rate of the fuel and multiply that quantity by the effective heat of combustion (ΔH_c).

$$\dot{Q} = \Delta H_c \dot{m}$$

Where:
$\dot{Q}$ = heat release rate (kW)
ΔH_c = heat of combustion (kJ/kg)
$\dot{m}$ = mass burning rate (kg/s)

NOTE: *Even though it is common to use the term* mass loss, *the mass is not magically or mystically lost from our universe because this would be a violation of the conservation of mass rule. Rather, mass is converted and used within the reaction or taken up with the buoyant plume. The term refers more to the mass that is lost from the object burning.*

However, the average burning flux ($\dot{m}''$) is most commonly determined experimentally and recorded for free burn fire tests. Thus, it is simpler to calculate the heat release rate by multiplying the burning flux by the fuel surface area and the heat of combustion, as follows:

$$\dot{Q} = A_F \Delta H_c \dot{m}''$$

Where:
$\dot{Q}$ = heat release rate (kW)
A_F = fuel surface area (m^2)
ΔH_c = heat of combustion (kJ/kg)
$\dot{m}''$ = mass burning flux (kg/m^{2}s)

The mass burning flux ($\dot{m}''$) is the loss of mass over time from the fuel's surface when it is undergoing combustion. The mass loss rate ($\dot{m}$), often referred to as the burning rate, is the rate at which the fuel is losing mass while undergoing combustion and is usually expressed in grams per second (g/s) (see Table 9.3). There are slight differences between fuel being evolved from the fuel (mass loss rate) compared to the fuel that is actually being combined with oxygen and undergoing combustion (burning rate), but the distinction will not be further evaluated in this text. Typically, the burning rate ($\dot{m}$) is not constant and depends greatly on the orientation, surface area–to–mass ratio, and various other fire spread issues. Therefore, it is important to know the area that is undergoing this vaporization to be able to better calculate the mass burning flux ($\dot{m}''$)

The mass being lost is in the gaseous form, so the heat of gasification must be reached at the fuel's surface for gases to be evolved. This means that the amount

TABLE 9.3	Mass Loss Rates for Common Fuels

FUEL	$\dot{m}''$ (g/m²-s)
Liquified propane	100–130
Liquified natural gas	80–100
Benzene	90
Butane	80
Hexane	70–80
Xylene	70
JP-4	50–70
Heptane	65–75
Gasoline	50–60
Acetone	40
Methanol	22
Polystyrene (granular)	38
Polymethyl methacrylate (granular)	28
Polyethylene (granular)	26
Polypropylene (granular)	24
Rigid polyurethane foam	22–25
Flexible polyurethane foam	21–27
Polyvinyl chloride (granular)	16
Corrugated paper cartons	14
Wood crib	11

of heat flux returning to the surface from the flame must overcome the heat losses from the fuel's surface. In other words, the fuel must be supplied with enough heat for continued vaporization, or pyrolysis; otherwise, combustion ceases.

$$\dot{m}'' = \frac{\dot{Q}''_{flame} - \dot{Q}''_{loss}}{L_g}$$

Where:
$\dot{m}''$ = mass loss flux (kg/m² · s)
$\dot{Q}''_{flame}$ = heat flux from flame to the surface (kW/m²)
$\dot{Q}''_{loss}$ = heat flux loss from the surface (kW/m²)
L_g = heat of gasification (kJ/kg)

To make the calculation simpler, under many conditions, you can assume that the burning of the fuel is at a steady state:

$$\dot{m}'' = \frac{\dot{q}''}{L_g}$$

Where:

$$\dot{m}'' = \text{mass loss flux (kg/m}^2 \cdot \text{s)}$$
$$\dot{q}'' = \text{net heat flux to the surface (kW/m}^2\text{)}$$
$$L_g = \text{heat of gasification (kJ/kg)}$$

The term *heat of gasification* is used to describe the amount of energy that is required to produce a unit mass of flammable vapor from a combustible that is initially at ambient temperatures (Drysdale, 2008). Heats of gasification for many materials have been experimentally determined; some are listed in Table 9.4.

TABLE 9.4	Flammability Parameters			
COMBUSTIBLES	L_v **(kJ/g)**	$\dot{Q}''_{flame}$ **(kW/m^2)**	$\dot{Q}''_{loss}$ **(kW/m^2)**	$\dot{m}''_{ideal}$ **(g/m^2-s)**
FR phenolic foam (rigid)	3.74	25.1	98.7	11[a]
FR polyisocyanurate foam (rigid, with glass fibers)	3.67	33.1	28.4	9[a]
Polyoxymethylene (solid)	2.43	38.5	13.8	16
Polyethylene (solid)	2.32	32.6	26.3	14
Polycarbonate (solid)	2.07	51.9	74.1	25
Polypropylene (solid)	2.03	28.0	18.8	14
Wood (Douglas fir)	1.82	23.8	23.8	13[a]
Polystyrene (solid)	1.76	61.5	50.2	35
FR polyester (glass-fiber-reinforced)	1.75	29.3	21.3	17
Phenolic (solid)	1.64	21.8	16.3	13
Polymethylmethacrylate (solid)	1.62	38.5	21.3	24
FR polyisocyanurate foam (rigid)	1.52	50.2	58.5	33
Polyurethane foam (rigid)	1.52	68.1	57.7	45
Polyester (glass-fiber-reinforced)	1.39	24.7	16.3	18
FR polystyrene foam (rigid)	1.36	34.3	23.4	25
Polyurethane foam (flexible)	1.22	51.2	24.3	32
Methyl alcohol (liquid)	1.20	38.1	22.2	32
FR polyurethane foam (rigid)	1.19	31.4	21.3	26
Ethyl alcohol (liquid)	0.97	38.9	24.7	40
FR plywood	0.95	9.6	18.4	10[a]
Styrene (liquid)	0.64	72.8	43.5	114
Methylmethacrylate (liquid)	0.52	20.9	25.5	76
Benzene (liquid)	0.49	72.8	42.2	149
Heptane (liquid)	0.48	44.3	30.5	93

[a]Charring materials. $\dot{m}''_{ideal}$ taken as the peak burning rate.

FIGURE 9.3 Vertical orientation versus horizontal orientation and impact of burning rates.

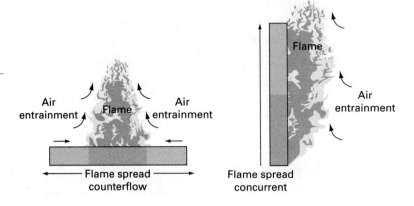

Area

The fuel's surface area that is undergoing combustion has a major impact on the heat release rate. As more surface area of the fuel burns, a greater amount of mass is exposed to heating and therefore more mass may be lost to allow for greater combustion. Several variables affect the exposed fuel's surface area, including the orientation, surface area–to–mass ratio, geometry, density, surface roughness, and thickness. All of these variables also control the flame spread along the fuel's surface. Flame spread is a series of piloted ignitions across a fuel's surface, with the flames serving as the source of heat flux and the piloted ignition source. More discussion regarding flame spread will be introduced in Chapter 11. For the heat release rate, a horizontal orientation versus a vertical orientation for the exact same material greatly affects the flame spread rate and exposed surface area, which in turn affects the heat release rate (see Figure 9.3). In addition, the ignition sequence can greatly affect the amount of surface area initially exposed, which influences the heat release rate of the fuel (especially within the incubation period).

EXAMPLE 9.1

Assume that a slab of PMMA (0.12 m²) is burning and is weighed as it burns. The average mass loss flux is 30 g/m²s. The complete heat of combustion for polymethyl methacrylate (PMMA) has been listed as 25 kJ/g. Assume that the combustion efficiency is 60 percent and estimate the resulting energy release rate.

Analysis:

$$\dot{Q} = \chi \cdot \dot{m}'' \cdot A_f \cdot \Delta H_c = (0.12 \text{ m}^2)\left(25\frac{\text{kJ}}{\text{g}}\right)(0.6)(30 \text{ g/m}^2\text{s}) = 54 \text{ kW}$$

MEASUREMENT OF HEAT RELEASE RATE

Early measurement techniques used to rely on the base calculations from above to assist in determining an item's heat release rate. Fuels were placed on a load cell (scale), and the mass loss was monitored during the combustion reaction. This mass loss rate was multiplied by the theoretical heats of combustion, and an estimated heat release rate was obtained. Several issues with this method caused unreliable results. If the fuel consisted of different materials, several different heats of combustion had to be accounted for during the calculations. For example, a polyurethane couch is typically constructed of exterior upholstery, a polyurethane foam interior, and a wood frame. Each of these fuels have different heats of combustion, not to mention that it would simply be impossible to know exactly which material was burning at certain times during the test.

Greater accuracy in determining the amount of energy emitted is achieved through measuring oxygen consumption (oxygen used in the reaction) rather than relying on assuming a theoretical heat of combustion of the fuel and measuring its mass loss. As mentioned previously, each gram of oxygen reacted yields approximately 13.1 kJ of energy, regardless of the fuel burning (Table 9.2). In the 1980s, oxygen consumption calorimeters were developed and have become the standardized method of testing a fuel's heat release rate ever since. Oxygen consumption calorimetry is based on the premise that air in normal atmospheric conditions has an oxygen concentration of approximately 21 percent. Consequently, if the combustion gases (i.e., smoke, etc.) could be collected into a duct and then the oxygen concentration of these gases measured, the amount of oxygen consumed or converted during the reaction can be estimated. Knowing the oxygen used during the reaction would in turn allow the heat release rate to be calculated. Because most of the oxygen is not completely consumed in the reaction, most calorimeters also measure CO and CO_2 gases.

Many types of oxygen consumption calorimeters, all based on this same principle, have been developed for different purposes and applications. The most commonly used small-scale calorimeter is known as the cone calorimeter, which derives its name from the cone-shaped heater used in the apparatus (see Figure 9.4). Dr. Vytenis Babrauskas developed this calorimeter in the early 1980s (Babrauskas, 2008). The cone calorimeter measures the heat release rate for a 0.1 m × 0.1 m specimen size.

The cone calorimeter is used only for small samples, which may not be suitable for thick or complex objects. Therefore, an intermediate-scale calorimeter (ICAL) was developed. The standard test specimen is approximately 1 m × 1 m (see Figure 9.5).

As mentioned previously, it is often difficult to understand the behavior of a complex item by looking solely at an individual component. Thus, for large furniture items, it is easier and better to burn the complete object under a large exhaust hood. The test apparatus used for larger tests is known as an open air calorimeter or furniture calorimeter. Room fire tests are used frequently when it is anticipated that room effects will exacerbate the burning of the object. Finally, larger-scale calorimeters, such as the industrial calorimeters used by Factory Mutual, test rack storage arrays, and the U.S. Bureau of Alcohol, Tobacco and Firearms (ATF) lab uses large calorimeters for mocking-up full-scale houses and rooms for incendiary cases (see Figure 9.6).

A detailed discussion of the test apparatus and methods is beyond the scope of this book. Babrauskas (2008) and Janssens (2008) provide an excellent overview of oxygen consumption calorimetry and heat release rate test methods.

FIGURE 9.4
(a) Schematic of cone calorimeter,
(b) photograph of University of North Carolina—Charlotte's cone calorimeter.
Source: Babrauskas (2008).

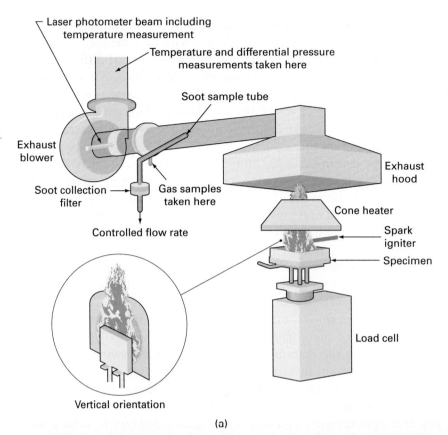

Laser photometer beam including temperature measurement

Temperature and differential pressure measurements taken here

Soot sample tube

Exhaust blower

Soot collection filter

Gas samples taken here

Controlled flow rate

Exhaust hood

Cone heater

Spark igniter

Specimen

Load cell

Vertical orientation

(a)

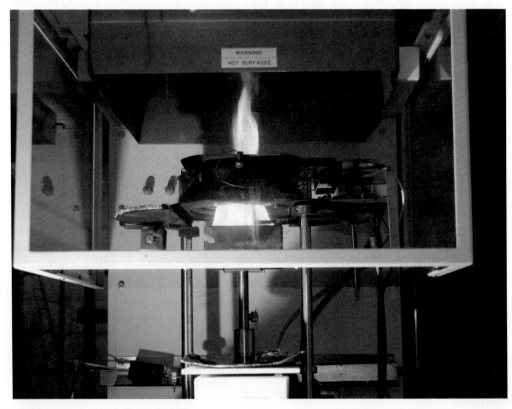

(b)

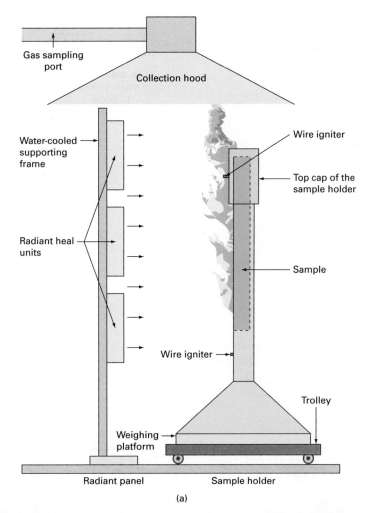

Gas sampling port

Collection hood

Water-cooled supporting frame

Radiant heal units

Wire igniter

Top cap of the sample holder

Sample

Wire igniter

Trolley

Weighing platform

Radiant panel

Sample holder

(a)

FIGURE 9.5
(a) Schematic and (b) photograph of University of North Carolina—Charlotte's intermediate scale calorimeter (ICAL).
Source: Babrauskas (2008).

(b)

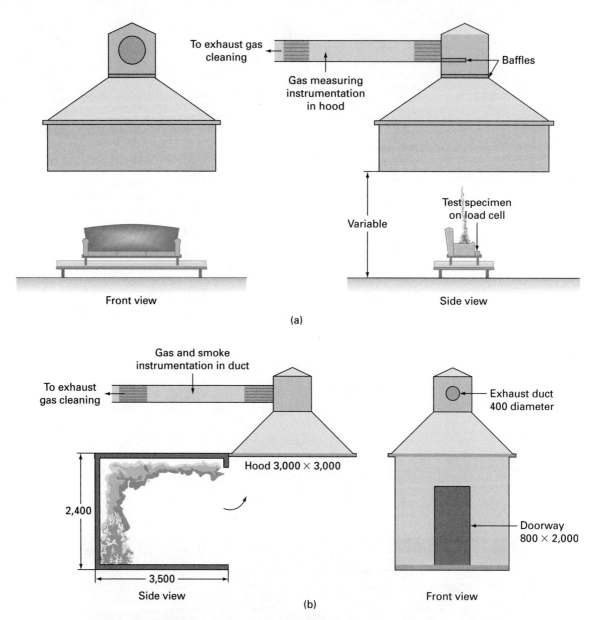

FIGURE 9.6 (a) Schematic of a furniture calorimeter, (b) schematic of a room HRR test calorimeter. (measurement values are in millimeters)
Source: Babrauskas (2008).

HEAT RELEASE RATE CURVES

The best approach for determining the exact heat release rate for a specific fuel is to use one of the testing devices discussed above and obtain a heat release rate curve. Consequently, it is equally important for the fire safety professional to understand how to interpret this data and implement the results for use in practical applications.

During the oxygen consumption calorimetry tests, the fuel burns and a heat release rate curve develops. The curve maps out the release of energy over the duration of the fuel burning. The x-axis of the graph is the time element of the curve representing the duration of the fuel burning; it is typically given in seconds. The

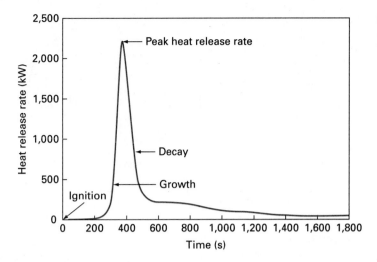

FIGURE 9.7 Example heat release rate graph.

y-axis of the graph is the heat release rate, typically given in kilowatts (kW). The curve typically starts from ignition (0 s, 0 kW) and steadily increases in heat release rate (HRR), similar to an exponential growth curve. Eventually the curve will come to its highest level (highest kW), which is referred to as the peak heat release rate for the fuel. Many fuels have multiple spikes or smaller peaks, but the highest value is recorded as the peak. Values listed in tables, like Table 9.1, represent that fuel's peak heat release rate.

For example, the heat release rate graph of an upholstered chair with polyurethane foam padding and weighing approximately 25 pounds is shown in Figure 9.7. The chair is constructed with a wood frame, polyurethane foam interior, and an upholstered exterior (see Figure 9.8). The fuel reaches its peak heat release rate of 1 MW (1,000 kW) in about 250 seconds and quickly starts to decrease, but around 300 seconds, the heat release rate increases and peaks once more to approximately 950 kW. After this second spike, the heat release decreases to approximately 500 kW at 400 seconds and then spikes again at approximately 600 kW. Slowly, the heat release rate decays until it nears zero at 900 seconds. The reason for multiple peaks is that, as mass was lost from the chair, additional fuels (and possibly different types of fuels) that were previously protected from

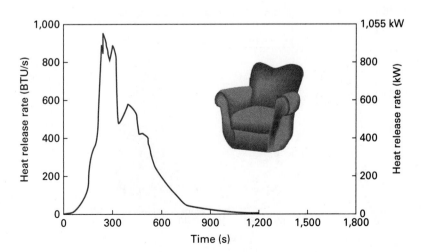

FIGURE 9.8 Heat release rate graph of upholstered chair with polyurethane foam padding.
Source: Data derived from Klote, J., and Milke, J. (2002).

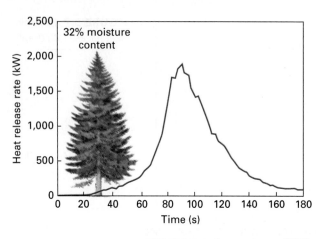

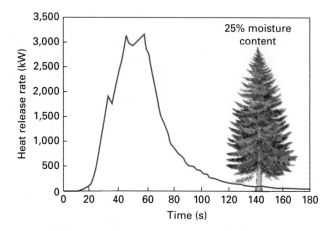

FIGURE 9.9 Comparison of HRR curves of Christmas trees with different moisture content: 32 percent moisture content (left) versus 25 percent moisture content (right).
Source: Data derived from Stroup, DeLauter, Lee, and Roadarmel (1999).

the heat were exposed and began to add to the energy released. Fuels that have multiple materials are expected to have multiple peaks associated with the burning of the different materials over the duration of the fire.

Similar fuels may have different heat release rate curves due to a multitude of factors, including orientation, method of ignition (i.e., single point or entire area), and material properties (i.e., moisture content). For example, the National Institute of Standards and Technology (NIST) performed tests to analyze the safety of Christmas trees. They burned similar types of Christmas trees, but one tree had a moisture content of 32 percent, while the other had a moisture content of 25 percent (Stroup, DeLauter, Lee, and Roadarmel, 1999) (see Figure 9.9). The peak heat release rate for the 32 percent moisture content tree was 2,000 kW, while the peak HRR for the 25 percent moisture content tree was 3,000 kW. You can also see in heat release rate graphs that the speed at which the fire grew to its peak is significantly different. The lower moisture content tree released its energy much faster. In addition, Madrzykowski (2008) also performed HRR tests of dry pine trees (see Figure 9.10). He performed five tests, four in the vertical

FIGURE 9.10
Comparison of HRR curves of pine Christmas trees with different orientations.
Source: Data derived from Madrzykowski (2008).

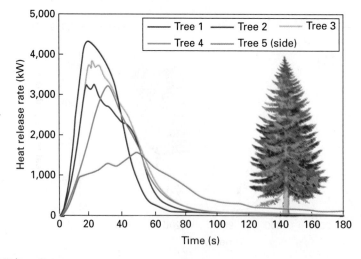

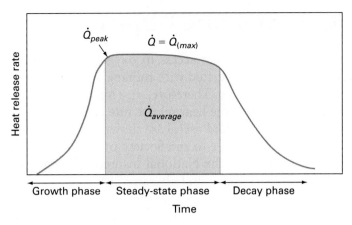

FIGURE 9.11 A simple design fire curve for a compartment fire.
Source: Data derived from Karlsson, B., and Quintiere, J. (2000).

orientation and one in the horizontal orientation (test 5). The HRR of the tree lying on its side is much lower, but it occurs over a longer duration. As you can see from these examples, some fuels, depending on their properties or how they are ignited, can quickly reach their peak HRR, while other similar types of fuels may burn more slowly and not reach as high a peak HRR. However, it is important that the same fuel theoretically contains the same amount of potential energy and therefore produces the same amount of energy under the curve. These variables affect the way that energy can be released and how fast the energy is released.

So far, the heat release rate graphs have been for fuel items. The heat release rate curves are also collected and used to quantify and analyze compartment fires. A compartment fire typically has a growth phase associated with its heat release rate, which will be slowed or stopped only if either the fuel or the ventilation is limited (see Figure 9.11). Assume that there is a large enough fuel item that is capable of causing the room to transition into full-room involvement. The heat release rate curve steadily grows until the compartment transitioned into full-room involvement (the transition would be fast if it happened via flashover, slow if not). Either way, the fire is now limited in growth by the amount of oxygen that can enter the volume and mix with the available fuel to support combustion. This phase of the fire is called the steady-state or steady-burning phase because the heat release rate plateaus and remains steady until the fuel inside the compartment is exhausted. Once either the fuel or oxygen is depleted, the heat release rate begins to decrease or decay. A heat release rate curve for a compartment fire is commonly referred to as the fire growth curve. A **fire growth** curve represents the different phases in heat release rate (ignition, growth, steady state, decay) over the duration of the entire fire (i.e., HRR over time).

fire growth
■ The initiation, growth, and decay of a fire's heat release rate over the duration of the fire.

As a fire grows inside a compartment, the driving factors are the fuel, its burning rate, and the impact of the surrounding surfaces. When a fire begins to develop within a compartment, heat is transferred to the interior wall and ceiling linings. Portions of this heat are reflected back into the compartment. As the heated gases begin to develop in the upper layer of the compartment, heat is being radiated back to the fuels within the lower layer, including the fuel that is undergoing combustion. This reflected heat and radiant heat from the upper layer can influence the burning rate of the fuel item and must be considered by the fire safety professional.

Heat Release Rate of Some Objects

Many fire safety professionals do not have direct access to oxygen consumption calorimeters and even if they did, the costs associated with running a test for all the fuels that they come across would be prohibitive. Therefore, they must rely on published research of similar fuels, if available. Some heat release rate curves of everyday items that the fire safety professional may find useful are shown in Figures 9.12 through 9.31. For a more complete listing, refer to the Society of Fire Protection Engineers (SFPE) *Engineering Handbook* and the National Institute of Standards and Technology's website Building and Fire Research Laboratory (BFRL).

Once a basic heat release rate curve is obtained, the application of this data to fire safety problems is the next step. This data can be utilized for basic calculations (i.e., radiant heat flux calculations) as well as for computer fire modeling for the analysis of different fire scenarios. It can also be employed for fuel comparison purposes.

FIGURE 9.12 HRR of televisions.
Source: Data derived from Babrauskas (2008).

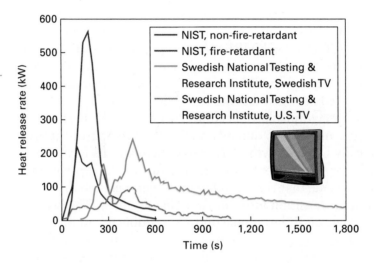

FIGURE 9.13 HRR of cars.
Source: Data derived from Klote, J., and Milke, J. (2002).

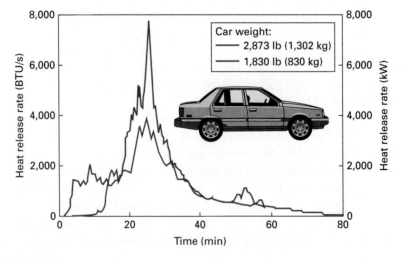

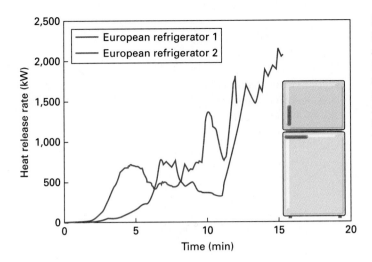

FIGURE 9.14 HRR of refrigerators (*Note:* Applicable only to European, not North American, refrigerators.)
Source: Data derived from Babrauskas (2008).

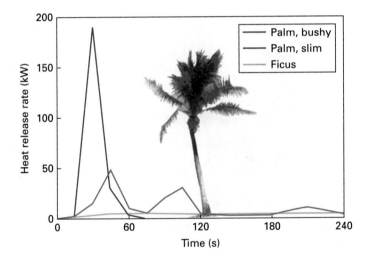

FIGURE 9.15 HRR of plastic house plants.
Source: Data derived from Babrauskas (2008).

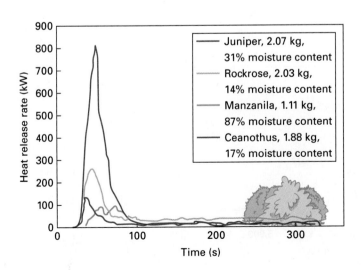

FIGURE 9.16 HRR for bushes.
Source: Data derived from Babrauskas (2008).

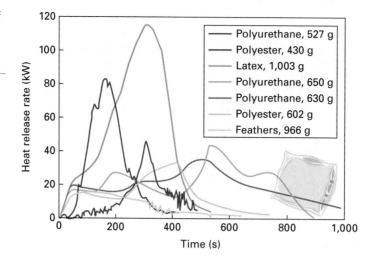

FIGURE 9.17 HRR of pillows.
Source: Data derived from Babrauskas (2008).

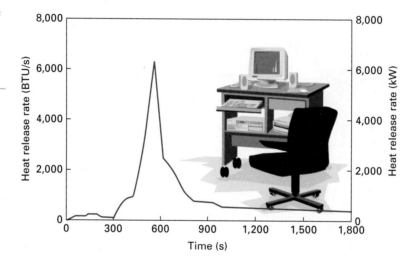

FIGURE 9.18 HRR of workstations.
Source: Data derived from Klote, J., and Milke, J. (2002).

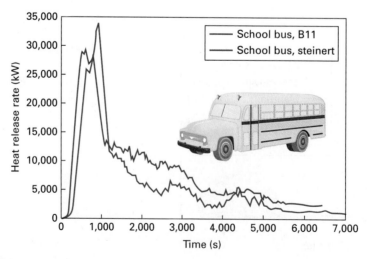

FIGURE 9.19 HRR of school buses.
Source: Data derived from Babrauskas (2008).

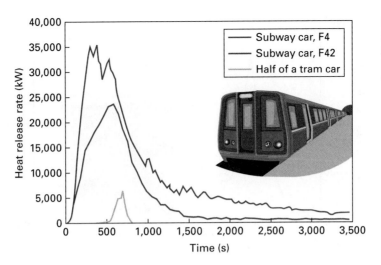

FIGURE 9.20 HRR of subway cars.
Source: Data derived from Babrauskas (2008).

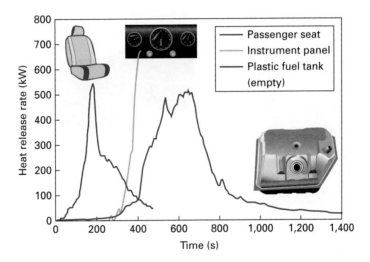

FIGURE 9.21 HRR of larger components from a passenger vehicle.
Source: Data derived from Babrauskas (2008).

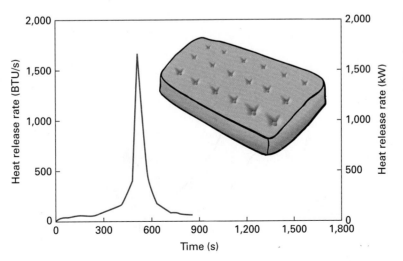

FIGURE 9.22 HRR of mattresses.
Source: Data derived from Klote, J., and Milke, J. (2002).

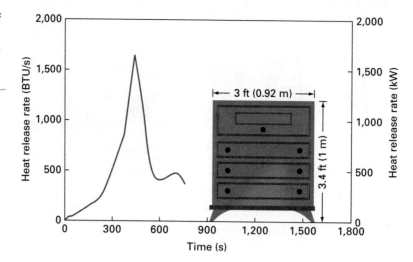

FIGURE 9.23 HRR of wooden dresser.
Source: Data derived from Klote, J., and Milke, J. (2002).

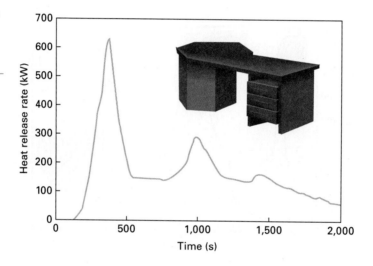

FIGURE 9.24 HRR of wooden desk.
Source: Data derived from Babrauskas (2008).

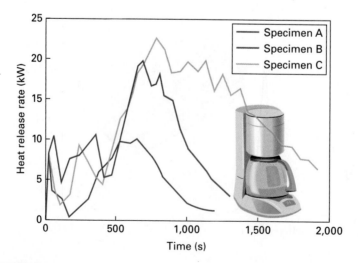

FIGURE 9.25 HRR of coffee makers.
Source: Data derived from Babrauskas (2008).

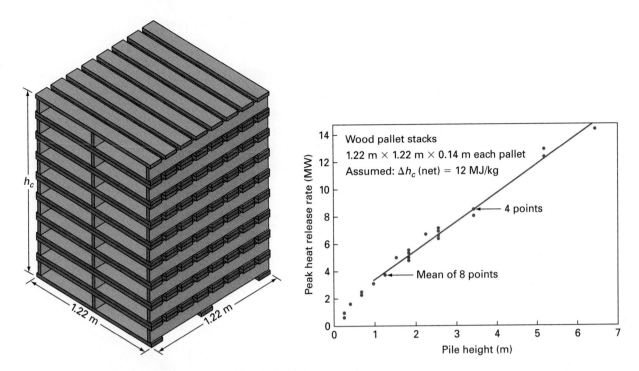

FIGURE 9.26 HRR of pallets.
Source: Data derived from Babrauskas (2008).

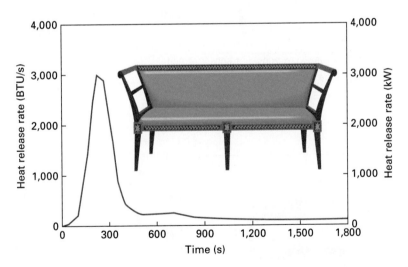

FIGURE 9.27 HRR of polyurethane sofa.
Source: Data derived from Klote, J., and Milke, J. (2002).

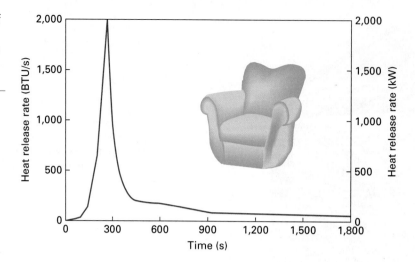

FIGURE 9.28 HRR of polyurethane chair.
Source: Data derived from Klote, J., and Milke, J. (2002).

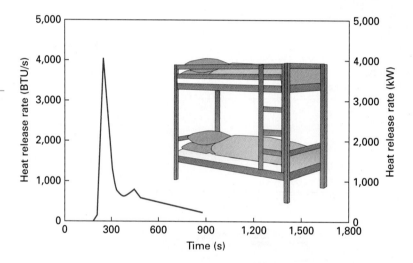

FIGURE 9.29 HRR of bunk bed.
Source: Data derived from Klote, J., and Milke, J. (2002).

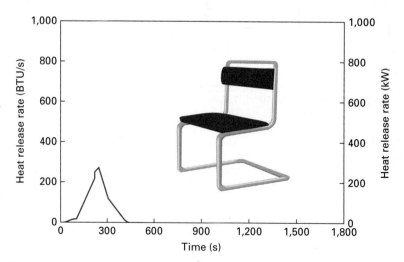

FIGURE 9.30 HRR of metal frame chair with polyurethane foam.
Source: Data derived from Klote, J., and Milke, J. (2002).

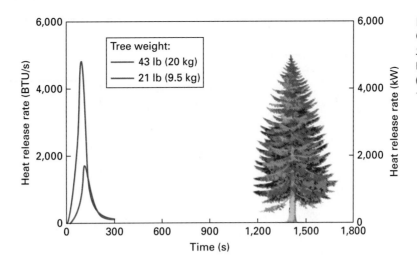

FIGURE 9.31 HRR of Christmas tree. *Source:* Data derived from Klote, J., and Milke, J. (2002).

Methods of Applying Heat Release Rate Curves

SIMPLIFIED SHAPES FOR FIRE GROWTH CURVES

One means of implementing a heat release rate curve of the various single-fuel items in an analysis is to assume that the heat release rate curve is a simplified shape. A review of the different heat release rate curves provided in Figures 9.12 through 9.31 reveals that many of the curves form a simple shape. For instance, the curves for Christmas trees and upholstered furniture form a basic triangular curve shape, while the curves for pallet and crib fires form a basic square shape (see Figures 9.32 and 9.33). The heat release rate curve for many fuel items can be simplified by obtaining or assuming a peak heat release rate (Q_{max}), and a base time (t_b) for the growth, steady state, and decay stages of the fire. Once this data is compiled, a heat release rate curve can be created and implemented in fire models.

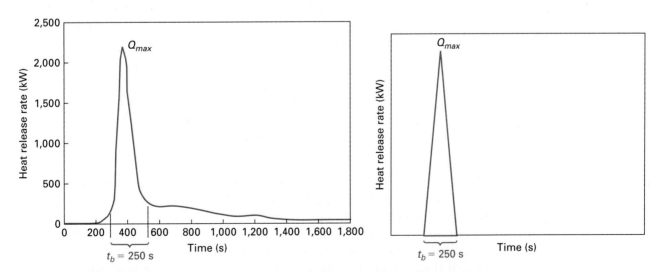

FIGURE 9.32 Heat release rate curves: simplified triangular curve shape, for example, upholstered furniture, Christmas trees.

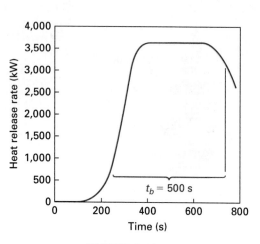

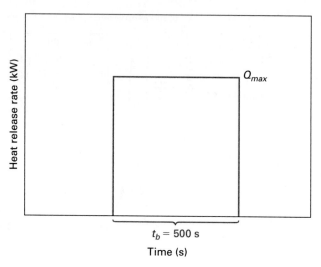

FIGURE 9.33 Heat release rate curves: simplified square curve shape, for example cribs, pallets, pool fires.

This method can often be utilized when the exact curve of the fuel cannot be obtained, but test data is available for a similar fuel. The base time of the comparison fuel and a peak heat release rate may provide substantially similar results and can be incorporated into your analysis.

t^2 FIRE GROWTH CURVES

Another method for identifying fire growth rates is to compare the items to categorize the fire growth rates of various fuel items burning. Since the first use of heat release rate tests, they have been analyzed to determine if there is a pattern to the growth rates. Based on thousands of test data collected, patterns of data plots, or four ideal curves, started to emerge. Many common fuel items burn at similar growth rates represented by an idealized parabolic equation that is approximately proportional to the time squared (t^2).

$$\dot{Q} = \alpha(t - t_o)^2$$

Where:

$\dot{Q}$ = heat release rate of fire (kW)
α = fire growth coefficient (kW/s^2)
t = time after ignition (s)
t_o = effective ignition time or incubation time (s)

The fire growth rate curves have been established by identifying the time each fuel item takes to reach a 1 megawatt (MW) fire and by identifying fire growth coefficients to standardize the curves (see Table 9.5). Four fire growth coefficients have been found to represent all of the fuel items burned, including slow, medium, fast, and ultrafast fires (see Figure 9.34).

One of the uncertainties with all of the fire growth rate methods is that the early stages of the fire are often uneven and slow growing. These rates depend greatly on the ignition sequence and fuel orientation. This initial growth period is

TABLE 9.5	Characteristic Times to Reach 1 Megawatt (MW) for t^2 Fires

COMMODITY	t_1 (s)
Wood pallets, stacked 1½ ft high	155–310
Wood pallets, stacked 5 ft high	92–187
Wood pallets, stacked 10 ft high	77–115
Wood pallets, stacked 16 ft high	72–115
Mail bags, filled, stored 5 ft high	187
Cartons, compartmented, stacked 15 ft high	58
Paper, vertical rolls, stacked 20 ft high	16–26
Cotton, polyester garments in 12-foot-high rack	21–42
"Ordinary combustibles" rack storage, 15–30 ft high	39–262
Paper products, densely packed in cartons, rack storage, 20 ft high	461
PE letter trays, filled, stacked 5 ft high on cart	189
PE trash barrels in cartons, stacked 15 ft high	53
PE bottles packed in compartmented cartons, 15 ft high	82
PE bottles in cartons, stacked 15 ft high	72
PE pallets, stacked 3 ft high	145
PE pallets, stacked 6–8 ft high	31–55
PU mattress, single, horizontal	115
PU insulation board, rigid foam, stacked 15 ft high	7
PS jars packed in compartmented cartons, 15 ft high	53
PS tubs nested in cartons, stacked 15 ft high	115
PS insulation board, rigid foam, stacked 14 ft high	6
PS bottles packed in compartmented cartons, 15 ft high	8
PP tubs packed in compartmented cartons, 15 ft high	9
PP and PE film in rolls, stakced 14 ft high	38
Distilled spirits in barrels, stacked 20 ft high	24–39

Source: Based on data from Heskestad (1983).

PE = polyethylene, PU = polyurethane, PVC = polyvinyl chloride, PS = polystyrene, PP = polypropylene, FRP = fiberglass-reinforced-polyester

often referred to as the incubation period, and it is observed on most heat release rate curves. The duration of this incubation period depends mostly on the ignition and fire spread sequence. This period lasts until the fire reaches a state of established burning or established growth. Established burning is the beginning portion of the parabolic fire growth curve that indicates the fire will no longer self-extinguish or continue to burn slowly (see Figure 9.35).

FIGURE 9.34 Fire
growth constants for
t^2 fires.

Source: Data derived from
Klote, J., and Milke, J.
(2002).

		NFPA 92B		NFPA 72
	$\alpha(\text{BTU/s}^3)$	$\alpha(\text{kW/s}^2)$	$t_g(\text{s})$	Range of t_g (s)
Slow	0.002778	0.002931	600	$t_g \geqslant 400$
Medium	0.01111	0.01127	300	$150 \leqslant t_g < 400$
Fast	0.04444	0.04689	150	$t_g < 150$
Ultrafast	0.1778	0.1878	75	N/A

The length of the incubation period is often ignored, and four idealized curves (slow, medium, fast, ultrafast) are created based on the multiplication of a fire growth coefficient by time squared (see Figure 9.36).

$$\dot{Q} = \alpha t^2$$

Where:

$\dot{Q}$ = heat release rate of fire (kW)
α = fire growth coefficient (kW/s^2)
t = time after ignition (s)

Not all fuels follow the time squared fire growth rate. Many fuel commodities may have faster growth rates depending on their configuration. Zalosh (2003) states that the initial growth rate for rack storage is more appropriately proportional to time cubed (t^3). Zalosh (2003) generated the following equation that depends on the number of tiers of storage to determine the heat release rate of the fire:

$$\dot{Q} = Nt^3$$

Where:

$\dot{Q}$ = heat release rate of fire (kW)
N = number of tiers of storage ($1 < N < 6$)
t = time after ignition (s)

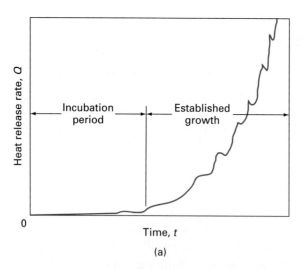

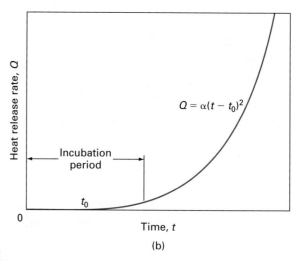

FIGURE 9.35 Fire growth curves: (a) typical HRR curve, (b) idealized parabolic curve.
Source: Data derived from Klote, J., and Milke, J. (2002).

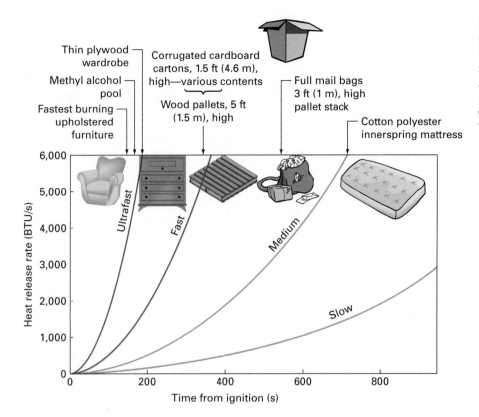

FIGURE 9.36 t^2 fire growth curves and the relation of t^2 fires to some fire tests.
Source: Data derived from Klote, J., and Milke, J. (2002).

Labels on figure: Thin plywood wardrobe; Corrugated cardboard cartons, 1.5 ft (4.6 m), high—various contents; Full mail bags 3 ft (1 m), high pallet stack; Methyl alcohol pool; Wood pallets, 5 ft (1.5 m), high; Cotton polyester innerspring mattress; Fastest burning upholstered furniture

Curves: Ultrafast; Fast; Medium; Slow

Y-axis: Heat release rate (BTU/s) — 0, 1,000, 2,000, 3,000, 4,000, 5,000, 6,000
X-axis: Time from ignition (s) — 0, 200, 400, 600, 800

Zalosh (2003) also investigated the initial fire growth rates of storage of special commodities, such as roll paper, and determined that these initial fire growth rates are proportional to time raised to the ninth power (t^9).

FUEL PACKAGE FIRE GROWTH CURVE

If multiple fuel items are located adjacent to each other (for example, a polyurethane foam sofa next to a chair), and if one item were ignited so that it would inevitably radiantly ignite the second item, it would be necessary to evaluate the influence of both items burning as a fuel package. Essentially, the first fuel ignited (i.e., the polyurethane sofa) would be expected to follow its heat release rate curve, assuming ventilation does not change. The heat release rate curve for the second fuel item would also be expected to follow its normal curve, but the second fuel item would not begin releasing its energy until it had become ignited. Therefore, Klote and Milke (2002) developed a method to account for multiple fuel items. The heat release rate curves need to be known or estimated. The point at which the second item will ignite can be calculated by using the point source radiant heat transfer calculation (it is discussed in more detail in Chapter 11, including its limitations):

$$\dot{q}'' = \frac{X_r \dot{Q}}{4\pi r^2}$$

Where: $\dot{q}''$ = energy radiated per unit time per unit area (W/m^2) or (kW/m^2)

$$X_r = \text{fraction of energy radiated relative to the total energy released}$$
$$\dot{Q} = \text{heat release rate (W or kW)}$$
$$4\pi r^2 = \text{surface area of sphere, where } r\text{(m) is the radial distance to the target fuel (m}^2\text{)}$$

This equation allows the user to determine the heat flux imposed on an object at a distance radial from the first item burning. This formula can be rearranged to find the heat release rate that is required to reach the ignition point for the second item.

$$\dot{Q}_{r,i} = 4\pi r^2 \dot{q}_r''$$

Where:

$$\dot{q}'' = \text{energy radiated per unit time per unit area (W/m}^2\text{) or (kW/m}^2\text{)}$$
$$r = \text{radial distance from point source (m)}$$
$$\dot{Q}_{r,i} = \text{heat release rate needed for ignition (kW)}$$

As soon as the heat release rate required for ignition is found, then the first fuel's heat release rate curve can be placed on a timeline. When the first fuel reaches the calculated heat release rate required to ignite the second item, then the starting point for the second fuel's heat release rate curve can be added to the total fire growth curve. See Example 9.2 for further explanation of this method.

EXAMPLE 9.2

The fuel load in a large atrium consists of the polyurethane foam–filled sofas and chairs shown in Figure 9.37. The ceiling is sufficiently high so that successful sprinkler suppression is not anticipated. The HRR of the sofas is the same as that of Figure 9.27, and the peak HRR is 2,960 BTU/s (3,120 kW). The HRR of the chairs is the same as that of Figure 9.28, and the peak HRR is 2,010 BTU/s (2,120 kW). How many sofas and chairs make up the base fuel package, and what is the HRR of the base fuel package?

Part I: Initial Estimate of Base Fuel Package

Use a radiant flux for nonpiloted ignition of $\dot{q}''_{r,i} = 20 \text{ kW/m}^2$.

For the sofa, $\dot{Q}_r = X_r\dot{Q} = 0.3(3120 \text{ kW}) = 937 \text{ kW}$. From the heat release rate formula above, $\dot{Q}_{r,i} = 4\pi r^2 \dot{q}_{r,i}$, we can solve for the radius. We obtain a separation distance from the burning sofa of:

$$r_{SD} = \sqrt{\frac{\dot{Q}_r}{4\pi\dot{q}''_{r,i}}} = \sqrt{\frac{937 \text{ kW}}{4\pi\left(\dfrac{20 \text{ kW}}{\text{m}^2}\right)}} = 1.9 \text{ m}$$

This shows that a fire on sofa 1 would not be expected to ignite sofa 2, but it would be expected to ignite chair 1. Because fires are often off center, the center of the fire is taken as the + on the side near the chair. This is conservative because ignition of the chair would be sooner than if the center of the fire were farther away.

For the chair, $\dot{Q}_r = X_r\dot{Q} = 0.3(2120 \text{ kW}) = 636 \text{ kW}$.

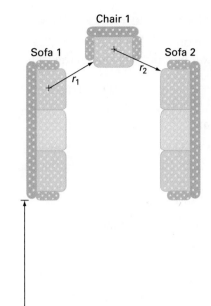

Chair 1

Sofa 1

Sofa 2

r_1

r_2

FIGURE 9.37
Arrangement of furniture in the atrium of Example 9.2.
Source: Data derived from Klote, J., and Milke, J. (2002), *Principles of Smoke Management*, Atlanta, GA: ASHRAE, Figure 2.30, page 26.

18 ft
(5.5 m)

7 ft
(2.1 m)

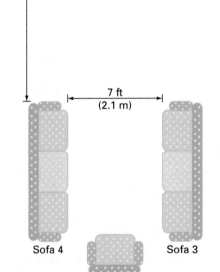

Sofa 4

Sofa 3

Chair 2

Note: $r_1 = r_2$ = 3.6 ft (1.1 m).

Again, the separation distance from the burning sofa can be calculated by the following:

$$r_{SD} = \sqrt{\frac{\dot{Q}_r}{4\pi \dot{q}_{r,i}''}} = \sqrt{\frac{636 \text{ kW}}{4\pi \left(\frac{20 \text{ kW}}{\text{m}^2}\right)}} = 1.6 \text{ m}$$

This shows that the fire of chair 1 would be expected to ignite sofa 2. Because sofas 3 and 4 are at least 18 feet (5.5 m) away from sofas 1 and 2, ignition of sofas 3 and 4 would not be expected. For now, the base fuel package is considered to consist of sofas 1 and 2 and chair 1.

Part II: Calculate the Heat Release Rate of the Base Fuel Package

In Figure 9.37, the distance from the center of the fire on sofa 1 is $r_1 = 3.6$ ft (1.1 m). The heat release rate that results in ignition at r_1 can be calculated from the following heat release equation:

$$\dot{Q}_{r,i} = 4\pi r^2 \dot{q}_r'' = 4\pi(1.1 \text{ m})^2 \left(\frac{20 \text{ kW}}{\text{m}^2} \right) = 309 \text{ kW}$$

This means that when the fire of sofa 1 reaches 293 BTU/s (309 kW), the chair would be expected to ignite. Because $r_1 = r_2$, ignition of sofa 2 is expected when the chair 1 fire also reaches 293 BTU/s (309 kW).

Calculations of the HRR are done graphically in Figure 9.38:

a. The HRR of sofa 1 is taken from Figure 9.27. The ignition time of chair 1 is determined at the intersection of the sofa 1 curve and 293 BTU/s (309 kW).
b. The HRR of chair 1 is taken from Figure 9.28.
c. The ignition time of sofa 2 is determined in a manner similar to step (a), and the HRR curve for sofa 2 is also taken from Figure 9.27.
d. The curves for sofas 1 and 2 and chair 1 are added to obtain the curve for the base fuel package.

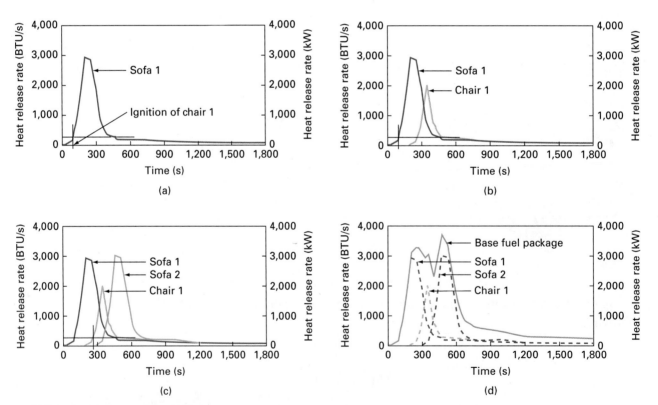

FIGURE 9.38 Graphic determination of the base fuel package of Example 9.2: (a) curve of sofa 1, and locate ignition point of chair 1; (b) curve of chair 1, transposed with sofa 1; (c) curve of sofa 1, chair 1, and ignition point of sofa 2; (d) base fuel package by adding the three previous curves.
Source: Data derived from Klote, J., Milke, J. (2002), *Principles of Smoke Management,* Atlanta, GA: ASHRAE, Figure 2.31, page 23.

Note that adding the HRR curves as done in step (d) assumes that the objects will burn as they would in free air under a calorimeter and neglects any effect of radiation from other burning objects.

Part III: Check the Base Fuel Package

Part III is a check to see if the base fuel package will ignite other materials. The highest peak of the HRR curve of Figure 9.38(d) is at 3,600 BTU/s (3,800 kW).

For the base fuel package, $\dot{Q}_r = \chi_r \dot{Q} = 0.3(3800 \text{ kW}) = 1140 \text{ kW}$. The separation distance from the base fuel package is:

$$r_{SD} = \sqrt{\frac{\dot{Q}_r}{4\pi \dot{q}''_{r,i}}} = \sqrt{\frac{1140 \text{ kW}}{4\pi \left(\dfrac{20 \text{ kW}}{\text{m}^2}\right)}} = 2.1 \text{ m}$$

The other fuel items in Figure 9.37 are 18 ft (5.5 m) from the base fuel package, so ignition of these items is not expected. So the base fuel package and its HRR curve can be used directly for a design analysis, or a simplified design HRR curve can be adapted from it. If there were fuel items within this separation distance, these items would have to be added to the base fuel package, and a new HRR curve would have to be determined.

Source: Klote, J., Milke, J. (2002). *Principles of Smoke Management.* American Society of Heating, Refrigerating and Air-Conditioning Engineers, Inc.: Atlanta, GA. WITH PERMISSION

Some Practical Applications

POOL FIRES

The heat release rate (HRR) of fires involving burning liquids is derived by calculating the potential energy of the fuel and the surface area exposed to air. When both are known, calculating a heat release rate is relatively simple. Pools may have physical restrictions that establish boundaries (i.e., dikes or reservoirs), but pools also include situations where liquids are spilled without restriction. Once the fuel is ignited, it is assumed that the flame spread across the fuel's surface is almost instantaneous. For those fuels with higher flashpoints and viscosities, the entire surface area may not ignite simultaneously, and the flame spread may need to be calculated. For more information on calculating this type of flame spread, refer to the work of Gottuk and White (2008).

The heat release rate of a pool fire can be calculated by the same HRR equation given above:

$$\dot{Q} = A_F \Delta H_c \dot{m}''$$

Where:

$\dot{Q}$ = heat release rate (kW)
A_F = fuel surface area (m^2)
ΔH_c = heat of combustion (kJ/kg)
$\dot{m}''$ = mass burning rate (kg/m^2-s)

As you can see from the equation above, the heat release rate depends greatly on the extent of the burning fuel surface area. The heat release rates of **pool fires** are no

pool fires
■ Fires involving horizontal fuel surfaces.

TABLE 9.6 — Large Pool Fire Burning Rate Data

MATERIAL	MASS LOSS RATE $\dot{m}''$ (kg/m²-sec)	HEAT OF COMBUSTION $\Delta H_{C,eff}$ (kJ/kg)	DENSITY ρ (kg/m³)	EMPIRICAL CONSTANT $k\beta$ (m⁻¹)
Cryogenics				
Liquid H₂	0.017	12,000	70	6.1
LNG (mostly CH₄)	0.078	50,000	415	1.1
LPG (mostly C₃H₈)	0.099	46,000	585	1.4
Alcohols				
Methanol (CH₃OH)	0.017	20,000	796	100[a]
Ethanol (C₂H₅OH)	0.015	26,800	794	100[a]
Simple Organic Fuels				
Butane (C₄H₁₀)	0.078	45,700	573	2.7
Benzene (C₆H₂)	0.085	40,100	874	2.7
Hexane (C₆H₁₄)	0.074	44,700	650	1.9
Heptane (C₇H₁₆)	0.101	44,600	675	1.1
Xylene (C₈H₁₀)	0.090	40,800	870	1.4
Acetone (C₃H₈O)	0.041	25,800	791	1.9
Dioxane (C₄H₈O₂)	0.018	26,200	1,035	5.4
Diethyl ether (C₄H₁₀O)	0.085	34,200	714	0.7
Petroleum Products				
Benzine	0.048	44,700	740	3.6
Gasoline	0.055	43,700	740	2.1
Kerosene	0.039	43,200	820	3.5
JP-4	0.051	43,500	760	3.6
JP-5	0.054	43,000	810	1.6
Transformer oil, hydrocarbon	0.039	46,400	760	0.7
Fuel oil, heavy	0.035	39,700	940–1,000	1.7
Crude oil	0.022–0.045	42,500–42,700	830–880	2.8
Solids				
Polymethylmethacrylate (C₅H₈O₂)ₙ	0.020	24,900	1,184	3.3
Polypropylene (C₃H₈)ₙ	0.018	43,200	905	100[a]
Polystyrene (C₈H₈)ₙ	0.034	39,700	1,050	100[a]
Miscellaneous				
561® Silicon Transformer Fluid	0.005	28,100	960	100[a]

[a] These values are to be used only for computation purposes; the true values are unknown.

different, except that the burning rate is not a set value with all surface areas (diameters). Much experimental work has been performed on pool fires, and it has been shown that the burning rate increases with an increasing diameter (0.2 m in diameter and greater) of the pool fire until the pool gets to a set diameter. At this point, a maximum average burning rate is achieved, which is known as the asymptotic burning rate ($\dot{m}''_\infty$). Two empirical constants have been developed to characterize this burning rate adjustment, which include the extinction-absorption coefficient of the flame (k) and a mean beam length corrector (β). To better account for this burning rate adjustment due to diameter, the following equation is used:

$$\dot{m}'' = \dot{m}''_\infty \cdot (1 - e^{-k\beta D})$$

Where:
$\dot{m}''_\infty$ = asymptotic mass burning rate (kg/m²-s) (values in Tables 9.3 and 9.6)
$k\beta$ = empirical constant (m^{-1})
D = diameter (m)

Reminder: The surface area of a pool fire is calculated by $A = \pi r^2$ or $A = \dfrac{\pi D^2}{4}$.

If the fuel is not a circle in shape, then an effective diameter for a noncircular pool fire can be calculated by the following equation:

$$D = \sqrt{\frac{4A}{\pi}}$$

Taking this adjustment and substituting it into the HRR, we arrive at:

$$\dot{Q} = A_F \Delta H_c \dot{m}''_\infty (1 - e^{-k\beta D})$$

EXAMPLE 9.3

A truck carrying kerosene and located 50 meters away from a warehouse wall collides with a utility pole and causes a kerosene spill. The kerosene spreads horizontally and results in a pool approximately 5.5 meters in diameter. This pool ignites. What is the HRR?

$$\dot{Q} = A_F \Delta H_c \dot{m}'' (1 - e^{-k\beta D})$$

Known: From Tables 9.1, 9.3, and 9.6:

$$k\beta = 3.5 \text{ m}^{-1}$$
$$D = 5.5 \text{ m}$$
$$\dot{m}'' = 0.039 \text{ kg/m}^2\text{-s}$$
$$\Delta H_c = 43{,}200 \text{ kJ/kg}$$

Assumptions: The entire pool surface area ignites instantly, and there is no flame spread.

Analysis: The surface area (A_F) must be determined prior to the equation being solved.

$$A = \frac{\pi D^2}{4} = \frac{\pi (5.5 \text{ m})^2}{4} = 23.76 \text{ m}^2$$

The heat release rate can now be calculated by substituting all values into the HRR equation:

$$\dot{Q} = A_F \Delta H_c \dot{m}'' (1 - e^{-k\beta D})$$

$$= (23.76 \text{ m}^2)\left(43{,}200\frac{\text{kJ}}{\text{kg}}\right)\left(0.039\frac{\text{kg}}{\text{m}^2 \text{ s}}\right)(1 - e^{-(3.5\text{m}^{-1})(5.5\text{m})})$$

Unit cancellation:

$$(\text{m}^2)(\text{kJ/kg})(\text{kg/m}^2\text{-s})((\text{m}^{-1})(\text{m})) = \text{kJ/s or kW}$$
$$\dot{Q} = 40{,}009.3 \text{ kW or } 40.01 \text{ MW}$$

SOLID FUELS: UPHOLSTERED FURNITURE

Dr. Vytenis Babrauskas developed a means to estimate the peak HRR of upholstered furniture by simplifying the expression with the use of small-scale calorimeter tests.

$$\dot{Q} = 0.63[\dot{Q}''_{bs}] \, [\text{mass factor}][\text{frame factor}][\text{style factor}]$$

Where:

$\dot{Q}$ = estimated peak full-scale HRR (kW)

$\dot{Q}''_{bs}$ = average rate of heat release over the first 180 seconds in a cone calorimeter test at 25 kW/m^2 irradiance (kW/m^2)

mass factor = combustible mass, in kg

frame factor = 1.66 for noncombustible frames, 0.58 for melting plastic, 0.30 for wood, or 0.18 for charring plastic

Style factor = 1.0 for plain, primarily rectilinear construction; 1.5 for ornate, convoluted shapes; or 1.25, or as appropriate, for intermediate shapes

Babrauskas performed hundreds of tests on upholstered furniture and thus developed this basic correlation. He found that by taking the cone calorimeter (small-scale test) results plus the mass of the furniture item in kg, the frame factor and its simple geometric shape (style factor) were the primary influences on estimating the peak HRR. However, because cone calorimeter tests are not always available for furniture items, Babrauskas developed a more general correlation:

$$\dot{Q} = 210[\textit{fabric factor}][\textit{padding factor}][\textit{mass factor}][\textit{frame factor}][\textit{style factor}]$$

Where:

fabric factor = 1.0 for thermoplastic fabrics (i.e., polyolefin melts prior to burning), 0.4 for cellulosic fabrics (i.e., cotton), or 0.25 for PVC or polyurethane film-type coverings (faux leather)

Padding factor = 1.0 for polyurethane foam or latex foam, 0.4 for cotton batting, 1.0 for mixed materials, or 0.4 for neoprene foam

While this correlation cannot be used for every solid, nor can small-scale tests be easily extrapolated into full-scale estimates for all fuels, it does provide an estimate for furniture items.

BURNING DURATION

Calculating the HRR of a fire is an important component in determining the fire's effect on various fire protection elements, structural integrity, ignition of secondary and tertiary fuels, as well as many other applications. Equally important is the duration of the fire. Some basic calculations can be performed to analyze the potential duration of a fire.

Let's use liquids, or pool fires, as an example. The duration of burning for pool fires can be easily calculated, assuming no interference by the lack of oxygen. The surface area of the liquid is the primary mechanism for limiting the rate at which the fuel will burn. For example, 5 gallons of gasoline, if placed in a container with only a 1- or 2-inch opening, will burn for a much longer time than taking the 5 gallons of gasoline and spilling it throughout a large warehouse. The burning duration (t_b) can be estimated by the following equation:

$$t_b = \frac{4V}{\pi D^2 \gamma}$$

Where:
V = volume of liquid (m^3)
D = pool diameter (m)
γ = regression rate (m/s)

The regression rate is the volumetric loss of liquid per surface area. In other words, the depth of the fuel decreases as the fuel is consumed from the volume of liquid. This value is obtained from both the material properties and experimental results of the liquid, and it can be calculated by using the following equation:

$$\gamma = \frac{\dot{m}''}{\rho}$$

Where:
$\dot{m}''$ = mass burning flux (kg/m^2-s)
ρ = density of fuel (kg/m^3)

EXAMPLE 9.4

Find the burning duration of the kerosene pool fire from Example 9.3. Assume that the truck was carrying 15 m^3, and all fuel entered into the pool fire.

Known: From Tables 9.1, 9.3, and 9.6:

$$\rho = 820 \text{ kg/m}^3$$
$$D = 5.5 \text{ m}$$
$$V = 15 \text{ m}^3$$
$$\dot{m}'' = 0.039 \text{ kg/m}^2\text{-s}$$

Assumptions: The entire pool surface area ignites instantly, and there is no flame spread. All fuel enters into the pool area and is consumed.

Analysis: First, the regression rate needs to be solved for:

$$\gamma = \frac{\dot{m}''}{\rho} = \frac{0.039 \dfrac{\text{kg}}{\text{m}^2\text{s}}}{820 \dfrac{\text{kg}}{\text{m}^3}} = 4.75 \times 10^{-5} \text{ m/s}$$

Next, all values can be substituted into the burning duration equation and solved:

$$t_b = \frac{4V}{\pi D^2 \gamma} = \frac{4 \cdot 15 \ \text{m}^3}{\pi (5.5 \ \text{m})^2 (4.75 \times 10^{-5} \ \text{m/s})} = \mathbf{13{,}274 \ s \ or \ 221 \ min}$$

FLAME HEIGHT

The flame height can also be calculated if you know the heat release rate of the burning fuel. Flame height provides an indication of the heat release rate in diffusion flames. The premise is that, if more fuel gases are present, a greater volume of space is required for the fuel/air ratio to become proper for combustion. Fuel gases near the flame's center, where the fuel/air ratio is too rich to support combustion, rise until they can mix with adequate oxygen for combustion to occur.

Many correlations have been developed for assessing the heat release rate and the corresponding flame height. Two, from Thomas, Webster, and Raftery (1961) and Heskestad (1983), are shown below:

Thomas

$$H_f = 42D \left(\frac{\dot{m}''}{\rho_a \sqrt{gD}} \right)^{0.61}$$

Where:

H_f = flame height (m)
D = diameter of the fire (m)
$\dot{m}''$ = mass burning flux (kg/m^2-s)
ρ_a = density of ambient air (kg/m^3)
g = gravitational acceleration (m/s^2)

EXAMPLE 9.5

Determine the flame height for the fire in Example 9.3. Recall that a truck carrying kerosene was 50 meters away from a warehouse wall when it collided with a utility pole and caused a kerosene spill. The kerosene spread horizontally and resulted in a pool approximately 5.5 meters in diameter. The fire's HRR was calculated as 40,009 kW.

$$H_f = 42D \left(\frac{\dot{m}''}{\rho_a \sqrt{gD}} \right)^{0.61}$$

Known: From Tables 9.1, 9.3, and 9.6:

$D = 5.5 \ \text{m}$
$\dot{m}'' = 0.039 \ \text{kg/m}^2\text{-s}$
$\rho_a = 1.20 \ \text{kg/m}^3$
$g = 9.81 \ \text{m/s}^2$

$$H_f = 42D \left(\frac{\dot{m}''}{\rho_a \sqrt{gD}} \right)^{0.61} = 42(5.5 \ \text{m}) \left(\frac{0.039 \dfrac{\text{kg}}{\text{m}^2 \text{s}}}{1.2 \dfrac{\text{kg}}{\text{m}} \sqrt{9.81 \dfrac{\text{m}}{\text{s}^2} (5.5 \ \text{m})}} \right)^{0.61} = 8.5 \ \text{m}$$

Heskestad

$$H_f = 0.23\dot{Q}^{2/5} - 1.02D$$

Where:

H_f = flame height (m)
$\dot{Q}$ = heat release rate (kW)
D = diameter of the fire (m)

EXAMPLE 9.6

Determine the flame height for the fire in Example 9.3. Recall that a truck carrying kerosene located 50 meters away from a warehouse wall collided with a utility pole and caused a kerosene spill. The kerosene spread horizontally and resulted in a pool approximately 5.5 meters in diameter. The fire's HRR was calculated to be 40,009 kW.

Known: From Tables 9.1, 9.3, and 9.6:

D = 5.5 m
$\dot{m}''$ = 0.039 kg/m²-s
ρ_a = 1.20 kg/m³
g = 9.81 m/s²
$H_f = 0.23\dot{Q}^{2/5} - 1.02\ D = 0.23(40{,}009\ \text{kW})^{2/5} - 1.02(5.5\ \text{m}) = 10.3\ \text{m}$

NFPA 921 (2004) Method

An adaptation of the Heskestad correlation is found in the *Guide for Fire and Explosion Investigations* (NFPA 921, 2004). This document references the following flame height correlation for use in assessing flame heights:

$$H_f = 0.174(k\dot{Q})^{0.4}$$

Where:

H_f = flame height (m)
$\dot{Q}$ = convective heat release rate (kW)
k = location of the flame in relation to walls
     1 = flame away from walls
     2 = flame is against one wall
     4 = flame is against two walls (in a corner)

EXAMPLE 9.7

What is the flame height of a 1,200 kW fire along a wall?

$$h_f = 0.174(k\dot{Q})^{0.4} = 0.174(2 \cdot 1200\ \text{kW})^{0.4} = \textbf{3.9 m}$$

NFPA 921 also reverses the calculation, assuming that investigators have a known flame height from either witnesses or a fire pattern left remaining on a wall. In this case, the heat release rate can be estimated with the following equation:

$$\dot{Q} = \frac{79.18(H_f)^{5/2}}{k}$$

Where:
$\dot{Q}$ = convective heat release rate (kW)
H_f = flame height (m)
k = location of the flame in relation to walls
   1 = flame away from walls
   2 = flame is against one wall
   4 = flame is against two walls (in a corner)

NOTE: *The use of this formula to analyze the flame height and calculate the heat release rate has several assumptions and limitations that need to be investigated further by the user. The user is encouraged to review the original study by Heskestad (1983) for accurate usage.*

EXAMPLE 9.8

What is the heat release rate of a flame that was reported to be 2 meters in height and was located in the center of the room?

$$\dot{Q} = \frac{(79.18)(2 \text{ m})^{5/2}}{1} = 447.9 \text{ kW}$$

Keep in mind that relying on flame height alone to determine the heat release rate could prove unreliable. Wider flames with multiple apexes indicate higher heat release rates than would be calculated from simply evaluating the uppermost apex. It is important for the user to investigate the original research regarding these formulas and correlations to ensure their proper usage for a given scenario.

RELATIONSHIP OF HEAT RELEASE RATE AND THE PLUME

The time it takes for sprinklers or heat detectors to activate is based on the temperature rise (ΔT) and the velocity (U) of the heated gases, which is a function of the HRR (see Figure 9.39). As pointed out above, the HRR is also the central factor in determining the flame height.

Once a fire growth rate (HRR over time) is established for the burning fuel, then the HRR as a function of time can be substituted into further correlations. Five plume correlations for gas temperatures (ΔT) and gas velocities (U) in the plume have been developed: Zukowski, Heskestad, Alpert ceiling jet correlation,

FIGURE 9.39 An idealization of the ceiling jet flow beneath a ceiling. *Source:* Data derived from Karlsson, B, and Quintiere, J. (2000).

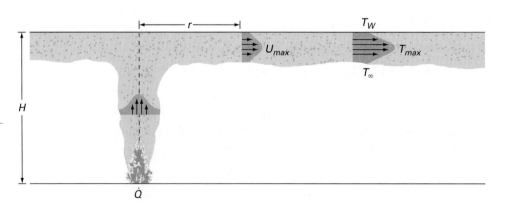

TABLE 9.7 List of the Primary Plume Correlations

CORRELATION	ΔT (TEMPERATURE RISE) (K)	U (GAS VELOCITY) (m/s)	VIRTUAL ORIGIN (Z_O)
Zukowski	$13.74\left[\dfrac{\dot{Q}_c^{2/5}}{Z}\right]^{5/3}$	$0.59\left[\dfrac{\dot{Q}_c}{Z}\right]^{1/3}$	N/A
Heskestad	$25\left[\dfrac{\dot{Q}_c^{2/5}}{Z-Z_o}\right]^{5/3}$	$1.0\left[\dfrac{\dot{Q}_c}{Z-Z_o}\right]^{1/3}$	$Z_o = 0.083\dot{Q}^{2/5} - 1.02D$ *Note:* $\dot{Q}$ is equal to total energy when calculating the virtual origin.
McCaffrey	$2.27\left[\dfrac{\dot{Q}^{2/5}}{Z}\right]^{5/3}$	$1.1\left[\dfrac{\dot{Q}}{Z}\right]^{1/3}$	N/A
Kung	$4.58\left[\dfrac{\dot{Q}_c^{2/5}}{Z^*-Z_o}\right]^{5/3}$	$1.29\left[\dfrac{\dot{Q}_c}{Z^*-Z_o}\right]^{1/3}$	$Z_o = 0.095\dot{Q}^{2/5} - Z_{oo}$ $Z_{oo} = 1.6$ m (2 tier) $\phantom{Z_{oo} = } 2.4$ m (3 and 4 tier) *Note:* Kung correlation used with large-scale arrays.
Alpert (ceiling jet correlation)	$16.9\dfrac{\dot{Q}^{2/3}}{H^{5/3}}$; $r/H < 0.18$	$0.95\left[\dfrac{\dot{Q}}{H}\right]^{1/3}$; $r/H < 0.15$	Near field: $r/H < 0.18$ Far field: $r/H > 0.18$
	$5.8\dfrac{(\dot{Q}/r)^{2/3}}{H}$; $r/H > 0.18$	$0.197\dfrac{(Q/H)^{1/3}}{(r/H)^{5/6}}$; $r/H > 0.15$	

Where:

- $\bullet$ = detector location
- r = radius (m) from centerline of plume to detector location
- H = height of ceiling above fuel surface (m)
- Z = flame height above fuel source (m)
- Z_o = virtual origin (m)
- D = effective diameter (m)
- $\dot{Q}_c$ = convective HRR (kW), approximately 60 to 80 percent of $\dot{Q}$. Example: $\dot{Q}_c = \dot{Q} * 0.7$
- $\dot{Q}$ = total energy released (kW)

McCaffrey, and Kung. Each correlation is listed in Table 9.7. One important concept that has yet to be covered that will be found with these correlations is the concept of *virtual origin* (see Figure 9.40).

Most plume correlations assume that the fire is a point source. However, when you have a fire that has a larger base area (i.e., pool fires or three-dimensional fires), the elevation of the heat source must be adjusted to account for these changes so that you can calculate the plume as a point source properly. For instance, if the area of the fire source is large compared to the HRR over that area,

FIGURE 9.40 Fire
plume with virtual
origin.
Source: Data derived from
Karlsson, B, and Quintiere,
J. (2000).

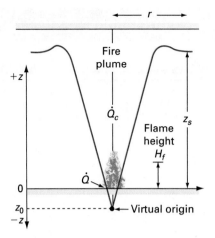

then the value of the virtual origin (Z_o) may have to be negative and lie beneath the fuel source to calculate it properly as a point source. If the area is small and has a high HRR, then Z_o may be positive, or located above the fuel source.

SPRINKLER AND HEAT DETECTOR ACTIVATION

Automatic sprinklers have thermal elements that are temperature-rated (either frangible bulbs or fusible links) that are set to fail at a specific temperature and allow water to flow. Now that plume correlations have been introduced, it is possible to calculate the temperature and velocity in the ceiling jet at the specific location of the sprinkler given the HRR. The only element that is not provided yet is the time it takes for the sprinkler to reach the activation temperature. Remember that the temperature in the gas layer is not the same as the temperature in the sprinkler element. The sprinkler has metal elements that conduct heat away from themselves, which slows the response of the sprinkler in reaching the activation temperature. The velocity and temperature difference at the specific sprinkler location has an impact on the activation of the sprinkler.

The delay in sprinkler temperature increase is known as the response time index (RTI) of the sprinkler. This is an experimentally measured sprinkler sensitivity for the specific type of sprinkler and accounts for the conductive heat exchange from the hot gases into the thermal element. The experiment is known as the plunge test and is recorded in units of m-s$^{1/2}$.

Therefore, the sprinkler activation correlation accounts for all of the elements discussed above, in the following equation:

$$T_d(t + \Delta t) = T_d(t) + \left[\frac{\sqrt{u_g(t)}}{\mathrm{RTI}} (T_g(t) + T_d(t))(\Delta t) \right]$$

Where:

T_d = temperature at the detector (°C)
t = time (s)
Δt = change in time (s)
u_g = ceiling jet velocity (m/s)
RTI = response time index (m-s$^{1/2}$)
T_g = ceiling jet gas temperature (°C)

All of the plume algorithms are an iterative process, meaning that these formulas are repetitive and adjust according to time. For instance, for every increase in time, the heat release changes, which in turn affects both the ceiling jet temperature and velocity. This change in temperature and velocity influences the temperature at the sprinkler head. Most heat detectors work on a very similar principle of temperature rise, which can also be calculated with the above correlation. For a more detailed explanation of plume correlations and their use in detector activation, refer to the SFPE *Handbook of Fire Protection Engineering* (2008).

EXAMPLE 9.9 CULMINATION OF PLUME EQUATIONS

Example 9.9 will tie in many of the fundamental concepts presented in this chapter, as well as drawing in information from other chapters throughout the text. It is easier to see these correlations applied to a real problem rather than listing them with a corresponding text discussion.

A dry pipe sprinkler system is being installed on a 6-meter high warehouse ceiling to protect three-tier-high (4.3 meters) storage of polystyrene cups in compartment cartons. There is a 90-second delay between sprinkler actuation and the arrival of water. The sprinkler links are 138°C, 168 (m-s)$^{1/2}$ RTI, situated at the optimum depth below the smooth ceiling.

1. Estimate an approximate flame height at the time of sprinkler activation and also at the time of sprinkler release. Also estimate the ceiling jet velocity and temperature at the nearest sprinkler at both these times (use Kung correlation).
2. Estimate the radiant heat flux at a commodity located 5 meters from the ignition site when the sprinklers open. Assume the flame to be a cylinder with a base equal to a diameter that encompasses four cartons.

Ultimately, the time it takes for the sprinkler to activate will be based on the temperature and the jet velocity of the heated gases, which is a function of HRR. The heat release rate is also the central factor in determining the flame height. Therefore, the best way to determine the approximate flame height at the time of sprinkler activation is to determine the convective HRR growth of such a commodity burning. An estimated fire growth can be obtained by referring to tests performed by Zalosh (2003) on polystyrene cups that were packaged in corrugated paper cartons on pallets arranged in two-, three-, four-, and five-tier storage. These tests found that the initial growth period can be characterized as heat release rates increasing by the third power, giving the following calculation:

$$Q_{con} = 0.0448N(t - t_o)^3; \text{ for: } t - t_o < 26 \text{ s}, 1 < N < 6$$

The fire growth rate (HRR over time) is provided in Figure 9.41. The first 26 seconds are based on the above correlation, but after 26 seconds, assume that the growth rate will follow a standard t^2 fire growth rate. Once a fire growth rate (HRR over time) is established for the burning commodity, then the HRR, as a function of time, can be substituted into the Kung correlation for gas temperatures and velocity.

FIGURE 9.41 Fire
growth curve, Kung
correlation.

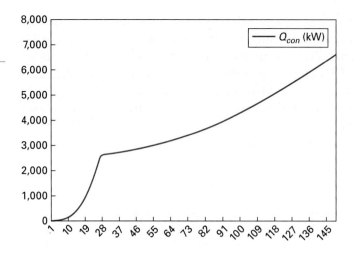

Kung:

$$\Delta T = 4.58 \left[\frac{\dot{Q}_c^{2/5}}{Z^* - Z_o} \right]^{5/3}$$

$$U = 1.29 \left[\frac{\dot{Q}_c}{Z^* - Z_o} \right]^{1/3}$$

$$Z_o = 0.095 \dot{Q}^{2/5} - Z_{oo}; \ Z_{oo} = 2.4 \ (3 \ \text{tier})$$

The next step is to determine the time required to activate the sprinkler. The sprinkler activation model is based on the following equation:

$$T_d(t + \Delta t) = T_d(t) + \left[\frac{\sqrt{u_g(t)}}{\text{RTI}} (T_g(t) + T_d(t))(\Delta t) \right]$$

The RTI has been given in the problem statement as 168 $(\text{m-s})^{1/2}$. The ceiling jet velocity (u_g) and gas temperature (T_g) is based on the HRR as a function of time calculated above by the Kung correlation. Therefore, the temperature at the detector (T_d) is also based on the changing velocity and gas temperature as a function of HRR and time. The activation temperature for the sprinkler has also been given in the problem statement as 138°C. Thus, when T_d is equal to 138°C, the sprinkler is expected to activate. The fire growth after 26 seconds is assumed to follow a modified t^2 fire growth curve and can be calculated in an Excel spreadsheet. Table 9.8 provides an overview of the Kung correlation calculations in conjunction with the HRR and detector temperature calculations. The detector temperature reaches the activation temperature of 138°C at 41 seconds, which corresponds to a convective heat release rate of 2,846.3 kW.

The approximate flame height can be calculated using the Heskestad correlation for cylindrical and conical flames and using the following equation:

$$H_f = 0.23 \dot{Q}^{2/5} - 1.02D$$

The diameter of the flame is based on the assumption that the flame is a cylinder with a base equal to a diameter that encompasses four cartons. Each carton is assumed to be the standard 1.22 m by 1.22 m. The area is therefore 1.4884 m^2 per carton. The area of four cartons, two wide and two deep (2.44 m by 2.44 m), is

TABLE 9.8 — Kung Correlation Summary

TIME (s)	Q_{con} (kW)	z_o	ΔT [K]	U [m/s]	T_D [k] (°C)
0	0	0	0	0	293.15 (20)
5	18	−2.1	3.40	2.17	293.21 (20.06)
10	144	−1.71	16.32	4.49	293.78 (27.7)
15	486	−1.27	46.09	7.05	296.08 (49.57)
20	1,152	−0.81	108.81	9.95	302.75 (96.96)
25	2,250	−0.32	244.15	13.38	320.5 (138.3)
30	2,699.78	−0.16	315.68	14.61	350.89 (77.74)
35	2,760.75	−0.14	326.27	14.77	379.7 (106.55)
40	2,831.1	−0.12	338.77	14.96	406.81 (133.6)
41	2,846.3	−0.11	341.52	15.00	412.06 (138.9)

assumed to be the configuration used. Therefore, the area is 5.954 m². The effective diameter for a noncircular pool fire can be calculated by the following equation (SFPE, 2008, pp. 3–285):

$$D = \sqrt{\frac{4A}{\pi}}$$

Therefore, the assumed effective diameter for this commodity array is $D = 2.75$ m.

The flame height at the time of sprinkler activation can then be approximated from the following:

$$H_f = 0.23(2846.3 \text{ kW})^{2/5} - 1.02(2.75) = \mathbf{2.73 \text{ m}}$$

The sprinkler system is a dry pipe system and has a delay of 90 seconds from activation of the sprinkler to the arrival of water. Therefore, it is expected that the HRR will continue to increase during this time.

Next, we want to determine what the heat release rate is at the time of actual water delivery, which would be at 131 seconds (41 s + 90 s). The formula used above for the fire growth rate may not be suitable past 26 seconds according to Zalosh (2003) and the SFPE *Handbook of Fire Protection Engineering* (SFPE, 2008, pp. 2–30). Most commonly, the fire growth rate is based on the second power of time, or t^2 fire growth curves. If we assume that the fire growth follows the t^2 growth rate and the 26-second t^3 rate, we come to the approximation shown in Figure 9.42.

The first 26 seconds of the curve in Figure 9.42 is based on a t^3 fire growth curve as established in Zalosh and SFPE for this type of commodity. The calculations have been calculated using an Excel spreadsheet. At 131 seconds, the time in which water would actually be flowing from the sprinkler, the convective HRR is approximated at 5,750.35 kW.

FIGURE 9.42 Fire
growth curve, Kung
correlation.

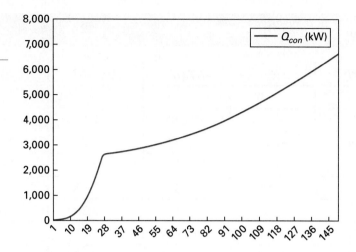

Given the HRR, the gas velocity and gas temperature can now be calculated using the Kung correlation.

$$\Delta T = 4.58 \left[\frac{5750.35 \text{ kW}^{2/5}}{(1.7 - 0.63)} \right]^{5/3} = 1313.3 \text{ K} => T_g = \textbf{1606.4 K } or \textbf{ 1333.3°C}$$

$$U = 1.29 \left[\frac{5750.35 \text{ kW}}{1.7 - 0.63} \right]^{1/3} = \textbf{22.6 m/s}$$

The approximate flame height can be calculated using the Heskestad correlation for cylindrical and conical flames and using the following equation:

$$H_f = 0.23 \dot{Q}^{2/5} - 1.02D$$

The diameter of the flame is still assumed to be $D = 2.75$ m. The flame height can then be approximated at the time of sprinkler flow and can be found from the following:

$$H_f = 0.23(5750.35 \text{ kW})^{2/5} - 1.02(2.75) = \textbf{4.53 m}$$

Once the flame height has been calculated, the heat flux can be approximated by using the following:

$$q'' = E\varphi\tau$$

Where:
$$\tau = \text{transmissivity} = 1$$
$$E = \epsilon\sigma T_f^4$$
$$\varphi = \text{view factor}$$

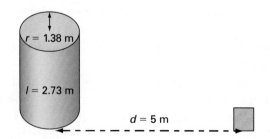

$$T_f = 1{,}190 \text{ K (Drysdale [1998], Table 2.10)}$$

$$F_{12} = \frac{1}{\pi D}\tan^{-1}\!\left(\frac{L}{\sqrt{D^2-1}}\right) + \frac{L}{\pi}\left[\frac{A-2D}{D\sqrt{AB}}\tan^{-1}\sqrt{\frac{A(D-1)}{B(D+1)}}\right.$$

$$\left. - \frac{1}{D}\tan^{-1}\sqrt{\frac{D-1}{D+1}}\right]$$

Where:

$$D = d/r;\ L = l/r;\ A = (D+1)^2 + L^2;\ B = (D-1)^2 + L^2$$
$$D = 5\text{ m}/1.38\text{ m} = 3.62;\ L = 2.73\text{ m}/1.38\text{ m} = 1.98;\ A = (3.62+1)^2$$
$$+ 1.98^2 = 25.26;\ B = (3.62-1)^2 + 1.98^2 = \mathbf{10.78}$$

$$F_{12} = \frac{1}{\pi 3.62}\tan^{-1}\!\left(\frac{1.98}{\sqrt{3.62^2-1}}\right)$$

$$+ \frac{1.98}{\pi}\left[\frac{25.6-2(3.62)}{3.62\sqrt{25.26*10.78}}\tan^{-1}\sqrt{\frac{25.26(2.62)}{10.78(4.62)}} - \frac{1}{3.62}\tan^{-1}\sqrt{\frac{2.62}{4.62}}\right]$$

$$F_{12} = \mathbf{0.0959}$$

E: $E = \epsilon\sigma T_f^4$

$\sigma = 5.67 \times 10^{-11}\text{ W/m}^2\text{K}^4$

$T_f = 1{,}190\text{ K}$

$\epsilon = 1 - \exp(-KL) = 1 - \exp(-5.3\text{ m}^{-1} * 0.7(2.75\text{ m})) = 0.9999$

$K = 5.3\text{ m}^{-1}$ (Drysdale [1998], Table 2.10)

$L = 0.7\,D$ (Drysdale [1998], Table 2.9)

$E = 0.9999(5.67 \times 10^{-11})(1190\text{ K})^4 = 113.69\text{ kW/m}^2$

q'':

$$q'' = 113.69\text{ kW/m}^2(0.0959) = \mathbf{10.9\text{ kW/m}^2}$$

Source: Industrial Fire Protection Engineering graduate course at Worcester Polytechnic Institute.

Summary

The heat release rate (HRR) is the single most important variable in a fire (Babrauskas and Peacock, 1991). The HRR provides fire safety professionals a way to quantify and discuss the complexity of fires, including their size, potential impact, and correlation with other variables (Icove and DeHaan, 2009). This chapter introduced you to the importance of understanding and recognizing heat release rates and their measurement.

Review Questions

1. Define heat release rate (HRR) and explain its importance, in your own words. Specifically, what elements affect the HRR of a single fuel package? What is the mathematical formula for determining the HRR, and what does each element of the equation represent? Write your answers in your own words. Provide all references.

   Determine the peak HRR of the following items in Review Questions 2–5 (provide references for your answers):

2. Bunk bed
3. Sofa
4. Mattress
5. Kiosk
6. Given the following graphs and information, discuss the hazard associated with this fuel. What is the peak HRR? Identify the *x*- and *y*-axes on each graph. Discuss what mass loss rate and the upper layer temperature represent. Specifically, look at the HRR graph and identify the different growth stages of this fuel at the different points denoted on the graph (i.e., what stage is this fire in at each particular point)?

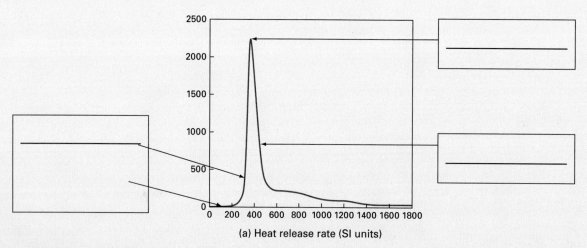

(a) Heat release rate (SI units)

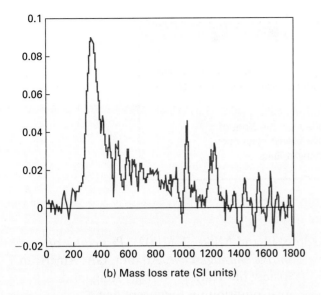

(b) Mass loss rate (SI units)

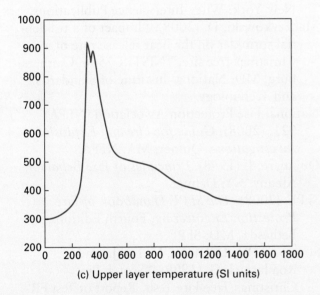

(c) Upper layer temperature (SI units)

Dynamics Tools (FDTs) spreadsheets to check your answer. Provide both your calculations by hand and a copy of the FDTs spreadsheet check.

a. 2 m^2 transformer oil fire
b. 2 m^2 wood pallets, pallet stack height is 1.5 m
c. Christmas tree with the following HRR curve:

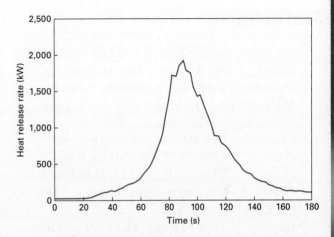

9. Why are there so many different flame height correlations? What is a correlation and why are they used for fire problems?

10. A truck carrying kerosene fuel flipped over while driving down the road. (a) What is the heat release rate for a 5.5-m diameter kerosene pool fire? Assume a combustion efficiency of 70 percent and 10 gallons of fuel spilled. (b) What is the flame height? (c) What would be the burning duration of this pool fire if 10 gallons were spilled? (d) What would the HRR be if there were 100 gallons spilled? (e) What would be the flame height? (f) What would the burning duration be for 100 gallons? (g) Is there a difference in HRR or flame height with the addition of volume? If not, why not? If so, why so?

7. What is the heat release rate for a 30-m diameter gasoline pool fire?

8. Estimate the flame height from the following burning objects. Show all three calculations by hand for each question. Then use Fire

References

American Society for Testing and Materials E1345 (2009). *Standard Test Method for Heat and Visible Smoke Release Rates for Materials and Products Using an Oxygen Consumption Calorimeter*. West Conshohocken, PA: ASTM International.

Babrauskas, V. (2008). "Heat Release Rates" (Section 3, Chapter 1) in *SFPE Handbook of Fire Protection Engineering*. Quincy, MA: SFPE/NFPA.

Babrauskas, V., and Peacock, R. (1991). "Heat Release Rate: The Single Most Important Variable in Fire Hazard." *Fire Safety Journal*. West Sussex, England: Elsevier Science Publishers.

Drysdale, D. (1998). *An Introduction to Fire Dynamics*. Chichester, West Sussex, Eng.: Wiley.

Drysdale, D. (2008). "Physics and Chemistry of Fire" (Section 1, Chapter 1). *Fire Protection Handbook*, Twentieth Edition. Quincy, MA: NFPA.

Gottuk, D., and White, D. (2008). "Liquid Fuel Fires" (Section 2, Chapter 15), in *SFPE Handbook of Fire Protection Engineering*. Quincy, MA: SFPE/NFPA.

Heskestad, G. (1983). "Luminous Heights of Turbulent Diffusion Flames." *Fire Safety Journal*, 5, 103–108.

Icove, D., and DeHaan, D. (2009). *Forensic Fire Scene Reconstruction*. Upper Saddle River, NJ: Brady Publishing.

Janssens, M. (2008). "Calorimetry." *SFPE Handbook of Fire Protection Engineering*, Fourth Edition. Bethesda, MD: SFPE.

Karlsson, B., and Quintiere, J. (2000). *Enclosure Fire Dynamics*. Boca Raton, FL: CRC Press.

Klote, J., and Milke, J. (2002). *Principles of Smoke Management*. Atlanta, GA: American Society of Heating, Refrigerating, and Air-Conditioning Engineers, Inc.

Kuvshinoff, B. (1977). *Fire Sciences Dictionary*. New York: Wiley-Interscience Publications.

Madrzykowski, D. (2008). "Impact of a residential sprinkler on the heat release rate of a Christmas tree fire," NISTIR 7506. Gaithersburg, MD: National Institute of Standards and Technology.

National Fire Protection Association (NFPA) 921. (2008). *Guide for Fire and Explosion Investigations*. Quincy, MA: NFPA.

Quintiere, J. (1998). *Principles of Fire Behavior*. Albany, NY: Delmar.

SFPE. (2008). *The SFPE Handbook of Fire Protection Engineering*, Fourth Edition. Bethesda, MD: SFPE.

Stroup, D. W., DeLauter, L., Lee J., and Roadarmel, G. (1999). "Scotch Pine Christmas Tree Fire Tests, Report of Test FR 4010." Gaithersburg, MD: National Institute of Standards and Technology.

Tewarson, A. (2008). "Generation of Heat and Gaseous, Liquid, and Solid Products in Fires" (Section 3, Chapter 4). *SFPE Handbook of Fire Protection Engineering*. Quincy, MA: SFPE/NFPA.

Thomas, P. H., Webster, C. T., and Raftery, M. (1961). "Some Experiments on Buoyant Diffusion Flames." *Combustion and Flame*, 5, 359–367.

Zalosh, R. (2003). *Industrial Fire Protection Engineering*. New York: Wiley.

10

Heat Transfer

OBJECTIVES

After reading this chapter, you should be able to:

- Define the concept of heat.
- Describe the manner by which convection and radiation affect the spread of a freely burning fire.
- Explain conduction, convection, and radiation heat transfer.
- Describe the concept of heat flux and its significance to the hazards of heat transfer from a fire.
- Perform simple mathematical computations in heat transfer.

Heat transfer is the most important concept to understand when analyzing fire phenomena. It is the driving factor for every aspect of fire development, including ignition, growth, detection, flashover, and suppression. When asked what factor in a fire has the greatest impact from an investigation, protection, prevention, or suppression perspective, the simple answer is heat. The transfer of energy in the form of heat causes fuels to begin pyrolyzing (vaporizing). When heat transfer is intense enough, for a long enough duration, a flammable mixture can form that allows for ignition. Heat is also the driving factor for flame spread across the initial fuel and/or radiant ignition of secondary fuels. Suppression occurs when the heat can be removed from the fuel's surface, which decreases the production of gaseous fuels, and the fire is extinguished. Some of the basic sources of heat consist of electrical (resistance, arcing, sparking), chemical reactions, mechanical (compression, friction), lightning, hot surfaces, open flames, and the sun.

heat
■ The form of energy that is transferred between two systems (or a system and its surroundings) by virtue of a temperature difference.

Heat is the transfer of energy based on a temperature difference between two objects or regions of a single object. The flow of energy is always in the direction from a high temperature to a lower temperature. Remember that temperature is simply a measure of the average kinetic energy of the molecules within a body and should not be confused with heat.

Heat can transfer between bodies in three different modes or processes, including conduction, convection, and radiation. Conduction and convection require an intervening medium (i.e., solid, liquid, or gas), while radiation requires no intervening medium.

Historical Background on Heat

It is easier to recognize the effects of heat or the work performed than it is to discuss the theory. For example, it is easier to recognize or feel a sunburn than it is to discuss the electromagnetic waves emitted by the sun traveling 92 million miles to the Earth and affecting your skin. It is also easier to recognize the increase in temperature when rubbing two solids together than it is to discuss the concept of friction and conductive heat transfer. In fact, this simple everyday observation inspired the first set of experiments to analyze heat. Count Rumford, an American-born British physicist, first analytically reviewed the relationship of energy and heat. He was a major general in the Bavarian army, where he supervised the boring of cannons. Count Rumford recognized that the process of boring barrels would produce great amounts of heat. To better measure this energy, he devised an experiment where he took a cannon barrel and placed it into a water bath and then measured the temperature of the water bath. This allowed him to analyze a known input of mechanical work converted to an increase in temperature, which is commonly known today as the mechanical equivalent of heat.

HEAT VERSUS TEMPERATURE

A material's temperature is simply the measurement of the amount of motion of the molecules or atoms. If the molecules within a material are moving very quickly, the temperature of this material will be high, and thus slow motion will equal a low temperature. Heat is the amount of energy transferred from one object to another due to a difference in temperature. Therefore, heat will not be transferred between two systems of the same temperature.

Conduction

Conduction is the transfer of energy in the form of heat by direct contact through the excitation of molecules and/or particles driven by a temperature difference. In other words, molecules with higher temperatures are more energetic and are constantly colliding with molecules that are less energetic (lower temperature), and a transfer of energy from the more energetic into the less energetic molecules must occur. This mode of heat transfer is most commonly recognized as occurring in solids because of the proximity of molecules and the ease for the vibration of one molecule to influence the kinetic energy of those neighboring molecules. Remember, however, that conductive heat transfer also occurs in liquids and gases, but it is typically not as obvious due to the distance between the molecules. The best example of conduction is the increase in temperature at the opposite end of the metal object that is being held in the flame of a candle (see Figure 10.1). At the molecular level, the portion of the metal object inside the candle flame is increasing in temperature, causing the excitation and increased vibration of those molecules, which in turn begin to affect the neighboring molecules and cause a chain reaction through the material until it reaches your fingers and causes a significant increase in temperature.

Conductive heat transfer in fire is prevalent during ignition, flame spread, activation of suppression and heat detection elements, heat transfer through compartment boundaries, spot fires in adjacent compartments, fire resistance, and fire patterns utilized for investigations (see Figure 10.2). It is possible to calculate the conductive heat transfer through a known material and thickness. This equation gives the amount of energy being transferred per time and is commonly known as a rate equation. The rate at which heat is transmitted through a material by conduction

conduction
- The transfer of energy in the form of heat by direct contact through the excitation of molecules and/or particles driven by a temperature difference.

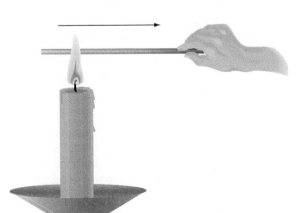

FIGURE 10.1 Candle flame with metal rod inserted into the flame. Heat transfer flows from the high-temperature region to the lower-temperature region.

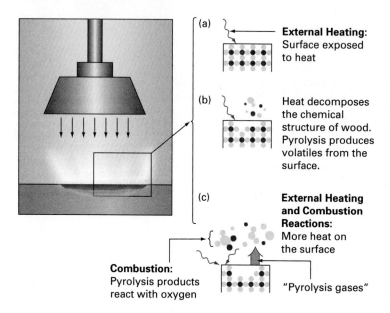

FIGURE 10.2 Heat being conducted into a common combustible.

(a)
External Heating: Surface exposed to heat

(b) Heat decomposes the chemical structure of wood. Pyrolysis produces volatiles from the surface.

(c) **External Heating and Combustion Reactions:** More heat on the surface

Combustion: Pyrolysis products react with oxygen

"Pyrolysis gases"

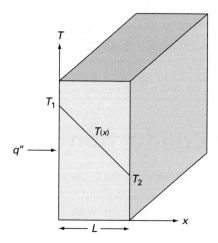

FIGURE 10.3 One-dimensional heat transfer by conduction (diffusion of energy).
Source: Data derived from Incropera, DeWitt, Bergston, and Lavine (2007).

thermal conductivity
■ The property of matter that represents the ability to transfer heat by conduction.

depends on the **thermal conductivity**, k (W/m · K) and the temperature difference. The thermal conductivity is a characteristic of the material (see Figure 10.3). This rate equation for conduction heat transfer is known as Fourier's law:

$$\dot{q}'' = -k\frac{dT}{dx}$$

Where:

$\dot{q}'' =$ heat flux (W/m²)
$k =$ thermal conductivity (W/m · K)
$dT/dx =$ temperature gradient across the material (K/m)

heat flux
■ The measure of the rate of heat transfer to a surface, expressed in kilowatts/m², kilojoules/m² · sec, or BTU/ft² · sec.

The **heat flux**, $\dot{q}''$ (W/m²) is the heat transfer rate in the x direction per unit area perpendicular to the direction of the transfer, and it is proportional to the temperature gradient, dT/dx, in this direction (Incropera, DeWitt, Bergman, and Lavine, 2007). The minus sign for the thermal conductivity is included because heat is transferred in the direction of decreasing temperature and will be rearranged later.

Remember that a dot above a symbol in an equation means that the symbol represents a value of that quantity per time, or a rate. Also, double prime marks (") after a symbol with a dot means that the quantity will be a flux, which means per unit area. In the formula for conduction heat transfer, q with a dot and double prime marks indicates that the answer for this formula will be given as a heat flux, or W/m².

The equation above can be simplified by making some basic assumptions, including the fact that heat is being conducted in only one dimension and is occurring at a steady state. The formula then becomes:

$$\dot{q}'' = -k\frac{T_2 - T_1}{L}$$

We can rearrange the formula to get rid of the minus sign before the k by switching the temperatures, as seen in the following equation:

$$\dot{q}'' = k\frac{T_1 - T_2}{L} = k\frac{\Delta T}{L}$$

Where:
$\dot{q}''$ = heat flux (W/m^2)
k = thermal conductivity (W/m · K)
T_1 = higher temperature (°C or K)
T_2 = lower temperature (°C or K)
ΔT = change in temperature (°C or K)
L = thickness of the material (m)

This formula can be expressed simply as a heat rate (W) by multiplying the heat flux by the exposed area, such as:

$$\dot{q} = \dot{q}'' \times A \ or \ \dot{q} = kA\frac{T_1 - T_2}{L} = kA\frac{\Delta T}{L}$$

EXAMPLE 10.1

The wall of an industrial furnace is constructed from 0.15-m-thick fireclay brick having a thermal conductivity of 1.7 W/m · K. Measurements made during steady-state operation reveal temperatures of 1,400 and 1,150 K at the inner and outer surfaces, respectively. What is the rate of heat loss through a wall that is 0.5 m × 1.2 m on a side?

Known: Steady-state conditions with prescribed wall thickness, area, thermal conductivity, and surface temperatures.

Find: Wall heat loss.

Schematic:

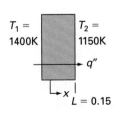
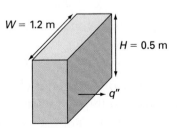

Assumptions:
1. Steady-state conditions
2. One-dimensional conduction through the wall
3. Constant thermal conductivity

Analysis: Because heat transfer through the wall is by conduction, the heat flux may be determined from Fourier's law.

$$\dot{q}'' = k\frac{\Delta T}{L} = 1.7\frac{W}{m \cdot K} \times \frac{(1,400 - 1,150\, K)}{0.15\ m} = 2,833\frac{W}{m^2}$$

The heat flux represents the rate of heat transfer through a section of unit area, and it is uniform across the surface of the wall. The heat loss through the wall of area $A = H \times W$ is then:

$$\dot{q} = \dot{q}'' \times A = 2,833\frac{W}{m^2} \times (0.5\ m \times 1.2\ m) = 1,700\ W$$

Comments: Note the direction of heat flow and the distinction between heat flux and heat rate.

In our discussion and use of the conduction heat transfer mode, we will also discuss other properties that affect the rate of heat transfer through a material. The most important of these properties are referred to as the material's thermophysical properties, which include the density (ρ), specific heat (c_p), and the thermal conductivity (k) (see Table 10.1). In pure heat transfer analyses, the ratio of the thermal conductivity to the heat capacity is an extremely important property; it is called the **thermal diffusivity (α)**, which is calculated as follows:

thermal diffusivity (α)
■ The ratio of thermal conductivity to the heat capacity, expressed as m^2/s.

$$\alpha = \frac{k}{\rho c_p}$$

Where: α = thermal diffusivity (m^2/s)
k = thermal conductivity (W/m $\cdot$ K)
ρ = density (g/m^3)
c_p = specific heat (J/g $\cdot$ K)

In other words, the thermal diffusivity is a measurement that compares the ability of the material to conduct energy through it versus its ability to store the energy. These three thermophysical properties for fire dynamics are typically more important when the product of the three properties are calculated and substituted into fire formulas, which is termed the **thermal inertia**.

thermal inertia
■ The properties of a material that characterize its rate of surface temperature rise when exposed to heat. It is related to the product of the material's thermal conductivity (k), its density (ρ), and its heat capacity (c).

$$\text{thermal inertia} = k \cdot \rho \cdot c_p$$

Where: k = thermal conductivity (W/m $\cdot$ K)
ρ = density (g/m^3)
c_p = specific heat (J/g $\cdot$ K)

The thermal inertia of a material characterizes its ability to conduct energy away from its surface and through its mass. More specifically, it provides a means of determining the temperature increase at the surface of a material, which is the principal issue when looking at ignition. The higher the thermal conductivity, the faster that energy can transfer away from the material's surface and into the mass of the material. Also, molecules packed more closely into a material (i.e., the greater the density) affect the material's ability to transfer the energy through its mass and away from the exposed surface. Therefore, the higher the thermal inertia,

	TABLE 10.1	**Thermophysical Properties of Common Materials**			

MATERIAL	THERMAL CONDUCTIVITY k(W/m-K)	DENSITY ρ(kg/m³)	SPECIFIC HEAT c_p(J/kg-K)	THERMAL INERTIA $k\rho c_p$ (W²s/m⁴K²)
Copper	387	8940	380	1.30×10^9
Steel	45.8	7850	460	1.65×10^8
Brick	0.69	1600	840	9.27×10^5
Concrete	0.8–1.4	1900–2300	880	2×10^6
Glass	0.76	2700	840	1.72×10^6
Gypsum plaster	0.48	1440	840	5.81×10^5
PPMA	0.19	1190	1420	3.21×10^5
Oak	0.17	800	2380	3.24×10^5
Yellow pine	0.14	640	2850	2.55×10^5
Asbestos	0.15	577	1050	9.09×10^4
Fiberboard	0.041	229	2090	1.96×10^4
Polyurethane foam	0.034	20	1400	9.52×10^2
Air	0.026	1.1	1040	2.97×10^1

Source: Data derived from Drysdale 1998, 33; and from Table 2.2, Thermal Characteristics of Common Materials Found at Fire Scenes, in Icove and DeHaan (2008), *Forensic Fire Scene Reconstruction*, Second Edition, Upper Saddle River, NJ: Brady/Prentice Hall.

the easier the heat can be transferred through the material's mass away from the surface and the less chance of ignition. For instance, metals are good conductors of thermal energy and can dissipate the energy through their mass faster, resulting in lower temperatures at the surface. Wood products have a low thermal conductivity and density; for that reason, wood has a low thermal inertia. Consequently, the energy imposed on the face of a piece of wood is dissipated from its surface slowly, which allows the surface temperature to rise quickly. Thus, greater pyrolysis results in ignition sooner than would occur in products with a higher thermal inertia. These properties are vital for ignition predictions and are covered in much more detail in Chapter 11.

Convection

Convection is the transfer of energy through a circulating fluid to or from a solid object (see Figure 10.4). Remember that fluids can be either gases or liquids, so the heat transfer that occurs between a heated metal pot to the cooler liquid inside the pot when cooking is one example of convection heat transfer. Another example, one more significant to fire safety professionals, is the heat

convection
■ Heat transfer by circulation within a medium such as a gas or a liquid.

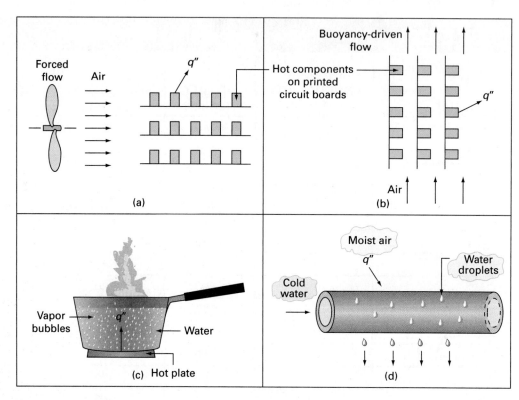

FIGURE 10.4 Convection heat transfer processes: (a) forced convection, (b) natural convection, (c) boiling, (d) condensation.
Source: Data derived from Incropera, DeWitt, Bergston, and Lavine (2007).

transfer between hot gases into cooler structural elements during a fire. Convection actually consists of two energy transfer mechanisms. First, an energy transfer occurs due to direct contact through the excitation of molecules, otherwise known as conduction. The second energy transfer mechanism is due to the bulk motion of the fluid. In other words, as the fluid flows past a solid object in the presence of a temperature difference between the fluid and the solid object, energy is transferred by the bulk motion of the fluid (**advection**) and due to the molecular excitation (conduction). A good example of the impact of bulk fluid motion is the difference in cooling a hot soup by letting it sit versus blowing on it (forced convection).

The primary mode of heat transfer during a combustion reaction is via convection. Approximately 70 percent of the energy liberated from a burning combustible is in the form of convection. This percentage changes, however, depending on the type of fuel and the ventilation conditions for the reaction. Table 10.2 provides a list of common fuels burning under well-ventilated situations. The table shows that, as a general rule, the majority of fuels liberate approximately 70 percent of their energy due to convection. Regardless of the exact percentage, it is important to note that the majority of the energy is transferred by convection.

The convection heat transfer process can also be calculated easily when the fluid properties and the temperatures of the fluid and the surface are known.

advection

■ One of the mechanisms that comprises the convection heat transfer mode dealing with the transport due to bulk fluid motion.

TABLE 10.2 — Chemical, Convective, and Radiative Heats of Combustion for Well-Ventilated Fires

MATERIAL	ΔH_T (kJ/g)	y_{CO_2} (g/g)	y_{CO}	y_{ch}	y_s	ΔH_{ch} (kJ/g)	ΔH_{con}	ΔH_{rad}
Common gases								
Methane	50.1	2.72	—	—	—	49.6	42.6	7.0
Ethane	47.1	2.85	0.001	0.001	0.013	45.7	34.1	11.6
Propane	46.0	2.85	0.005	0.001	0.024	43.7	31.2	12.5
Butane	45.4	2.85	0.007	0.003	0.029	42.6	29.6	13.0
Ethylene	48.0	2.72	0.013	0.005	0.043	41.5	27.3	14.2
Propylene	46.4	2.74	0.017	0.006	0.095	40.5	25.6	14.9
1,3-Butadiene	44.6	2.46	0.048	0.014	0.125	33.6	15.4	18.2
Acetylene	47.8	2.60	0.042	0.013	0.096	36.7	18.7	18.0
Common liquids								
Methyle alcohol	20.0	1.31	0.001	—	—	19.1	16.1	3.0
Ethyl alcohol	27.7	1.77	0.001	0.001	0.008	25.6	19.0	6.5
Isopropyl alcohol	31.8	2.01	0.003	0.001	0.015	29.0	20.6	8.5
Acetone	29.7	2.14	0.003	0.001	0.014	27.9	20.3	7.6
Methylethyl ketone	32.7	2.29	0.004	0.001	0.018	30.6	22.1	8.6
Heptane	44.6	2.85	0.010	0.004	0.037	41.2	27.6	13.6
Octane	44.5	2.84	0.011	0.004	0.038	41.0	27.3	13.7
Kerosene	44.1	2.83	0.012	0.004	0.042	40.3	26.2	14.1
Benzene	40.1	2.33	0.067	0.018	0.181	27.6	11.0	16.5
Toluene	39.7	2.34	0.066	0.018	0.178	27.7	11.2	16.5
Styrene	39.4	2.35	0.065	0.019	0.177	27.8	11.2	16.6
Hydrocarbon	43.9	2.64	0.019	0.007	0.059	36.9	24.5	12.4
Mineral oil	41.5	2.37	0.041	0.012	0.097	31.7	—	—
Polydimethyl siloxane	25.1	0.93	0.004	0.032	0.232	19.6	—	—
Silicone	25.1	0.72	0.006	0.008	—	15.2	12.7	2.5
Chemicals and solvents								
Tetrahydrofuran (C_4H_8O)	32.2	2.29	0.021	—	—	30.3	—	—
Phenol (C_6H_6O)	31.0	2.63	0.057	—	0.099	27.6	13.3	14.3
Acetonitrile (C_2H_3N)	29.6	2.04	0.025	—	0.026	29.0	23.0	6.0
Ethylisonicotate ($C_8H_9O_2N$)	26.3	2.37	0.029	—	0.142	24.3	12.8	11.5

Source: Data derived from Tewarson, 2008.

(continued)

TABLE 10.2	**Chemical, Convective, and Radiative Heats of Combustion for Well-Ventilated Fires (*continued*)**							

MATERIAL	ΔH_T (kJ/g)	y_{CO_2}	y_{CO} (g/g)	y_{ch}	y_s	ΔH_{ch}	ΔH_{con} (kJ/g)	ΔH_{rad}
Adiponitrile ($C_6H_8N_2$)	33.1	2.35	0.045	—	0.045	31.1	22.1	9.0
Hexamethylenediamine ($C_6H_{15}N_2$)	35.3	2.28	0.029	—	0.045	32.6	15.7	16.9
Toluenedisocyanate ($C_9H_6O_2N_2$)	23.6	1.77	0.052	—	0.141	19.3	11.1	8.2
Diphenylmethanedisocyanate MDI ($C_{15}H_{10}O_2N_2$)	27.1	0.95	0.042	—	0.154	19.6	13.7	5.9
Polymeric MDI ($C_{23}H_{19}O_3N_3$)	29.6	1.22	0.032	—	0.165	23.3	15.0	8.3
Isoproturon ($C_{12}H_{18}ON_2$)	32.8	1.70	0.056	—	0.115	23.9	14.0	9.9
3-Chloropropene (C_3H_5Cl)	23.0	0.75	0.076	—	0.179	10.8	6.9	3.9
Monochlorobenzene (C_6H_5Cl)	26.4	0.86	0.083	—	0.232	11.2	—	—
Dichloromethane (CH_2Cl_2)	6.0	0.11	0.088	—	0.081	2.0	—	—
1,3-Dichloropropene ($C_3H_4Cl_2$)	14.2	0.35	0.090	—	0.169	5.6	—	—
Ethylmonochloroacetate ($C_4H_7O_2Cl$)	15.7	1.24	0.019	—	0.138	14.1	10.1	4.0
Chloronitrobenzoic acid ($C_7H_4O_4NCl$)	15.9	0.39	0.057	—	—	4.4	—	—
Aclonilen ($C_{12}H_9O_3N_2Cl$)	19.9	0.68	0.063	—	0.186	7.0	—	—
2,6-Dichlorobenzonitrile (dichlobenil) ($C_7H_3NCl_2$)	17.8	0.39	0.068	—	—	4.3	—	—
Diuron ($C_9H_{10}ON_2Cl_2$)	20.3	0.76	0.080	—	0.159	10.2	7.7	2.5
Trifluoromethylbenzene ($C_6H_5CF_3$)	18.7	1.19	0.069	—	0.185	10.8	5.1	5.7
Metatrifluoromethylphenylacetonitrile ($C_9H_6NF_3$)	16.0	0.89	0.058	—	0.168	7.3	4.0	3.3
Tetramethylthiurammonosulfide ($C_6H_{12}N_2S_3$)	22.6	1.06	0.041	—	—	19.6	—	—
Methylthiopropionylaldehyde (C_4H_8OS)	25.0	1.62	0.001	—	0.005	23.8	18.8	5.0
Pesticides								
2,4-D acid (herbicide, $C_8H_6O_3Cl_2$)	11.5	0.50	0.074	—	0.163	4.5	3.0	1.5
Mancozeb ($C_4H_6N_2S_4Mn)_i(Zn_{0.4}$)	14.0	0.50	—	—	—	9.5	—	—
Folpel ($C_9H_4O_2NSCl_3$)	9.1	0.37	0.072	—	0.205	3.6	—	—
Chlorfenvinphos ($C_{12}H_{24}O_4Cl_3P$)	18.0	0.43	0.011	—	0.288	7.7	—	—
Chlormephos ($C_5H_{12}O_2S_2ClP$)	19.1	0.51	0.075	—	0.055	13.9	—	—
Natural materials								
Tissue paper	—	—	—	—	—	11.4	6.7	4.7
Newspaper	—	—	—	—	—	14.4	—	—
Wood (red oak)	17.1	1.27	0.004	0.001	0.015	12.4	7.8	4.6

MATERIAL	ΔH_T (kJ/g)	y_{CO_2}	y_{CO}	y_{ch}	y_s	ΔH_{ch}	ΔH_{con}	ΔH_{rad}
			(g/g)				(kJ/g)	
Wood (Douglas fir)	16.4	1.31	0.004	0.001	—	13.0	8.1	4.9
Wood (pine)	17.9	1.33	0.005	0.001	—	12.4	8.7	3.7
Corrugated paper	—	—	—	—	—	13.2	—	—
Wood (hemlock)	—	—	—	—	0.015	13.3	—	—
Wool 100%	—	—	—	—	0.008	19.5	—	—
Synthetic materials—solids (abbreviations/names in the nomenclature)								
ABS	—	—	—	—	0.105	30.0	—	—
POM	15.4	1.40	0.001	0.001	—	14.4	11.2	3.2
PMMA	25.2	2.12	0.010	0.001	0.022	24.2	16.6	7.6
PE	43.6	2.76	0.024	0.007	0.060	38.4	21.8	16.6
PP	43.4	2.79	0.024	0.006	0.059	38.6	22.6	0
PS	39.2	2.33	0.060	0.014	0.164	27.0	11.0	16.0
Silicone	21.7	0.96	0.021	0.006	0.065	10.6	7.3	3.3
Polyester-1	32.5	1.65	0.070	0.020	0.091	20.6	10.8	9.8
Polyester-2	32.5	1.56	0.080	0.029	0.089	19.5	—	—
Epoxy-1	28.8	1.59	0.080	0.030	—	17.1	8.5	8.6
Epoxy-2	28.8	1.16	0.086	0.026	0.098	12.3	—	—
Nylon	30.8	2.06	0.038	0.016	0.075	27.1	16.3	10.8
Polyamide-6	—	—	—	—	0.011	28.8	—	—
IPST	—	—	—	—	0.080	23.3	—	—
PVEST	—	—	—	—	0.076	22.0	—	—
Silicone rubber	21.7	0.96	0.021	0.005	0.078	10.9	—	—
Polyether ether ketone (PEEK-$CH_{0.63}O_{0.16}$)	31.3	1.6	0.029	—	0.008	17.5	—	—
Polysulfone (PSO-$CH_{0.81}O_{0.15}S_{0.04}$)	29.0	1.8	0.034	—	0.020	24.3	—	—
Polyethersulfone (PES-$CH_{0.67}O_{0.21}S_{0.08}$)	25.2	1.5	0.040	—	0.021	20.4	—	—
Polyetherimide (PEI-$CH_{0.68}N_{0.05}O_{0.14}$)	30.1	2.0	0.026	—	0.014	27.2	—	—
Polycarbonate (PC-$CH_{0.88}O_{0.13}$)	31.6	1.5	0.054	—	0.112	18.4	—	—
Polyurathane (flexible) foams								
GM21	26.2	1.55	0.010	0.002	0.131	17.8	8.6	9.2
GM23	27.2	1.51	0.031	0.005	0.227	19.0	10.3	8.7

(continued)

TABLE 10.2

Chemical, Convective, and Radiative Heats of Combustion for Well-Ventilated Fires (*continued*)

MATERIAL	ΔH_T (kJ/g)	y_{CO_2}	y_{CO}	y_{ch}	y_s	ΔH_{ch}	ΔH_{con}	ΔH_{rad}
		(g/g)				(kJ/g)		
GM25	24.6	1.50	0.028	0.005	0.194	17.0	7.2	9.8
GM27	23.2	1.57	0.042	0.004	0.198	16.4	7.6	8.8
Polyurethane (rigid) foams								
GM29	26.0	1.52	0.031	0.003	0.130	16.4	6.8	9.6
GM31	25.0	1.53	0.038	0.002	0.125	15.8	7.1	8.8
GM35	28.0	1.58	0.025	0.001	0.104	17.6	7.8	9.8
GM37	28.0	1.63	0.024	0.001	0.113	17.9	8.7	9.2
GM41	26.2	1.18	0.046	0.004	—	15.7	5.7	10.0
GM43	22.2	1.11	0.051	0.004	—	14.8	6.4	8.4
Polystyrene foams								
GM47	38.1	2.30	0.060	0.014	0.180	25.9	11.4	14.5
GM49	38.2	2.30	0.065	0.016	0.210	25.6	9.9	15.7
GM51	35.6	2.34	0.058	0.013	0.185	24.6	10.4	14.2
GM53	37.6	2.34	0.060	0.015	0.200	25.9	11.2	14.7
Polyethylene foams								
1	41.2	2.62	0.020	0.004	0.056	34.4	20.2	14.2
2	40.8	2.78	0.026	0.008	0.102	36.1	20.6	15.5
3	40.8	2.60	0.020	0.004	0.076	33.8	18.2	15.6
4	40.8	2.51	0.015	0.005	0.071	32.6	19.1	13.5

Source: Tewarson, 2008.

convection heat transfer coefficient
■ A quality that represents the ability of heat to be transformed from a moving fluid to a surface.

The mathematical expression is known as Newton's law of cooling. The fluid properties, the surface geometry, and the nature of the fluid motion affect how heat is transferred and is compiled into one variable, known as the **convection heat transfer coefficient,** h. Table 10.3 provides some typical values for convection coefficients. A detailed discussion of the convection coefficient is beyond the scope of this book. Incropera, DeWitt, Bergman, and Lavine (2007) provide an excellent overview of the experimental research into these coefficients. Two formulas are presented below to calculate the heat flux (W/m²); the first one is used if the fluid temperature is greater (i.e., hot gases flowing next to structural elements), while the second is used if the surface temperature is greater than the fluid (i.e., hot plate surface). The heat flux formula is given as:

$$\dot{q}'' = h(T_f - T_s) \quad or \quad \dot{q}'' = h(T_s - T_f)$$

TABLE 10.3   Typical Values for Convective Coefficients, *h*

FLUID CONDITION	$h(\text{W/m}^2\,°\text{C})$
Buoyant flows in air	5–10
Laminar match flame	–30
Turbulent liquid pool fire surface	–20
Fire plume impinging on a ceiling	5–50
2 m/s wind speed in air	–10
35 m/s wind speed in air	–75

Where: h = convection heat transfer coefficient (W/m² · K)
T_f = temperature of fluid (°C or K)
T_s = temperature of surface (°C or K)

If the area of the heat transfer is known, then the heat transfer (W) formula becomes:

$$\dot{q} = hA(T_f - T_s) \quad or \quad \dot{q} = hA(T_s - T_f)$$

Where: h = convection heat transfer coefficient (W/m² · K)
A = solid surface area (m²)
T_f = temperature of the fluid (°C or K)
T_s = temperature of the surface (°C or K)

There are two basic types of convection heat transfer: natural (free) or forced. Natural convection is typically associated with buoyancy forces, with no external forces acting to speed up or affect the heat transfer. The most common natural convection is found when water evaporates. Forced convection is typically associated with some mechanical means of increasing the convection process by applying a fan or some other external means of blowing air across a surface. Two examples of forced convection are when small fans are applied on the inside of a computer to more efficiently cool the interior electronics, or when you blow on hot coffee before taking a sip.

As mentioned previously, the bulk fluid movement affects the heat transfer capabilities. More specifically, the velocity and characteristics of the flow are driving factors of the heat transfer capabilities. When a fluid flows over a surface, a boundary layer develops at this interface (see Figure 10.5). The manner or characteristics of how this flow proceeds over the surface will be either laminar or turbulent. Laminar flows are highly structured and the particles seem to generally flow in a similar motion and direction. Turbulent flows are highly disorganized and have random three-dimensional motion. On longer surfaces, both types of flow may occur across the surface. In this case, there will also be a transition zone between the two flows, which has characteristics of both types of flows. The manner in which the fluid flows across the surface affects the heat transfer coefficient, which in turn affects the capabilities of heat transfer. Review the differences in the turbulent and laminar flow heat transfer coefficients in Table 10.3.

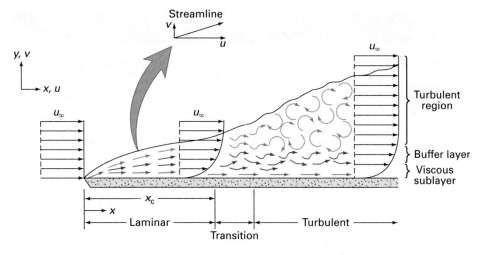

FIGURE 10.5 Velocity boundary layer development on a flat plate. *Source:* Data derived from Incropera, DeWitt, Bergston, and Lavine (2007).

EXAMPLE 10.2

Find the convection heat flux from a turbulent flame to a cold wall at 20°C. Estimate the h value from Table 10.3 for free convection as 5 W/m² · K or 5 W/m² · °C. The maximum time-averaged flame temperature is approximately 900°C.

$$\dot{q}'' = h(T_f - T_s)$$

$$\dot{q}'' = 5\frac{W}{m^2\,°C}(900°C - 20°C) = 4{,}400\frac{W}{m^2} \text{ or } 4.4\frac{kW}{m^2}$$

EXAMPLE 10.3 HEAT TRANSFER CALCULATIONS

Problem: A fire in a storage room in an automobile repair garage raises the temperature of the interior surface of the brick walls and ceiling over a period of time to 500°C (932°F or 773 K). The ambient air temperature on the outside of the brick wall to the garage is 20°C (68°F or 293 K). The thickness of the brick wall is 200 mm (0.2 m), and its convective heat transfer coefficient is 12 W/m²K. Calculate the temperature to which the exterior of the wall will be raised.

Suggested Solution—Conductive-Convective Problem: In this solution, the configuration is considered a plane wall of uniform homogenous material (brick) having a constant thermal conductivity, a given interior surface temperature of $T_s = 500°C$ (932°F or 773 K), and exposed to an external garage ambient air temperature of $T_a = 20°C$ (68°F or 293 K). The external surface temperature, T_E, of the wall is unknown. This temperature might be important later if combustible materials come into contact with this wall surface.

Because the steady-state heat transfer rate of conduction through the brick wall is equal to the heat transfer rate of convection passing into the ambient air outside the garage,

$$\frac{\dot{q}}{A} = \frac{T_s - T_E}{\frac{\Delta x}{K_a}} = \frac{T_E - T_a}{\frac{1}{h_G}}$$

$$\frac{773\ K - T_E}{\dfrac{0.2\ m}{0.69\ W/mK}} = \frac{T_E - 293\ K}{\dfrac{1}{12\ W/m^2K}}$$

$$2665 - 3.448\ T_E = 12.00\ T_E - 3516$$

$$T_E = \frac{6181}{15.45} = 400\ K = 126°C(258°F)$$

Where:

$\dot{q}$ = conduction heat transfer rate (J/s or W)

$\dot{q}''$ = conductive heat flux (W/m^2)

h = convective heat transfer coefficient (W/m^{2}K) (brick to air)

k = thermal conductivity of material (W/m-K) (through brick)

A = area through which the heat is being conducted (m^2)

$\Delta T = (T_S - T_E)$ wall face temperature differences between the warmer T_s and cooler T_E

Suggested Alternative Solution—Electrical Analog: Assuming that the brick maintains its integrity during the fire, determine the steady-state temperature of the wall's surface in the garage (example modified from Drysdale 1998, p. 72). A one-dimensional electrical circuit analog is used in this solution because we can look at the way heat transfers through a material in the same way that current flows through resistors.

Solutions by electrical analogs are often used. The steady-state heat transfer solution is best illustrated by a direct-current electrical circuit analogy, where the temperatures are voltages ($V = T$), and the thermal resistances of heat convection and conduction are represented as electrical resistances ($R = 1/h_b$ and $R = L_n/k_n$, respectively). The heat flux, $\dot{q}''$, whose units are W/m^2, is analogous to the current; that is $I = \dot{q}''$.

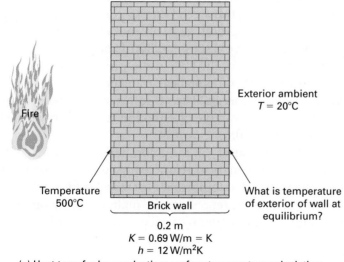

(a) Heat transfer by conduction: surface temperature calculation

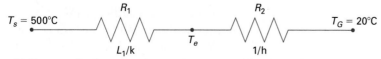

(b) Heat transfer by conduction: surface temperature using electrical analog

The problem can be represented as a steady-state electrical circuit analog, where R_1 and R_2 are the thermal resistances of the brick wall and inner air of the garage.

Temperature in storage room	$T_S = 500°C$ (773 K)
Temperature in garage	$T_G = 20°C$ (293 K)
Thermal conductivity	$k = 0.69$ W/mK
Thickness of the brick	$L = 200$ mm (0.2 m)
Coductive thermal resistance	$R_1 = \Delta L/k = (0.2 \text{ m})/(0.69 \text{ W/mK})$ $= 0.290 \text{ m}^2\text{K/W}$
Convective heat transfer coefficient	$h = 12$ W/m^{2}K
Convective resistance	$R_2 = 1/h = (1)/(12 \text{ W/m}^2\text{K})$ $= 0.083 \text{ m}^2\text{K/W}$
Total heat flux	$\dot{q}'' = (T_S - T_G)/(R_1 + R_2) = 1287 \text{ W/m}^2$ $T_E = 773 \text{ K} - (\dot{q}'')(R_1)$
Exterior wall temperature T_E	$= 773 \text{ K} - 374 \text{ K}$ $= 399 \text{ K}$ $T_E = 399 \text{ K} - 273 = 126°C (258°F)$

Convection heat transfer is the primary mode of heat transfer early in a compartment fire. It is also the reason for preheating fuels overhead as ceiling jets and an upper layer begin to form, which may result in ignition of secondary fuels in a compartment.

Radiation

radiation
■ Heat transfer by way of electromagnetic energy.

All matter that is above absolute zero temperature emits energy as thermal radiation. Thermal **radiation** is the transfer of energy through electromagnetic waves. All states of matter (i.e., liquid, solid, gas) can emit energy as radiation, which becomes important later when we discuss radiant heat transfer from an upper hot gas layer and from the flame itself. Unlike the previous modes of heat transfer (conduction and convection), radiant heat transfer does not require a material medium for the transfer of energy. Therefore, energy transfer by radiation can occur in a vacuum. The best example of radiation is the energy transferred by the sun to the Earth (see Figure 10.6).

The energy emitted by matter via radiation depends strongly on its temperature. The fundamental mathematical relationship, which is based on the Stefan-Boltzmann law, illustrates the importance of the surface temperature of the matter. The following equation represents the highest energy output for this matter given the temperature and assuming an ideal radiator or blackbody:

$$E = \sigma T_s^4$$

Where:

E = energy radiated per unit time per unit area from an ideal radiator (W/m^2)

σ = Stefan-Boltzmann constant = 5.67×10^{-8} (W/m$^2 \cdot$ K^4) or 5.67×10^{-11} (kW/m$^2 \cdot$ K^4)

T_s = absolute temperature of surface (K)

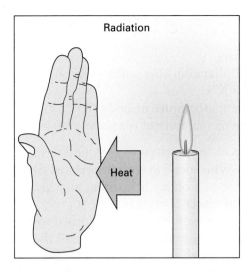

FIGURE 10.6 Radiant heat transfer travels in all directions.
Source: Adapted from International Fire Service Training Association (IFSTA), *Essentials of Firefighting*, Fifth Edition (2008).

Notice that temperature is raised to a fourth power, which means that a slight change in temperature will drastically change the energy emitted. For example, assume that the temperature of a hot plate is 773 K (500°C) and that the energy radiated is 20,244 W/m². If the temperature of the hot plate is raised only 25°C (798 K), the energy emitted has been increased to 22,993 W/m².

A **blackbody** is a perfect absorber and emitter of energy. However, the actual heat flux emitted by real objects is less than that of a blackbody and is related to a blackbody by the objects' emissivity, ε. The emissivity is a radiative property of the surface that relates the efficiency of the surface to emit energy compared to a blackbody. Emissivity values range between 0 and 1 (1 being a blackbody). To account for real surfaces emitting energy, the following formula takes into account the emissivity:

blackbody
- A perfect absorber and emitter of radiation.

$$E = \epsilon\sigma T_s^4$$

Where:

E = energy radiated per unit time per unit area (W/m²)

ϵ = emissivity

σ = Stefan-Boltzmann constant = 5.67×10^{-8} (W/m² · K⁴)

T_s = absolute temperature of the surface (or body if a gas) (K)

Based on the above equations, the heat flux emitted from a surface to its surroundings can be calculated, as long as the temperatures are provided and the emitter is completely exposed to the surroundings. This formula is then rewritten with the common notation of $\dot{q}''$ in place of E to give the following:

$$\dot{q}'' = \epsilon\sigma(T_s^4 - T_{sur}^4)$$

Where:

$\dot{q}''$ = energy radiated per unit time per unit area (W/m^2)

ϵ = emissivity

σ = Stefan-Boltzmann constant = 5.67×10^{-8} (W/m$^2 \cdot$ K^4)

T_s = absolute temperature of body (K)

T_{sur} = absolute temperature of the surroundings (K)

The above equation assumes that the heat is being transferred to surroundings that encompass the entire high-temperature object. When dealing with fire-related situations, however, it is more common to find that one object is burning and heat is being transferred to another fuel item at a set distance away from the source. And the second fuel item is typically situated such that only one side is exposed to the source of heating. Therefore, the exposed fuel item receives only a portion of the radiant heat transfer from the burning item, based on the configuration of the burning item in relation to that of the exposed fuel item. This circumstance, along with the distance between the two objects, is known as the **configuration factor** (also known as the *view factor*). Radiation, similar to light waves, does not bend around objects and lessens in intensity as distance increases away from the source. Therefore, for an item to be exposed to radiation from a burning object, that item must be in view of the burning object. Also, the closer the second item is to the burning object, the more intense the exposure. For example, if a wall is placed between an object burning and a target fuel, the wall absorbs the energy and protects or shields the target fuel from exposure to radiant heating. Because most radiant heat transfer issues relate to this type of problem, the calculations must account for the fact that the second fuel item will receive only a fraction of the total energy being emitted by the burning fuel. The following equation accounts for the fraction of energy by multiplying the previous equation by a configuration factor (F_{12}):

$$\dot{q}'' = \epsilon \sigma T_s^4 F_{12}$$

Where:

$\dot{q}''$ = energy radiated per unit time per unit area (W/m^2)

ϵ = emissivity

σ = Stefan-Boltzmann constant = 5.67×10^{-8} (W/m$^2 \cdot$ K^4)

T_s = absolute temperature of emitting body (K)

F_{12} = configuration factor (depends on orientation and distance)

To determine the configuration factor, the geometry of both the emitting object and the absorbing object must be known, as well as the angles and distance between the two objects. Figure 10.7 shows an example of a common method for calculating a configuration factor. To solve for the configuration factor (F_{12}), simply take the dimensions of the hot surface or emitter and the distance between the two objects, and substitute those into the equation provided.

$$X = \frac{a}{c}$$

$$Y = \frac{b}{c}$$

configuration factor
■ The fraction of the radiation leaving surface *I* that is intercepted by surface *j*. Also commonly referred to as the *view factor*.

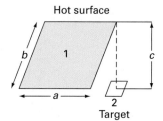

Hot surface

FIGURE 10.7
Example of a configuration factor for a rectangle emitter in relation to a target fuel.

$$F_{12} = \frac{1}{2\pi}\left(\frac{X}{\sqrt{1+X^2}}\tan^{-1}\frac{Y}{\sqrt{1+X^2}} + \frac{Y}{\sqrt{1+Y^2}}\tan^{-1}\frac{X}{\sqrt{1+Y^2}}\right)$$

Typically, when calculating the heat transfer from a flame plume, a simple geometric shape can be utilized to represent the flame in order to model the view factor or configuration factor. The most commonly utilized shapes are rectangles and cylinders. For a large compilation of radiation heat transfer configuration factors, Dr. John Howell at the University of Texas at Austin has cataloged many different geometries, which can be accessed via the website http://www.me.utexas.edu/~howell/. The Society of Fire Protection Engineers' (SFPE) *Handbook of Fire Protection Engineering* (2008) also provides a list of common configuration factors. * Use Radians with this calculation.

EXAMPLE 10.4

A truck carrying kerosene located 50 m away from a warehouse wall collides with a utility pole. Kerosene spills and forms a pool 5.5 m in diameter. This pool ignites, and the flame is approximately 10.33 m in height.

a. Assuming this flame is rectangular in shape, with a base of 4.5 m, what is the configuration factor?
b. What is the resulting heat flux to the wall, if we assume that the flame temperature is 1,600 K and the emissivity, ϵ, is 0.997?

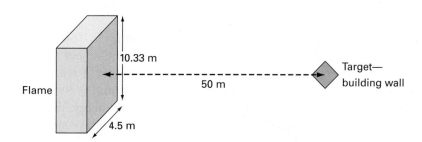

a. $X = \dfrac{a}{c} = \dfrac{10.33 \text{ m}}{50 \text{ m}} = 0.2066$

$Y = \dfrac{b}{c} = \dfrac{4.5 \text{ m}}{50 \text{ m}} = 0.09$

$F_{12} = \dfrac{1}{2\pi}\left(\dfrac{X}{\sqrt{1+X^2}}\tan^{-1}\dfrac{Y}{\sqrt{1+X^2}} + \dfrac{Y}{\sqrt{1+Y^2}}\tan^{-1}\dfrac{X}{\sqrt{1+Y^2}}\right)$

$$F_{12} = \frac{1}{2\pi}\left(\frac{0.207}{\sqrt{1 + 0.207^2}}\tan^{-1}\frac{0.09}{\sqrt{1 + 0.207^2}} + \right.$$

$$\left.\frac{0.09}{\sqrt{1 + 0.09^2}}\tan^{-1}\frac{0.207}{\sqrt{1 + 0.09^2}}\right) = 0.006$$

b. $\dot{q}'' = \epsilon\sigma T_s^4 F_{12} = (0.997)\left(5.67 \times 10^{-11}\frac{kW}{m^2K^4}\right)(1600\ K)^4(0.006) = 2.2\ \frac{kW}{m^2}$

A simplified method for calculating radiant heat flux is to assume that the heat source is a point source (see Figure 10.8). With this assumption, there is no need to calculate the configuration factor between the two objects. This is an acceptable means of calculating radiant heat flux as long as the distance from the source to the target fuel is greater than double the diameter of the flame ($R > 2D$). The formula is:

$$\dot{q}'' = \frac{X_r\dot{Q}}{4\pi R^2}$$

Where:

$\dot{q}''$ = energy radiated per unit time per unit area (W/m^2 or kW/m^2)

X_r = fraction of energy radiated relative to the total energy released

$\dot{Q}$ = heat release rate (W or kW)

$4\pi r^2$(m^2) = surface area of sphere, where r is the radial distance to the target fuel

The variable X_r relates to the fraction of energy radiated relative to the total energy released during the combustion reaction. Remember that, of the total energy liberated during a combustion reaction, convection accounts for approximately 70 percent. The other 30 percent, as a rule of thumb, is emitted as radiant heat transfer. This percentage changes, however, depending on the type of fuel and the ventilation conditions for the reaction. Table 10.2 provides a list of common fuels in a well-ventilated situation, which shows that, as a rule, the majority of fuels liberate approximately 30 percent of their energy due to radiation.

Radiant heat transfer is emitted by the object burning in 360°, which means that radiant heat transfer occurs outward from the burning object and is spherical in shape. The $4\pi r^2$ portion of the equation represents the surface area of a sphere. This is commonly referred to as the inverse square law and is usually discussed in many physical science fields that deal with force and energy. The law reveals that, as energy moves away from its source, it must spread out over an area that is

FIGURE 10.8

Schematic of radiant heat transfer from a point source of a fire plume to a target fuel.

Source: Adapted from Icove and DeHaan (2008), *Forensic Fire Scene Reconstruction*, Second Edition.

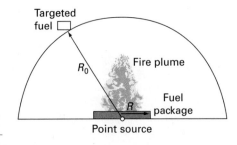

proportional to the square of the distance from the source. This means that, as radiation passes through a unit area, it is inversely proportional to the square of the distance from the source.

EXAMPLE 10.5

Refer to Example 10.4. Calculate the point source radiant heat flux to the brick wall. Assume that the heat release rate is 40,000 kW.

Point source: flame ◄- ► Target: building wall

50 m

$$\dot{q}'' = \frac{X_r \dot{Q}}{4\pi R^2} = \frac{(0.3)(40,000 \text{ kW})}{4\pi(50 \text{ m})^2} = 0.4 \frac{\text{kW}}{\text{m}^2}$$

Radiant heat transfer is the primary means of igniting local combustibles and spreading a fire to secondary and tertiary fuel items surrounding the initial fuel burning. Radiant heat transfer also becomes the most prevalent heat transfer mechanism during a compartment fire when the enclosure is nearing flashover and is, in fact, the driving factor that causes flashover. The radiant heat transfer coming from the flame plume typically causes the radiant ignition of the other fuel items. Because radiant heat transfer is greatly affected by the temperature of the emitting object, it is important to note that typical flame plume temperatures range from 1,190 to 1,600 K. The radiant heat transfer coming from the upper hot gas layer during a compartment fire is the cause for flashover, the rapid ignition of all radiant view combustibles within the lower layer. Typical smoke layers are generally at temperatures ranging from 1,100 to 1,500 K (Tien, Lee, and Stretton, 2009).

It is important to list once again the variables or factors that affect radiant heat transfer, including the temperature of the emitter, shielding items that prevent electromagnetic waves from impinging on second fuels, the distance between the two objects, emissivity, and the view factor.

Relationship to the Flame

All three modes of heat transfer occur during a combustion reaction. The best example is the candle flame (see Figure 10.9). When an open flame is placed to the wick, wax moves up through the wick by capillary action, which allows a greater surface area–to–mass ratio of the wax, thus allowing the flame to easily vaporize the fuel for ignition.

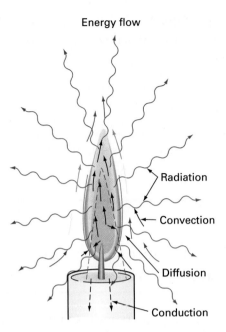

FIGURE 10.9 Candle flame diagram showing all three heat transfer mechanisms: radiation (wavy arrows) carries about 25 to 30 percent of the total energy of combustion equally in all directions from the flame. Roughly 4 percent of the radiation that feeds back to the candle melts the wax to fuel the flame. The melted wax moves upward through the wick by capillary action. Vaporized there by intense radiation from the surrounding reaction zone, the wax molecules go into cracking and combustion reactions that lead ultimately to oxidation of their constituent carbon atoms to form carbon dioxide and water.
Source: Lyons, J. (1985), *Fire*, New York: Scientific American Books.

Conductive heat transfer is the mechanism that causes the excitation of molecules within the fuel (wax), which allows the molecules to break down and melt. This liquefied wax moves up the wick and, in the presence of continued heating, begins to evolve vapors. As soon as these vapors are within their flammable range, ignition occurs, and sustained combustion commences if sufficient heating continues. Ignition of this fuel causes an increase in temperature and convection heat transfer to occur, which in turn drives the buoyancy forces of the plume. Convection heat transfer is the reason your hand begins to hurt when it is placed over the flame.

Radiant heat is emitted in 360° from the combustion reaction, preheating nearby fuels and, more important, radiating heat back to the fuel surface (wax and wick). These events allow the evolution of more vapors, and the chain reaction thus continues.

Summary

Most fire safety professionals state that the most important concept to understand for them to perform their duties is heat. Heat is the driving factor for every aspect of fire development, including ignition, growth, detection, flashover, and suppression. Three different modes of heat transfer were discussed in this chapter, including conduction, convection, and radiation. Conduction and convection require an intervening medium (i.e., solid, liquid, or gas), while radiation requires no intervening medium. Assumptions can be made to assist the fire safety professional in calculating or predicting heat transfer for different applications.

Review Questions

1. The transfer of energy through a solid medium is
   a. convection
   b. conduction
   c. radiation
   d. none of the above
2. Heat energy is transferred from the
   a. cold body to the warm body
   b. warm body to the cold body
3. If the flow of a fluid, liquid or gas, is straight, the flow is said to be
   a. turbulent
   b. eddy
   c. laminar
   d. lateral
4. The transfer of heat energy from the warm body to the colder body through a gaseous medium, is an example of
   a. convection
   b. conduction
   c. radiation
   d. energy transfer in a true blackbody
5. What is the inverse square law? How does it relate to radiation wavelengths? Where does the denominator $(4\pi r^2)$ come from in the radiation formula?
6. Explain in your own words the three modes of heat transfer. Give examples of each, provide mathematical formulas, and explain their importance in fire dynamics.
7. Consider steady-state conditions for one-dimensional conduction in a plane wall having a thermal conductivity $k = 50$ W/m · K and a thickness of $L = 0.25$ m, with no internal heat generation.

Determine the heat flux and the unknown quantity for each case. Sketch the temperature distribution and indicate the direction of the heat flux (adapted from Incropera, DeWitt, Bergman, and Lavine, 2007).

CASE	T_1(°C)	T_2(°C)	dt/dx (K/m)
1	50	−20	
2	−30	−10	
3	70		160
4		40	−80
5		30	200

8. A fire in a room rapidly raises the temperature of the surface of the walls and maintains that temperature at 1,000°C for a prolonged period. On the other side of one wall is a large warehouse, whose ambient temperature is normally 25°C. If the wall is solid brick and 250 mm thick, and it retains its integrity,

what is the steady-state temperature of the surface of the wall on the warehouse side? Assume that the thermal conductivity of the brick is independent of the temperature and the convection heat transfer (h) = 12 w/m^{2}K (adapted from Drysdale, 1998).

9. Find the heat flux due to radiation from a gasoline pool fire whose flame is 3 m tall with a 1 m diameter, and it is 3.5 m away from a target fuel. Assume that $\epsilon = 0.98$ with a flame temperature of 1,300 K. Solve this problem by determining the configuration factor and assume a point source.

References

Cengel, Y., and Boles, M. (2006). *Thermodynamics: An Engineering Approach*, Fourth Edition. New York: McGraw Hill.

Drysdale, D. (1998). *An Introduction to Fire Dynamics*. Chichester, West Sussex, Eng.: Wiley.

Icove, D., and DeHaan, J. (2008). *Forensic Fire Scene Reconstruction*, Second Edition. Upper Saddle River, NJ: Brady/Pearson.

Incropera, F., Dewitt, D., Bergman, T., and Lavine, A. (2007). *Introduction to Heat Transfer,* Fifth Edition. Danvers, MA: Wiley.

Lyons, J. (1985). *Fire.* New York: Scientific American Books.

NFPA 921 (2008). *Guide for Fire and Explosion Investigations*. Quincy, MA: National Fire Protection Association.

Quintiere, J. (1999). *Principles of Fire Behavior.* Albany, NY: Delmar.

SFPE (2008). *The Handbook of Fire Protection Engineering*. Bethesda, MD: The Society of Fire Protection Engineers.

Tewarson, A. (2008). "Generation of Heat and Gaseous, Liquid, and Solid Products in Fires" (Section 3, Chapter 4) in *SFPE Handbook of Fire Protection Engineering.* Quincy, MA: SFPE/NFPA.

Tien, C., Lee, K., and Stretton, A. (2009). "Radiation Heat Transfer" (Section 1, Chapter 4) in *SFPE Handbook of Fire Protection Engineering*, Fourth Edition. Quincy, MA: Society of Fire Protection Engineers, NFPA.

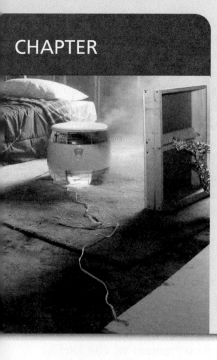

CHAPTER

11
Ignition

KEY TERMS

autoignition temperature, *p. 206*

concurrent flow flame spread, *p. 216*

counterflow flame spread, *p. 218*

ignition temperature, *p. 206*

minimum ignition energy (MIE), *p. 207*

piloted ignition temperature, *p. 206*

spontaneous ignition, *p. 209*

surface area–to–mass ratio, *p. 214*

thermal inertia, *p. 213*

thermally thin solid, *p. 219*

thermally thick solid, *p. 219*

thermal runaway, *p. 223*

OBJECTIVES

After reading this chapter, you should be able to:

- Explain the difference between piloted and autoignition.
- Define the concept of ignition.
- Understand the concept of ignition temperature for solids.
- Use formulas to predict the ignition time of solids.

PEARSON

myfirekit™

For additional review and practice tests, visit **www.bradybooks.com** and click on MyBradyKit to access book-specific resources for this text!

As has been shown in previous chapters, the three legs of the triangle (fuel, heat, and oxidizer) must be present and in sufficient quantity for ignition to occur. So far, we have discussed the different states of matter and how most solids and liquids require heat to be added for fuel to be produced in the gaseous state and thus for combustion to commence. Ignition is defined as the "initiation of combustion" (International Organization for Standardization [ISO], n.d.). Combustion is referred to as a self-sustained, high-temperature oxidation reaction (Babrauskas, 2003). Therefore, ignition is the process of initiating self-sustained combustion (National Fire Protection Association [NFPA] 921, 2008). This chapter will focus on ignition fundamentals for solids, liquids, and gases. It will also introduce the concept of flame spread on these fuels. At first glance, it may seem strange to see ignition correlated with flame spread; as will be pointed out, however, flame spread is a series of piloted ignitions across the fuel.

Regardless of the fuel's physical state, ignition can occur only if the fuel gases surrounding the fuel have reached their flammable limit. Many texts about ignition and flammability hazards express the importance of an ignition temperature. While ignition temperature is a good benchmark for a fuel, there are much more important items to look at relating to the heat source, the thermophysical properties of the fuel, and the duration of the heat exposure. When discussing solid fuels, the piloted ignition temperature is simply a measurement of the surface temperature required for the given fuel to produce gases (pyrolyze) at a sufficient rate to produce a flammable mixture. Some liquids with flash points lower than the ambient temperature may already be evolving sufficient vapors for the flammable limit to be reached without an external heat source. Gaseous fuels, once within their flammable limits, require only a minor amount of energy to begin the reaction, known as the minimum ignition energy.

Ignition temperature is the minimum temperature required for ignition of fuel gases under specific test conditions. The **piloted ignition temperature** is the temperature required to reach a fuel's lower flammable limit so that it can be ignited in the presence of a piloted ignition source (i.e., flame, spark). **Autoignition temperature** (also known as *autogenous ignition temperature*) describes the temperature at which oxidation reactions initiate within fuel/air mixtures without a piloted heat source. At the point where the fuel/air mixture is proper for ignition, the *minimum ignition energy* must also be present, and it is typically measured in millijoules (mJ).

> Note: *Some fuels, including some forms of carbon and magnesium, may ignite and burn directly in the solid form. However, the majority of fuels must first evolve sufficient vapors and/or gases for ignition to occur and combustion to be sustained, and these vapors are the primary concern for our discussion in this text.*

Sidebar definitions

ignition temperature
- Minimum temperature at which a substance will ignite under specific test conditions.

piloted ignition temperature
- Minimum temperature at which a substance will ignite in the presence of a pilot source (i.e., pilot flame, spark, hot surface).

autoignition temperature
- The lowest temperature at which a combustible material ignites in air without a spark or flame. Also known as *autogenous ignition temperature*.

Fire Ignition Statistics

The most common ignition sources for fires in the United States between 1994 and 1998 were reported to be cooking equipment (i.e., stove, oven) and heating equipment (i.e., space heaters, chimney, fireplace) (Babrauskas, 2003). The most common ignition sources for fires in Sweden between 1982 and 1991 were reported as lightning and electrical (Babrauskas, 2003). It is important to review the most common ignition sources of fires prior to discussing the variables that affect the very issue. For a more exhaustive review of ignition statistics and ignition sources, see *Ignition Handbook* by Dr. Vytenis Babrauskas (2003).

Ignition Energy

Most ignition sequences begin when sufficient heat energy contacts a fuel gas in the presence of oxygen where concentrations are sufficient to support combustion. Energy required to initiate a flame is defined and/or measured in multiple ways, including piloted ignition temperature, autoignition temperature, and **minimum ignition energy (MIE)**.

MINIMUM IGNITION ENERGY

The most reactive mixture for all gases is slightly fuel-rich or stoichiometric and therefore requires the lowest energy for ignition (see Table 11.1). Most of the common fuel gases have a minimum ignition energy of 0.25 to 0.3 milijoules. This energy, though minuscule, must be present in an area where the fuel/air mixture is within the flammable range.

PILOTED IGNITION

As discussed throughout the fuel chapters, a solid or liquid fuel must generate fuel gases or vapors at a sufficient rate to form a flammable mixture. Solid fuels typically

minimum ignition energy (MIE)
■ The minimum quantity of heat energy that should be absorbed by a substance to ignite it.

TABLE 11.1	Minimum Ignition Energies for Common Fuel Gases	
MINIMUM IGNITION GAS/VAPOR	**ENERGY (mJ)**
Acetone	1.15
Acetylene	0.02
Benzene	0.20
Butadiene	0.13
n-Butane	0.25
Ethane	0.24
Heptane	0.24
Hexane	0.24
Hydrogen	0.02
Methane	0.26
Methyl alcohol	0.14
Methyl ethyl ketone	0.53
n-Pentane	0.22
Propane	0.25

MATERIAL	$k\rho c$ $(kW/m^2K)^2s$	T_{ig} (°C)	$\dot{q}''_{Critical}$ (kW/m^2)
Plywood, plain (0.635 cm)	0.46	390	16
Plywood, plain (1.27 cm)	0.54	390	16
Plywood, FR (1.27 cm)	0.76	620	44
Hardboard (6.35 mm)	1.87	298	10
Hardboard (3.175 mm)	0.88	365	14
Hardboard (gloss paint), (3.4 mm)	1.22	400	17
Hardboard (nitrocellulose paint)	0.79	400	17
Particleboard (1.27 cm stock)	0.93	412	18
Douglas fir particleboard (1.27 cm)	0.94	382	16
Fiber insulation board	0.46	355	14
Polyisocyanurate (5.08 mm)	0.020	445	21
Foam, rigid (2.54 cm)	0.030	435	20
Foam, flexible (2.54 cm)	0.32	390	16
Polystyrene (5.08 cm)	0.38	630	46
Polycarbonate (1.52 mm)	1.16	528	30
PMMA type G (1.27 cm)	1.02	378	15
PMMA polycast (1.59 mm)	0.73	278	9
Carpet #1 (wool, stock)	0.11	465	23
Carpet #2 (wool, untreated)	0.25	435	20
Carpet #2 (wool, treated)	0.24	455	22
Carpet (nylon/wool blend)	0.68	412	18
Carpet (acrylic)	0.42	300	10
Gypsum board, (common) (1.27 mm)	0.45	565	35
Gypsum board, FR (1.27 cm)	0.40	510	28
Gypsum board, wall paper	0.57	412	18
Asphalt shingle	0.70	378	15
Fiberglass shingle	0.50	445	21
Glass reinforced polyester (2.24 mm)	0.32	390	16
Glass reinforced polyester (1.14 mm)	0.72	400	17
Aircraft panel, epoxy fiberite	0.24	505	28

TABLE 11.2 | Ignition Properties

Source: Data derived from Quintiere, 1998.

TABLE 11.3	Typical Piloted Ignition Temperatures for Common Fuels
FUEL	**TYPICAL PILOTED IGNITION TEMPERATURES**
Liquids	−40 to 400°C (−40 to 752°F)
Common Ignitable Liquids	−40 to 90°C (−40 to 194°F)
Heavy liquids (i.e., oils)	205 to 320°C (400 to 600°F)
Solids	270 to 450°C (518 to 842°F)
Non-fire-retardant plastics	270 to 360°C (518 to 680°F)
Wood-based products	330 to 375°C (626 to 707°F)
Fire-retardant solids	400°C or more (752°F)

pyrolyze or melt, then vaporize; liquid fuels typically vaporize when exposed to heat. The minimum temperature at which a flammable mixture is formed and can be ignited by the introduction of an external pilot ignition source (i.e., flame, spark) is known as the piloted ignition temperature. Many variables influence a fuel's ability to increase to a temperature sufficient to produce a flammable mixture. Therefore, the precise prediction or calculation of ignition is difficult at best. Because the principles of these variables can be accounted for and approximated, a good engineering estimation can be performed for different fuels. Tables 11.2 and 11.3 show the temperature ranges for ignition, based on these principles and experimental methods.

AUTOIGNITION (AUTOGENOUS IGNITION)

Autoignition temperature is the minimum temperature to which a flammable mixture must be raised to initiate combustion without any external pilot ignition source (i.e., spark, pilot flame) (Babrauskas, 2003) (see Table 11.4). In other words, the autoignition temperature relates to the ambient temperature that provides enough heat transfer to the fuel, which in turn allows the combustion reaction to occur without other energy sources. The ignition of the fuel gases without a pilot source is driven by the heat transfer characteristics of the chemical kinetics, a subject that is outside the scope of this text. Autoignition temperature has also been referred to as self-ignition temperature or **spontaneous ignition**, but for this text, autoignition is the form of ignition without a piloted ignition source. Spontaneous ignition temperature means something altogether different and is discussed later in this chapter.

spontaneous ignition
■ Initiation of combustion of a material by an internal chemical or biological reaction that has produced sufficient heat to ignite the material.

DIFFERENCES AND SIMILARITIES BETWEEN IGNITION CONCEPTS

When the minimum ignition energy is introduced to a small portion of a combustible fuel/air mixture, combustion initiates. The flame itself then becomes the pilot that serves to ignite the remaining volumes of the mixture. Ambient temperature of the entire mixture is inconsequential for piloted ignition, yet it is critical for autoignition.

TABLE 11.4	Autoignition Temperatures for Common Fuels	
PLASTIC	**TEMPERATURE (°C)**	**TEMPERATURE (°F)**
Polyethylene	365* 488	910
Polyisocyanate	525	977
Polymethylmethacrylate	310* 467	872
Polypropylene	330* 498	928
Polystyrene	360* 573	1,063
Polytetrafluoroethylene	660	1,220
Polyurethane foam (flexible)	456–579	850–1,075
Polyurethane foam (rigid)	498–565	925–1,050
Polyvinyl chloride	507	944
FUEL		
Acetone	465	869
Benzene	498	928
Diethyl ether	—	—
Ethanol (100 percent)	363	685
Ethylene glycol	398	748
Fuel oil #1 (kerosene)	210	410
Fuel oil #2	257	495
Gasoline (low octane)	280	536
Gasoline (100 octane)	456	853
Jet fuel (JP-6)	230	446
Linseed oil (boiled)	206	403
Methanol	464	867
n-pentane	260	500
n-hexane	225	437
n-heptane	204	399
n-octane	206	403
n-decane	210	410
Petroleum ether	288	550
Pinene (alpha)	255	491
Turpentine (spirits)	253	488

Source: DeHaan. J. (2007), *Kirk's Fire Investigation*, Sixth Edition, Table 5.3, p. 127, combined with Table 4.2, p. 69.
*These entries come from Babrauskas (2003).

Energy of the Ignition Source

When discussing ignition, it is important to understand the differences between heat and temperature. Many ignition sources are considerably higher in temperature than the reported ignition temperatures of most solids and liquids; however, these fuels will not immediately ignite upon exposure to these high temperatures (see Table 11.5). The heat source must remain exposed to the fuel surface to allow enough energy to be transferred into the fuel and thus allow its ignition temperature to be reached. This concept can be illustrated by running your finger quickly through a candle flame. Note that the skin on your finger will not be burned because very little heat energy is transferred to your finger. Consequently, the material properties of the fuel have a significant influence on the time to ignition.

The ability of a substance to conduct heat into itself and absorb that energy has a significant influence on that substance's ability to rise in temperature. For example, when a person steps out of bed and places one foot on a rug and another foot on

TABLE 11.5	Reported Burning and Sparking Temperatures of Selected Ignition Sources	
	TEMPERATURE	
SOURCE	°C	°F
Flames		
Benzene[a]	920	1690
Gasoline[a]	1026	1879
JP-4[b]	927	1700
Kerosene[a]	990	1814
Methanol[a]	1200	2190
Wood[c]	1027	1880
Embers[d]		
Cigarette (puffing)	830–910	1520–1670
Cigarette (free burn)	500–700	930–1300
Mechanical sparks[e]		
Steel tool	1400	2550
Copper–nickel alloy	300	570

Source: Data derived from NFPA 921, 2008.
[a]From Drysdale, *An Introduction to Fire Dynamics.*
[b]From Hagglund, B., Persson, L. E. (1976), Heat Radiation From Petroleum Fires, National Defence Research Inst., Stockholm, Sweden, FOA Report C20126-D6(A3).
[c]From Hagglund, B., Persson, L. E. (1974), Experimental Study of the Radiation From Wood Flames, National Defence Research Inst., Stockholm, Sweden, FOA Report C4589-D6(A3).
[d]From Krasny, J. (1987), *Cigarette Ignition of Soft Furnishings—A Literature Review with Commentary,* NBSIR 87-3509; National Bureau of Standards, Gaithersburg MD.
[e]From NFPA *Fire Protection Handbook,* 15th ed., Section 4, p. 167.

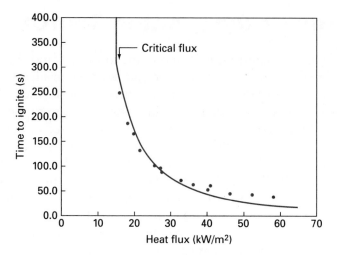

FIGURE 11.1 Radiant ignition times for an asphalt shingle.

Source: Data derived from Quintiere (1998).

a tile floor, which foot feels colder? Both feet are at the same temperature, and both the tile and the rug are at the same temperature, so what explains the perceived temperature difference? The foot that rests on the tile feels cooler due to the material property of the tile versus the rug. The rug is an insulator and does not allow heat to transfer easily from your foot into its mass. The tile is a conductor of the energy, dissipating the energy from your foot into its mass. The ability of energy to be dissipated throughout the fuel's mass is a thermophysical property that retards the temperature increase at the surface, which in turn prevents the item from readily raising in temperature. This same property influences ignition of fuels.

To overcome the energy losses by conduction into the fuel, convection losses to the surrounding atmosphere, radiation losses, and any energy loss associated with chemical changes to evolve vapors (i.e., heat of gasification and heat of vaporization), the temperature of the fuel must be raised sufficiently. The temperature is fuel dependent and is known as the ignition temperature for that particular fuel. Therefore, a heat source must overcome the thermal properties of the fuel, a situation known as the material's critical heat flux. If this minimum heat flux is not reached, the fuel will never ignite because its thermal properties will continue to dissipate the energy, never allowing ignition. Fuels that are relatively easy to ignite have critical heat fluxes between 7 and 10 kW/m^2, while most fuels have a critical heat flux between 10 and 20 kW/m^2. Figure 11.1 shows the radiant ignition times of an asphalt shingle. You can see in this figure that the time to ignition infinitely rises as the curve approaches the critical heat flux. In other words, any heat flux less than approximately 15 kW/m^2 (specific to this fuel) will never be sufficient to raise the temperature of the fuel and cause ignition. Also, it should be evident that the higher the heat flux being imposed on a fuel's surface, the less the time required for ignition.

Material Properties

Several issues influence whether ignition occurs and the time it takes to ignite the fuel, including the magnitude of the heat source, the material properties, and the duration of the heat exposure. The material properties of the fuel influence both ignition and flame spread. Flame spread is a series of piloted ignitions, so the same properties that affect the initial ignition of a fuel also drive the flame spread along the surface. These material properties include thermal conductivity, specific heat, and density. In addition, fuel characteristics, such as geometry and orientation, are extremely important factors when discussing flame spread.

THERMAL INERTIA

Thermal properties that affect the rate of heat transfer through a material play a significant role in the ignition and spread of fire on a solid fuel. The most important of these properties include the density (ρ), specific heat (c_p), and the thermal conductivity (k) (see Table 11.6). Thermal conductivity is a measurement that compares the ability of the material to conduct energy through it versus its ability to store the energy. In fire dynamics, the product of these three thermophysical properties is typically used as the **thermal inertia**.

$$\text{thermal inertia} = k \cdot \rho \cdot c_p$$

Where:

k = thermal conductivity (W/m · K)
ρ = density (g/m^3)
c_p = specific heat (J/g · K)

The thermal inertia of a material characterizes its ability to conduct energy away from its surface and through its mass. More specifically, it provides a means of determining the temperature increase at the surface of a material, which is the principal issue when examining ignition. The higher the thermal conductivity, the faster energy can transfer away from the material's surface and into the mass of the material. Also, the closer the molecules are packed into a material (i.e., the greater the density) also affects its ability to transfer the energy through its mass

thermal inertia
■ The properties of a material that characterize its rate of surface temperature rise when exposed to heat. These properties are related to the product of the material's thermal conductivity (k), its density (ρ), and its heat capacity (c).

TABLE 11.6	Thermal Properties of Select Fuels			
MATERIAL	THERMAL CONDUCTIVITY (k) (W/(mK))	DENSITY (ρ) (kg/m^3)	HEAT CAPACITY (C) (J/(kg-K))	THERMAL INERTIA ($k\rho c$) (W^2 · s/k^2 m^4)
Copper	387	8940	380	1.31E + 09
Concrete	0.8–1.4	1900–2300	880	1.34E + 06 – 2.83E + 06
Gypsum plaster	0.48	1440	840	5.81E + 05
Oak	0.17	800	2380	3.24E + 05
Pine (yellow)	0.14	640	2850	2.55E + 05
Polyethylene	0.35	940	1900	6.25E + 05
Polystyrene (rigid)	0.11	1100	1200	1.45E + 05
Polyvinylchloride	0.16	1400	1050	2.35E + 05
Polyurethane[a]	0.034	20	1400	9.52E + 02

Source: Data derived from Drysdale, 1998.
[a]Typical values and properties vary with temperature.

and away from the exposed surface. Low-density fuels (i.e., polyurethane foams) cannot dissipate the energy as quickly through their mass and thus the temperature increases easily. This increase in temperature allows the fuel to ignite and spread flame at a much faster rate.

The higher the thermal inertia, the easier heat can be transferred through the material's mass away from the surface and the less chance of ignition. For instance, metals are good conductors of thermal energy and can dissipate the energy through their mass faster, resulting in lower temperatures at the surface. Wood products, on the other hand, have a low thermal conductivity and density. For that reason, wood has a low thermal inertia. Energy imposed on the face of a piece of wood dissipates from its surface slowly, allowing the temperature at the point of heating to rise quickly, which supports pyrolysis and then ignition.

Surface Area–to–Mass Ratio

surface area–to–mass ratio
■ The ratio of surface area to total mass of an object. Thin objects have more surface per unit of mass than do thick objects comprised of like matter. Also known as *surface to mass ratio*.

The **surface area–to–mass ratio** (commonly called the surface to mass ratio) is an important concept that must be considered when analyzing ignition, flame spread, and heat release rate. As the surface area increases (ratio increases), the fuel particles become smaller or more finely divided. When heat is applied to a solid or even a liquid, the heat is conducted into that material. If a large mass exists behind the surface, the material has a greater ability to dissipate the energy away from its surface. The ability of this material to conduct energy away from its surface relates to its thermal properties (i.e., thermal conductivity, specific heat) and the density of that material (i.e., how closely or loosely packed the fuel particles are). (These topics are discussed later in the chapter.) If the material has been cut or divided out of the large mass, then its surface area–to–mass ratio has increased. When the same heat as before is now applied to this lesser mass, the energy cannot dissipate as much due to the lack of mass for the heat to be conducted into. Therefore, the energy is able to increase the temperature at the surface at a much quicker rate, which results in ignition of the material at a much faster rate. This is the principle that explains why atomized liquids can ignite when the same pool of that liquid cannot, and sawdust can ignite while the wood log may not.

For example, take a wood log and apply a match. Would you expect the wood to ignite? Most likely not because the energy imparted by the flame is being dissipated throughout the large mass of the wood and does not allow the temperature to increase to the point to cause pyrolysis. However, when we cut or sand this wood log, wood shavings and fine particles may be collected. If we apply the same type of match to the wood shavings and fine particles, do you expect ignition to occur now? Bear in mind, we have not changed the thermal properties or the density of the wood, nor the temperature or energy imparted by the flame. The only thing that has changed is the surface area–to–mass ratio. More likely than not, the wood shavings will ignite. This time, the match is exposed to less mass, which means that there is not as much mass to allow for heat to be conducted away and dissipated. Because the energy cannot be dissipated, the temperature of the wood can increase, enabling the pyrolysis process.

The same principle applies when trying to ignite a material on a bulk surface area versus an edge. For example, when trying to ignite a piece of paper, it is

FIGURE 11.2 Flat, concave, and convex fuel surfaces with an imposed heat flux.

much easier to ignite the paper on its edge rather than in the center. The material properties of the paper do not change from the center of the paper to its edge. The only item that changes is the ability of the energy to be dissipated easily throughout the material. The center of the paper allows heat to be conducted in every direction away from the heat source, preventing a fast temperature rise. The edge of this paper has fewer directions that the heat can be conducted into the remaining mass, which delays the energy dissipation from its surface.

GEOMETRY

The shape of the fuel surface affects the heat transfer into the fuel. Flat surfaces and concave surfaces reflect much more of the imposed heat flux compared to a convex surface (see Figure 11.2). The convex surface has a greater ability to absorb the reflected heat into its remaining fuel surfaces and thereby retain more of the imposed heat flux. This aspect is most pronounced when examining flame spread and ignition for surfaces that follow these geometric shapes, and especially when analyzing fires in corners and along edges of a fuel surface (see Figure 11.3). It has been shown experimentally that a corner retains more of the energy imposed and results in faster ignition and a much faster flame spread. The heat being released is absorbed by additional fuels within the corner rather than being emitted to the surroundings. The feedback from a flame is bounded by two surfaces, and radiation feedback increases.

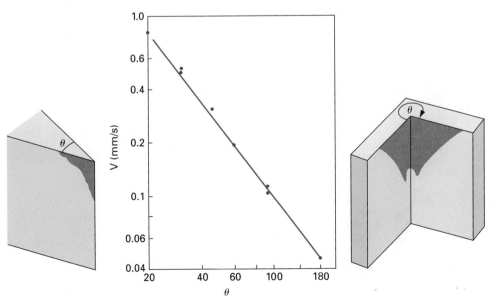

FIGURE 11.3 Corner and edge influence on flame spread.
Source: Data derived from Drysdale, D. (1998).

DENSITY

Density was discussed previously as an internal material property; however, the density of the fuel array or composite is also very important. The same principle applies as the internal density for the fuel. The lower-density items cannot dissipate energy as quickly and are more susceptible to ignition and a faster flame spread. For example, a phone book consists of hundreds of lightweight papers bound together, creating a high-density fuel. When it is closed, the book is not easily ignited and does not allow flames to spread quickly through the material due to its high density and the ability of the fuel to dissipate the imposed energy to the remaining mass. However, if the pages of the book were ripped out, crumpled up, and placed into a pile, the flame spread would be extremely fast. Two properties are affecting this flame spread. First, the low-density pile of crumpled pages does not allow the imposed energy to dissipate to the remaining mass. Second, the pockets of air allow for fuel and air to mix, which allows for better combustion within the fuel array.

ORIENTATION

concurrent flow flame spread
■ The flame spread direction is the same as the gas flow or wind direction.

The orientation of solid fuels has a major impact on how the fuel behaves once ignited. A fuel oriented vertically and ignited at the base has its virgin fuel heated by the flowing heated gases by direct contact via convection and conduction heat transfer, as well as a much greater radiant heat view factor associated with the parallel fuel and flame source (see Figure 11.4). This heat transfer quickly preheats the virgin fuel and increases the rate of flame spread. The flow of the heated gases, and thus the convective heat transfer, is flowing concurrently and aiding in the flame spread, which is known as **concurrent flow flame spread**. This method of flame spread is also seen with wind-aided flame spread and certain inclined angles (see Figure 11.5).

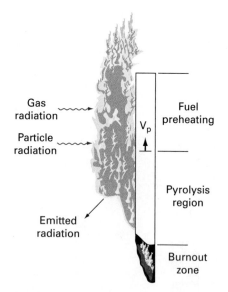

FIGURE 11.4 Vertical orientation of a fuel and its associated heat transfer.

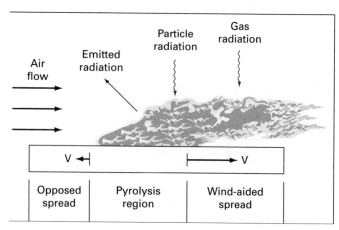

FIGURE 11.5 Wind-aided flame spread.

A fuel oriented horizontally has only the radiant feedback from the flame itself (approximately 30 percent of the total heat release rate) and its flame is spreading against the natural flow of air entrainment (see Figure 11.6). The flow of air entraining into this combustion reaction is going against the spreading flame and is known as **counterflow flame spread**. Counterflow flame spread is much slower due to the limited heat transfer imposed back to the virgin fuel (see Figure 11.7).

What happens when the orientation of the fuel is not exactly vertical or horizontal? The position of the fuel anywhere between the horizontal plane and the vertical plane, known as an inclined plane, also influences the flame spread

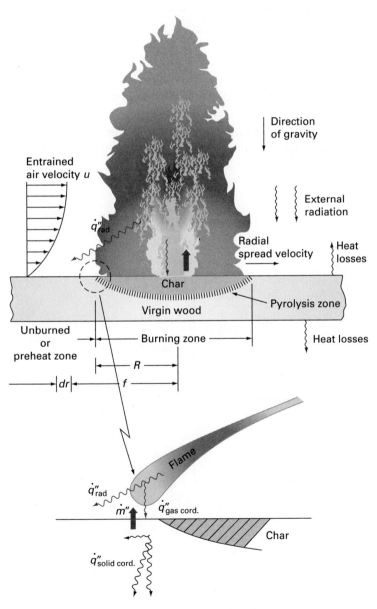

FIGURE 11.6 Flame spread on a horizontal surface.
Source: NFPA 921 (2008).

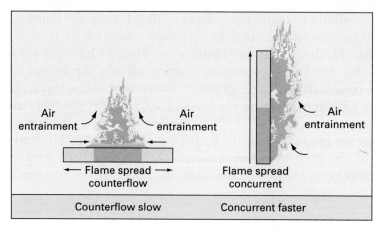

FIGURE 11.7 Counterflow and concurrent flow flame spread.

capabilities (see Figure 11.8). It has been reported that a shift from counterflow flame spread to concurrent flame spread occurs when the angle of the incline is increased past 25° (Drysdale, 1998). In other words, there is not a significant increase in flame spread between a fuel sitting horizontally to one that is inclined between 0° and 25°. However, once the fuel is inclined at an angle greater than 25°, flame spread is influenced.

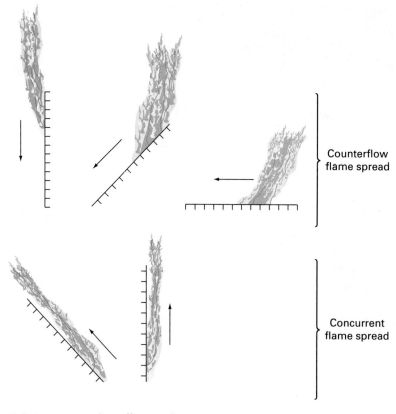

FIGURE 11.8 Incline effects on flame spread.
Source: Data derived from Drysdale, D. (1998).

Time to Ignition Calculations for Solid Fuels

It has been shown in this chapter that the ignition temperature for the fuel must be reached for a solid fuel to emit gaseous fuel within its flammable limit before ignition will occur. It has also been shown that the temperature increase for the fuel depends on its material properties and the duration of exposure of a heat source. This discussion can be represented mathematically. The ability for the fuel to transfer energy into itself also depends on whether the fuel is thermally thin or thermally thick, which provides the basis for modeling time to ignition.

The classification of a material as thermally thin corresponds to an object's thermal properties. Thermally thin does not necessarily relate to the thickness of the solid, but rather to its decreased ability to absorb heat energy, causing a quick transfer of heat through its mass. Typically, thermally thin materials are also thin in physical thickness, but not always. **Thermally thick solids** are solids that absorb heat energy more readily and do not transfer it as quickly through the material. In other words, **thermally thin solids**, when exposed to a heat flux, have a slight rise in temperature incurred to the entire mass of the fuel due to the fast transfer and dissipation of the imposed heat flux. In contrast, a thermally thick material, when exposed to a heat flux, experiences a substantial increase in temperature on the side facing the imposed heat flux. The opposite side incurs a slower increase in temperature.

thermally thick solid
- A solid that, while heated from one face, shows a negligible temperature rise at its opposite face. This characteristic is not simply a property of the material; it also depends on the time of exposure and the heat flux.

thermally thin solid
- A solid that, while heated from one face, shows a back-face temperature that is nearly identical to the temperature of the heated face. This characteristic is not simply a property of the material; it also depends on the time of exposure and the heat flux.

IGNITION OF THERMALLY THIN SOLIDS

As a heat flux is imposed on one side of the fuel's surface, a temperature increase begins to occur based on the severity and duration of the heat exposure as well as the fuel's material properties. Therefore, a mathematical expression has been created to account for this temperature increase:

$$T = T_\infty + \frac{\dot{q}''t}{\rho c l}$$

Where:

T = temperature of the solid (°C or K)
T_∞ = initial temperature (°C or K)
$\dot{q}''$ = imposed heat flux (kW/m^2)
ρ = density (g/m^3)
c = specific heat (J/g · K)
l = thickness (m)
t = time (s)

The above equation can be rearranged to solve for the time to ignition by assuming that the temperature of the solid has reached its ignition temperature (experimentally determined; see Table 11.6). However, as previously discussed, if the solid can dissipate the heat faster than the temperature can increase to this ignition temperature, ignition will not occur. Consequently, for a fuel to ignite, a minimum critical heat flux must be applied. The critical heat flux depends on the different material properties for different fuels (see Table 11.6). Figure 11.9 reviews this concept graphically. A low heating rate or imposed heat flux may never increase the surface temperature of the solid sufficiently to reach the ignition temperature. A critical

FIGURE 11.9
Temperature rise and
heating rate for
ignition.
Source: Data derived from
Quintiere (1998).

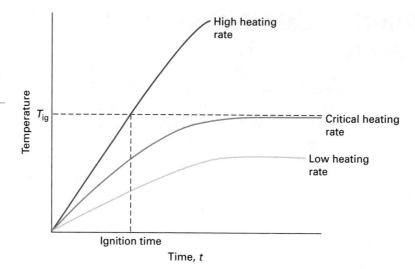

heat flux or critical heat rate is the minimum for ignition to occur and may require an extended duration for the solid to reach its ignition temperature. On the other hand, a fast heating rate or high heat flux ignites the fuel quickly.

$$t_{ig} = \rho c l \frac{(T_{ig} - T_\infty)}{\dot{q}''}$$

Where: T_{ig} = ignition temperature of the solid (°C or K) (see Table 11.6)
T_∞ = initial temperature (°C or K)
$\dot{q}''$ = imposed heat flux (kW/m²)
ρ = density (g/m³)
c = specific heat (kJ/kg · K)
l = thickness (m)
t_{ig} = time to ignition (s)

EXAMPLE 11.1

How much time does it take to ignite a cottonlike material with a density (ρ) of 0.59 g/cm³, $c = 0.35$ cal/g-K, thickness (l) = 1 mm, and ignition temperature of 300°C. Assume an initial temperature (T_∞) of 20°C with an imposed heat flux ($\dot{q}''$) of 12 kW/m².

Analysis: We must first convert all of the properties to the same units to be able to calculate the time to ignition.

$$\rho = 0.59 \frac{g}{cm^3} * \left(\frac{100\ cm}{1\ m}\right)^3 * \frac{1\ kg}{1000\ g} = 590 \frac{kg}{m^3}$$

$$c = 0.35 \frac{cal}{gK} * \left(\frac{1000\ g}{1\ kg}\right)\left(\frac{4.18\ J}{1\ cal} * \frac{1\ kJ}{1000\ J}\right) = 1.463 \frac{kJ}{kg\ K}$$

$$l = 1\ mm * \left(\frac{1\ m}{1000\ mm}\right) = 0.001\ m$$

Now that all variables are in similar units, we can calculate the time to ignition by substituting all the values into the equation:

$$t_{ig} = \rho c l \frac{(T_{ig} - T\infty)}{\dot{q}''} = \left(590 \frac{kg}{m^3}\right)\left(1.463 \frac{kJ}{kgK}\right)(0.001\ m)\frac{(300 - 20°C)}{12 \frac{kW}{m^2}}$$

$$= 20.1\ \text{seconds}$$

Note: Colors correspond to units that cancel each other.

IGNITION OF THERMALLY THICK SOLIDS

Thermally thick solids have a very similar formula used to determine their time to ignition:

$$t_{ig} = C(k\rho c)\left[\frac{(T_{ig} - T\infty)}{\dot{q}''}\right]^2$$

Where:
$C = \pi/4$ or 2/3 (see below)
t_{ig} = time to ignition (s)
T_{ig} = ignition temperature of the solid (°C or K) (see Table 11.6)
$T\infty$ = initial temperature (°C or K)
$\dot{q}''$ = imposed heat flux (kW/m^2)
ρ = density (g/m^3)
c = specific heat (kJ/kg · K)
k = thermal conductivity (W/m · K)

Specific heat, thermal conductivity, and the ignition temperature of the fuel are typically determined experimentally and are listed in Table 11.6. The C term is a constant that is somewhat dependent on heat flux (Thomas, 1995); C is equal to $\pi/4 = 0.785$ when the ideal case of no surface heat loss occurs and $C = \frac{2}{3} = 0.667$, assuming heat losses. The ignition calculations presented above have limitations and assumptions. One of the main assumptions is that the fuel has constant properties throughout its mass, which may or may not be appropriate for a given scenario. It is the user's responsibility to ensure the proper use of the calculation. See Fire Dynamics Tools (FDTs) for a more exhaustive list of assumptions and limitations (Iqbal and Salley, 2004).

EXAMPLE 11.2

Determine the time to ignite a piece of ½-inch particleboard subject to a flame heat flux of 25 kW/m^2 with a starting temperature of 25°C.

Properties:

$$T_{ig} = 412°C$$
$$T\infty = 25°C$$
$$k\rho c = 0.93\ (kW/m^2K)^2 s$$
$$C = \frac{\pi}{4} = 0.785\ or = \frac{2}{3} = 0.667$$

Assume:

$$C = \frac{\pi}{4} = 0.785$$

Analysis:

$$t_{ig} = C(k\rho c)\left[\frac{(T_{ig} - T\infty)}{\dot{q}''}\right]^2 = (0.785)\left(0.93\left(\frac{kW}{m^2K}\right)^2 s\right)\left[\frac{(412 - 25°C)}{25\frac{kW}{m^2}}\right]^2$$

$$= 174.9 \text{ s}$$

Assume:

$$C = \frac{2}{3} = 0.667$$

Analysis:

$$t_{ig} = C(k\rho c)\left[\frac{(T_{ig} - T\infty)}{\dot{q}''}\right]^2 = (0.667)\left(0.93\left(\frac{kW}{m^2K}\right)^2 s\right)\left[\frac{(412 - 25°C)}{25\frac{kW}{m^2}}\right]^2$$

$$= 148.6 \text{ s}$$

Note: Colors correspond to units that cancel each other.

Comments: The true answer may be between these upper and lower bounding values. This is why it is always important to use these values as general benchmarks, and not as absolutes.

Ignitability and Flammability Testing

Some calculations have been provided to assess the flammability hazards for solids. However, the primary means for determining a fuel's flammability hazard is still based on standardized testing. Many of these tests (i.e., flashpoint, heat release rate) have been presented in other chapters, but there are many standardized tests used to determine the relative hazard of a fuel by monitoring flame spread or ignitability. There are literally dozens of flammability tests, and they will not be covered in this text. For a more exhaustive review of these tests, see the National Fire Protection Association's *Fire Protection Handbook* (2008) or the specific standards that regulate the testing (i.e., the American Society for Testing and Materials [ASTM], Underwriters' Laboratories [UL], International Organization for Standardization [ISO]). Some of the more common tests and their standardizing bodies include the following:

1. Cone calorimeter (NFPA, ASTM, and ISO)
2. Fire propagation apparatus (NFPA and ASTM)
3. Lateral ignition and flame spread (LIFT) apparatus (ASTM and ISO)
4. Steiner tunnel test (ASTM)
5. Radiant panel test (ASTM)

6. Radiant panel flooring test (ASTM)
7. Vertical burn test (UL and ASTM)
8. Fabric and textile flammability tests (nearly all orientations) (ASTM, Code of Federal Regulations [CFR], and NFPA)

Spontaneous Ignition

Spontaneous ignition is brought on by self-heating. Self-heating is a process that is commonly associated with chemicals or reactions that result exothermically. Self-heating is most commonly due to organic materials, including animal and vegetable fats and oils, that react with oxygen to emit heat. Sometimes referred to as biological activity, such as that found in haystacks, decomposition oxidation can also result in self-heating. Some of the substances prone to self-heating are listed in Table 11.7.

Self-heating may often occur with these types of substances, but it rarely results in ignition. Spontaneous ignition requires the self-heating process to proceed to what is known as thermal runaway. **Thermal runaway** is commonly described as an instability that occurs when heat generation exceeds heat loss within the material (NFPA 921, 2008). In other words, for a self-heating process to increase to a point where ignition can occur, the self-heating process must have a significant increase in the rate of temperature rise, which continues to ignition.

thermal runaway
■ Self-heating that rapidly accelerates to high temperatures.

TABLE 11.7	Common Substances Prone to Self-Heating		
MICROBIOLOGICAL HEATING	**OXIDATIVE HEATING**	**MOISTURE-INDUCED HEATING**	**OTHER CHEMICAL PROCESSES**
Bagasse (sugar can residue)	Activated carbon	Chlorinated oxidizers	Monomers
Compost	Coal (low-rank)	Calcium oxide (lime)	Nitrocellulose
Grains	Cotton	Cotton bales	Peroxides such as lauroyl peroxide, and benzoyl peroxide
Hay (moist)	Foam rubber	Dry paper rolls	
Mulch (moist)	Metal filings and powders	Insulating boards (dried) (lingo-cellulosic)	
Pecans	Particleboard	Potassium phosphide	
Sewage sludge	Peat	Wool bales	
Soya beans	Sawdust		
Walnuts	Wood chips		

Source: Data derived from Zalosh, 2003.

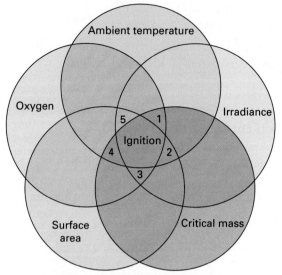

FIGURE 11.10 Conditions required for spontaneous ignition to occur in materials capable of self-heating. *Source:* NFPA 921 (2008).

(1) Insufficient surface area.
(2) Insufficient oxygen.
(3) Ambient temperature too low.
(4) Insufficient insulation—heat radiated away.
(5) Insufficient material.

Many variables influence spontaneous ignition, including the ambient temperature, availability of oxygen, moisture content, humidity, critical mass, surface area, and irradiance (see Figure 11.10). In essence, heat has to be insulated within the material to allow for thermal runaway, but there also has to be enough access to the interior portions for oxygen to enter into the reaction. Similarly, the surface area and the mass must be in the proper form for ignition to occur. Finally, when self-heating does progress to the pyrolyzing of the fuel and sufficient vapors formed, the MIE for those vapors must be present.

Summary

Ignition is the process of initiating self-sustained combustion. Even though it has been shown that many terms and different mechanisms affect ignition, there are a few basic concepts that relate to all fuels. Basically, the fuel (regardless of its state of matter) has to be within its flammable limits, and a minimum ignition energy must be present for ignition to occur. The same properties that affect ignition have also been shown to affect flame spread.

Review Questions

1. List five common products that are capable of self-heating.
2. Under what conditions can a liquid burn if its temperature is below its flash point as stated in reference materials?
3. If a pilot flame is present, how intense must a thermal radiative flux be to ignite wood? How does this flux compare to the intensity of the sunlight?
4. Why is upward flame spread over a vertical surface more rapid than downward flame spread?
5. How much time does it take to ignite a cotton-like material with a density (ρ) of 0.65 g/cm^3, c = 0.37 cal/g-K, thickness (l) = 1.2 mm, and an ignition temperature of 300°C. Assume an initial temperature (T_∞) of 20°C with an imposed heat flux (q'') of 18 kW/m^2.
6. Explain how pyrolysis of solid fuels is involved in ignition.
7. What is the difference between piloted ignition temperature and autoignition temperature?
8. What does the term *minimum ignition energy* mean?
   a. The temperature required to cause pyrolysis in fuels
   b. The energy required to initiate combustion in liquids
   c. The energy required to initiate combustion in vapors or gases
   d. The energy released at the start of combustion of solids
9. The surface area–to–mass ratio affects burning rate:
   a. Only when the fuel is oriented vertically
   b. Only when the fuel is oriented horizontally
   c. When it is low; lower surface area–to–mass ratio fuel items undergo pyrolysis more easily and ignite sooner
   d. When it is high; higher surface area–to–mass ratio fuel items undergo pyrolysis more easily and ignite sooner
10. Thermally thin fuels tend to ignite:
    a. More readily than thermally thick fuels because they produce more fuel gases
    b. Less readily than thermally thick fuels because the surface does not heat as quickly
    c. The same as thermally thick fuels
    d. None of the above
11. A vertical sheet of cotton fabric, 0.5 mm thick, is immersed in a stream of hot air at 200°C. Calculate how long it will take to reach 100°C if the heat transfer coefficient is h = 20 W/m^{2}K and ρ and c are 300 kg/m^3 and 1,400 J/kg, respectively. Assume the initial temperature to be 25°C (adapted from Drysdale, 1998).

References

Babrauskas, V. (2003). *Ignition Handbook.* Issaquah, WA: Fire Science Publishers.

DeHaan, J. (2007). *Kirk's Fire Investigation,* Sixth Edition. Upper Saddle River, NJ: Prentice Hall/Brady.

Drysdale, D. (1998). *An Introduction to Fire Dynamics,* Second Edition. New York: Wiley.

Fire Safety—Vocabulary (ISO 13943). (2008). Geneva: International Organization for Standardization.

Hagglund, B., and Persson, L. (1974). *Experimental Study of the Radiation from Wood Flames.* FOA Report C4589-D6 (A3). Stockholm, Sweden: National Defence Research Institute.

Hagglund, B., and Persson, L. (1976). *Heat Radiation from Petroleum Fires.* FOA Report C20126-D6 (A3). Stockholm, Sweden: National Defence Research Institute.

Iqbal, N., and Salley, M. (2004). *Fire Dynamics Tools—Quantitative Fire Hazard Analysis Methods for the U.S. Nuclear Regulatory Commission Fire Protection Inspection Program.* NUREG-1805. Washington, DC: Nuclear Regulatory Commission.

International Organizaton for Standardization (ISO). (n.d.). Jersey City, NJ: ISO.

Krasny, J. (1987). *Cigarette Ignition of Soft Furnishings—A Literature Review with Commentary.* NBSIR 87-3509. Gaithersburg, MD: National Bureau of Standards.

National Fire Protection Association (NFPA). (2008). *Fire Protection Handbook,* Twentieth Edition. Quincy, MA: NFPA.

NFPA 921 (2008). *Guide for Fire and Explosion Investigations.* Quincy, MA: NFPA.

Quintiere, J. (1998). *Principles of Fire Behavior.* Albany, NY: Delmar.

Thomas, P. (1995). "The Growth of Fire-Ignition to Full Involvement," in G. Cox (Ed.), *Combustion Fundamentals of Fire.* London, UK: Academic Press.

Zalosh, R. (2003). *Industrial Fire Protection.* New York: Wiley.

12
Enclosure Fire Dynamics

KEY TERMS

air entrainment, *p. 233*

backdraft, *p. 228*

ceiling jet, *p. 232*

established burning, *p. 228*

flameover, *p. 247*

flashover, *p. 238*

fuel-controlled fire, *p. 231*

full-room involvement, *p. 237*

neutral plane, *p. 234*

upper layer, *p. 233*

ventilation-controlled fire, *p. 246*

OBJECTIVES

After reading this chapter, you should be able to:

- Describe the stages of an enclosure fire.
- Identify components of fire dynamics that affect the growth of a fire inside a compartment.
- Differentiate among the rapid transitional events.
- Compare and contrast the mechanisms that result in rapid fire progression.
- Predict fire behavior in an enclosure fire.

PEARSON
myfirekit

For additional review and practice tests, visit **www.bradybooks.com** and click on MyBradyKit to access book-specific resources for this text!

Fire safety professionals must achieve a solid theoretical knowledge of fire behavior, specifically enclosure fire behavior, to perform their duties effectively. In general, enclosure fire behavior is the study of the chemical and physical mechanisms controlling a fire within a compartment or room. Statistics and historical data prove that enclosure fires are the most dangerous to human life. For example, in 2007, 80 percent of all civilian fire deaths in the United States resulted from fires inside residential structures (Hall and Cote, 2008). While the public may hear more about major conflagrations and large-loss fires, the truth is that the most dangerous place to be is in our own homes. Thus, fire safety professionals must truly understand and grasp all the components of enclosure fire behavior to succeed at their primary mission of saving lives.

Fire dynamics in enclosures change drastically throughout a fire's progression, including the possible various phenomena that may occur during the development of such a fire. Therefore, suppression personnel may misread the conditions unless they fully comprehend anticipated fire development in enclosure fires. Failure to understand differences in enclosures based on size and contents leads to fatalities and injuries of responders. Three distinct and commonly misunderstood phenomena—flameover, **backdraft**, and flashover—are the most dangerous because of the dramatic changes that occur rapidly throughout the compartment. This chapter will provide a review of the current theoretical understanding of the phenomena, along with methods of practical application.

The purpose of this chapter is to provide a deeper understanding of enclosure fire dynamics to assist you in better assessing fire behavior inside an enclosure. The major division between the types of fire dynamics discussed in the chapter is based on the availability of oxygen or lack thereof for combustion.

backdraft
■ Limited ventilation during an enclosure fire can lead to the production of large amounts of unburned pyrolysis products. When an opening is suddenly introduced, the inflowing air forms a gravity current and begins to mix with the unburned pyrolysis products, creating a combustible mixture of gases in some part of the enclosure. Any ignition source, such as a glowing ember, can ignite this combustible mixture, which results in an extremely rapid burning of gases and/or pyrolysis products forced out through the opening, in turn creating a fireball outside the enclosure.

Ignition

This chapter will also focus on the development of enclosure fires and will assume that ignition has occurred. A fire may follow one of two paths after ignition has occurred: It can self-extinguish, or it may grow and become a great threat to life and property. Obviously, this chapter focuses on the fires that develop and become a great threat to life and property due to their relative danger and importance.

The initial combustion process (commonly referred to as the incipient phase) after ignition has many variables that influence fire behavior and is generally very difficult to predict. However, the fire may reach a state when oxygen, fuel, and radiant feedback are sufficient to allow combustion to continue unless some outside force (for example, hose stream application) acts on it. This state is commonly referred to as established burning. **Established burning** is simply a benchmark in the development of enclosure fires distinguishing the point where the fire will not self-extinguish, typically quantified by a standard heat release rate or flame height. It is an important point of reference because the ability to predict fire growth increases once it is assumed that the fire will not self-extinguish. A flame size of 20 kW is most often accepted as established burning for most residential and commercial structures (Fitzgerald, 2004). Drysdale (1998) reports that experimental tests on upholstered furniture were much more predictable after 50 kW and suggested that this value be considered established burning for upholstered furniture.

established burning
■ The point in an oxidation reaction when a piloted flame is unnecessary to sustain pyrolysis. Radiant heat transfer from the fuel package is sufficient for continued pyrolysis.

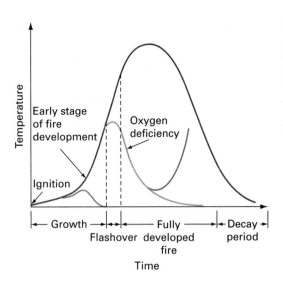

FIGURE 12.1
Idealized fire growth curve featuring different types of fire behavior.
Source: Adapted from Bengtsson (2001).

Ignition and the progression to established burning are the beginning stages of the fire growth curve (see Figure 12.1). The fire growth curve is a line graph that represents the heat release rate over the duration of the fire. The initial stages of the fire are at the bottom left-hand of the curve at the zero time mark. All fires initially begin with a moderate rising slope that steadily increases as the fire progresses through the duration of its fire growth. Eventually, in the progression of every fire, the fuel becomes exhausted or the available oxygen is not present to support combustion, and the flattening of the curve and the eventual decay represents this.

Growth

Typically, as an enclosure fire begins, only one fuel item is burning. Therefore, the rate at which the fire develops depends greatly on the initial growth stage. The driving force for the initial fire growth is primarily by flame spread across the first item burning. The secondary force affecting fire growth in the initial stages can be flame spread to nearby fuels by direct contact or possibly by radiant ignition. Therefore, orientation and fuel arrangement are extremely important in the initial stages of the enclosure fire. Orientation of the fuel greatly affects the rate of flame spread. For instance, flames spread at a faster rate along a vertical fuel surface than a horizontal fuel surface due to the combination of both radiant and convective heating. If a fire were to start at the base of a curtain, the flame spread would be substantially faster than if that same fabric were ignited while lying flat on a table (see Figure 12.2). A rapid flame spread contributes to an increase of the fire's area, which in turn increases the heat release rate. Additionally, the arrangement of fuels has a significant impact on the progression of the initial fire. The closer fuels are located to each other, the greater the potential for multiple items being ignited earlier in the growth of the fire. All of these variables combine to form the initial stages of the fire growth curve.

The controlling or limiting factor in the early stage of fire growth is the amount of fuel available in a physical state capable of supporting combustion and

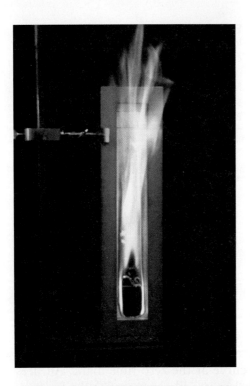

FIGURE 12.2
HRR differences due to fuel geometry.
Source: Photo by authors.

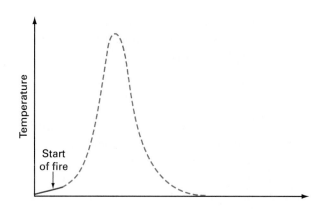

FIGURE 12.3 Initial growth curve of fire. *Source:* Adapted from Bengtsson (2001).

is commonly referred to as a **fuel-controlled fire** (see Figure 12.3). Essentially, the heat release rate is limited by the availability of fuel in the gaseous form capable of supporting combustion. The availability of fuel in the gaseous form is limited in turn by the heat feedback from the developing flame. At this point, an abundant source of available air (oxygen) is typically available to support combustion and ventilation is not a limiting factor in transforming the fuels' potential energy into kinetic energy.

As the fire spreads across the fuel's surface and possibly involves other fuels, the heat release rate increases. The greater the surface area involved, the greater the amount of energy released by the burning fuels. Steadily, the heat release rate increases and radiant heat transfer from the flame plume intensifies, allowing for greater potential of igniting additional nearby fuels. The fire continues to burn as long as sufficient fuel and oxygen are present.

fuel-controlled fire
■ A fire in which the heat release rate and growth rate are controlled by the characteristics of the fuel, such as quantity and geometry, and in which adequate air for combustion is available.

PLUME FORMATION

As a fuel undergoes combustion, it releases by-products of combustion in the form of carbon dioxide (CO_2), water (H_2O), carbon monoxide (CO), smoke, light, heat, and other products of incomplete combustion (HCN, etc.). These by-products of combustion are elevated in temperature compared to the surrounding air. The temperature elevation causes the combustion gases to increase in volume and become less dense compared to the surrounding air. This creates buoyancy forces between the two fluids (heated gases and air) and causes the higher-temperature, less-dense gases to rise in a fire. The upward-moving gases and smoke are commonly referred to as the thermal plume, convective column, or fire plume. The fire plume continues to rise until the temperature of the smoke and gases become similar to the surrounding air and/or until the plume cannot overcome the opposing forces.

One of the more influential opposing forces present during this buoyant flow is the viscous drag between the rising convective column of fluid (fire plume) and the stagnant fluid (air). As the upward-flowing fluid (fire plume) interacts with the nonmoving fluid (air), a resistance or drag force begins to affect this movement and serves to slow its progression, similar to friction force between one moving solid material over another.

At this point in the fire, the buoyant fire plume is primarily transferring heat energy by convection. The flame plume is preheating nearby fuels by radiant heat but, at this point, is not the primary means of heat transfer.

CEILING JET

Restricting the vertical movement of the buoyant plume is known as stratification. The stratification of the fire plume is commonly seen when you are sitting at a campfire and watching the smoke rise to a certain height, then spreading outward. Another example of stratification of a thermal plume is seen as the mushroom cloud associated with the atomic bomb. Stratification becomes an important aspect of fire protection in high spaces, such as atria, when temperatures may be significantly higher at upper elevations and may cause stratification of the plume before it can reach a ceiling where smoke and/or heat detectors are typically placed. A fire burning in the open (campfire), away from walls and a ceiling, has momentum and will continue to rise until these variables (i.e., temperature difference, viscous drag) cause it to stratify. However, a fire inside an enclosure involves a ceiling that intersects the natural progression of this buoyant flow. The ceiling does not stop the buoyant force of the thermal plume; rather, it redirects it beneath the ceiling. The thermal plume, while redirected, does not lose all of its buoyancy force and continues to spread outward 360 degrees from the centerline of the plume unless intersected by another barrier. It is important to note that the thermal plume, a fluid, follows the path of least resistance. The path of least resistance when intersected by a ceiling is laterally beneath that ceiling. The redirected flow of the buoyancy-driven thermal plume under a ceiling is known as a **ceiling jet** (see Figure 12.4). It is important to note that not all fires have sufficient power or heat release rate to create a high enough temperature difference for a plume to reach the ceiling. In addition, the ceiling height, shape, and relationship to the base of the fire are important factors to the development of a ceiling jet.

The ceiling jet is also a very important aspect to sprinkler and detector activation (see Figure 12.5). Ultimately, the time it takes for a sprinkler to activate is based on the temperature, velocity of the heated gases flowing by the sprinkler

ceiling jet
■ A relatively thin layer of flowing hot gases that develops under a horizontal surface (e.g., ceiling) as a result of plume impingement and the flowing gas being forced to move horizontally.

FIGURE 12.4
Ceiling jet formation.

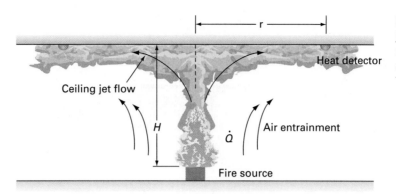

FIGURE 12.5 Ceiling jet formation in relationship to detector activation.

and/or detector, and the time it takes the detector or sprinkler to reach its activation setting (known as a response time index). Therefore, the detector and sprinkler activation is a function of the heat release rate, the height of the ceiling, and the distance from the axis of the fire plume (radius).

UPPER LAYER DEVELOPMENT

A ceiling jet is the natural progression of the buoyancy-driven thermal plume redirected beneath the ceiling, which continues to spread laterally below the ceiling until the temperature of the gases and particulates decreases to that of the surrounding air and/or the flow cannot overcome the opposing forces. In the case of an enclosure, a wall redirects the gases in a similar fashion as the ceiling. Remember that the thermal plume follows the path of least resistance. The path of least resistance for the flow, when intersected by a wall, is downward into the enclosure, yet still suspended high in the enclosure due to the differences in density. The collection of these gases by the ceiling and walls begins to form a relatively uniform layer throughout the upper portions of the compartment, which is commonly referred to as an **upper layer**. Once a uniform layer is formed, the additional flow of gases from the fire continues to collect in the upper layer, forcing the layer to descend from the ceiling. As the temperature and $\dot{Q}$ (heat release rate) is increased, the volume of the accumulated gases increases as well, forcing the layer to descend. Decreasing the temperature or $\dot{Q}$ causes the smoke layer to rise. The upper layer is transferring heat energy primarily by convection from the hot gases into the enclosure's lining materials and additional fuel surfaces. While convection is the primary mode of heat transfer at this point within the fire, it is important to recognize that the collected gases, smoke, and particulates are also emitting thermal radiation. The thermal radiation from the ceiling jet, the hot ceiling and walls, and the upper layer will have a major impact later in the fire.

upper layer
- A buoyant layer of hot gases and smoke produced by a fire in a compartment.

The lower layer consists of air at relatively cooler conditions. As the hot gases rise in the buoyant plume, the cooler air is pulled into the fire plume. This process is called **air entrainment**. Therefore, the lower layer is where the flow of air is being entrained into the fire plume.

air entrainment
- The process of air or gases being drawn into a fire, plume, or jet.

In simplest terms, fire scientists see the growth of a compartment fire by dividing the compartment into two stacked zones (see Figure 12.6). The first consists of an upper layer defined by the accumulation of combustion by-products

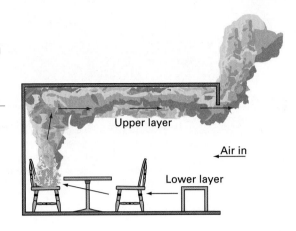

FIGURE 12.6
Elevation view of compartment showing layer formation.
Source: Adapted from Kennedy, P. (2003).

Upper layer

Air in

Lower layer

collecting, forming a layer, and banking down from the ceiling. The lower layer consists of relatively cool, oxygen-rich air. The formation of the upper layer, in turn, heats the ceiling and upper portions of the confining walls mostly by convection and conduction, creating additional fuel and products of combustion. The bottom of this ever-deepening upper layer represents the horizontal border or interface between the two layers. The lower layer remains relatively cool with the addition of entrained unheated air into the originating fire plume. Later in this chapter, we will collectively describe the soot, aerosols, and other combustion by-products as smoke.

VENTILATION OPENINGS

As discussed in Chapter 2 ("Math Review") of this textbook, when gases are elevated in temperature, there is an increase in volume. Gases and smoke produced by combustion are no different; therefore, the upper layer is filled with hot gases that have expanded. A pressure increase inside the compartment is associated with this increase in volume. The pressure differences between the outside and inside of the compartment drive the flow of smoke. Smoke follows the path of least resistance and is dependent on the pressure differences at the ventilation opening. The differences in pressure across the opening change along the height of the opening. The outside air pressure is a straight line, sloping with greater pressure near the ground. The increase in pressure near the ground is due to the increases in weight of the air column closer to the ground. This is the reason why cooler air will rush in by your feet and warm air by your head when you open the door to, let's say, a garage in the wintertime. The inside portion of the opening typically has a greater pressure at a greater height along the opening due to the expansion of the hot gases. There is a point along the opening where the internal and external pressures are equal, and this point is known as the neutral plane (see Figure 12.7). The **neutral plane** is the point along the ventilation opening where no flow occurs due to equality of internal and external pressures. In the lower region, it is possible for some cooler smoke to flow back into the compartment along with the air entrainment. The forces responsible for movement of smoke in a building are the buoyancy forces created by the fire, buoyancy forces arising from differences between internal and external ambient temperatures, and the effects of external wind and air movement (Drysdale, 1998).

neutral plane
■ The height above which smoke will or can flow out of a compartment. The height of zero pressure difference across a partition.

FIGURE 12.7 Diagram showing neutral plane.

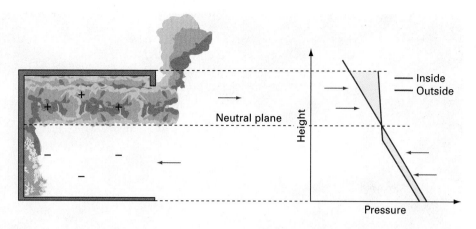

When the smoke descends below the soffit of a door or window, it may or may not flow out of the compartment depending on the pressure difference at the opening. If the pressure caused by the expanded gases exceeds the pressure outside, then the gases flow out of the compartment. However, if the outside pressure is greater than what is being produced by the expanding gases, the smoke continues to descend and remains inside the compartment. In either case, air is flowing into the compartment through the lower portions of the ventilation opening.

The same physics apply when fire service personnel choose to ventilate a structure on fire. If an opening (ventilation) is located low in the compartment, then only cooler gases from the lower layer are forced out of the compartment due to the expanding upper layer, while much cooler air is entering. However, if an opening (ventilation) is located high in the compartment, then the hot gases from the upper layer are forced out of the compartment by the higher pressures in the upper layer. This is the reason that fire service personnel attempt to ventilate a structure fire through the roof or highest portion of the building (via vertical ventilation). Fire service personnel must remember that their efforts modify ventilation rather than create ventilation and that combustion requires ventilation. Failure to change ventilation properly can adversely affect structure stability, occupants, and responders. Horizontal ventilation, especially at the same level of the fire, should be used with extreme caution. The creation of horizontal ventilation may allow cooler, fresher air to rush into the compartment and substantially increase the rate of burning.

In normal atmospheric conditions without wind, the smoke is expected to flow out of the compartment. The presence of wind and/or positive pressure ventilation may cause a higher pressure at the opening, in turn causing the smoke to stay within the compartment. In extreme cases, the smoke may be pushed into the back of the compartment and possibly forced out of other ventilation openings. This control of ventilation is what engineers design with air-handling systems and what suppression forces attempt to implement when placing air blowers for the practice known as positive pressure ventilation (PPV) (see Figure 12.8).

The smoke will continue to fill the compartment, but remember that no compartment is airtight. As long as the fire continues and the layer descends to near floor level, there is a positive pressure causing smoke to flow out of the compartment. If the fire becomes oxygen-deficient and begins to die out, however, then the pressure inside becomes less than that outside, allowing air to flow back into the compartment. This introduction of air often revitalizes the fire, causing the pressure to build

FIGURE 12.8
Pressure-driven flow
from compartment.
Source: Photo by authors.

once again and to push smoke out, which then causes the volume to become oxygen-deficient again and lowers the interior pressure. This constant circular combustion process looks like the building is breathing. Firefighters are warned against ventilation at this point because the introduction of air could cause the fire to build up rapidly and lead to a large fire or combustion at explosive speed.

BATHTUB ANALOGY

The collection of gases by the ceiling and the walls of the enclosure has been compared to the inverse of filling a bathtub with water (see Figure 12.9). This comparison, while not perfect, does provide a relatable practical example. The size of the water spigot, the water pressure, and the resulting flow relate to the heat release rate (power) of the fire. As both are fluids, the water compares to the smoke, heated gases, and incomplete combustion products of the fire. The bathtub volume is compared with the volume of the enclosure and its ability to contain

FIGURE 12.9
Bathtub analogy.
Source: Adapted from
Kennedy, P. (2003).

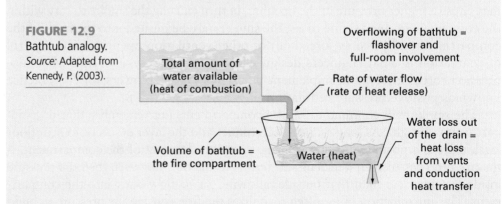

the fluid. The major visual difference is that the bathtub is upside down in comparison to the enclosure (because water is heavier than air and combustion by-products are lighter than air). The water, when turned on, will flow downward out of the spigot based on the force caused by the water pressure and gravitational pull. The water, when intersected by the bottom of the bathtub, will follow the path of least resistance and begin to spread outward. This flow of water continues outward until it is intersected by the sides of the bathtub and will then be redirected to fill the tub upward. This begins to form a uniform layer of water that rises in the bathtub (similar to the descending upper layer). The drain of the bathtub is similar to the openings to the outside from the enclosure. As the buildup of water begins to fill the volume, the filling rate is controlled by the flow from the spigot versus the flow out of the drain. This is similar to the buildup of gases and filling of the enclosure volume controlled by the heat release rate of the fire versus the rate of flow out through the ventilation.

Up to this point, most fires behave according to the discussion above from ignition to plume formation, including the effects of ventilation openings on the developing upper layer. From this point onward, however, a fire may develop in many ways depending heavily on the availability of oxygen and fuel. This textbook separates the remaining growth of an enclosure fire based on two broad categories: fuel-controlled and ventilation-controlled. The fuel-controlled or well-ventilated compartment fire behaves much differently from the ventilation-controlled or poorly ventilated compartment fire. The heat release rate in a fuel-controlled or well-ventilated compartment fire is controlled or limited by the availability of fuel in a state (vapor or gas within its flammable limits) capable of continued combustion and the rate at which that fuel is being released (mass loss rate). In the fuel-controlled case, the oxygen is abundant and is not the limiting factor for the heat release rate and fire growth. On the other hand, the heat release rate in a ventilation-controlled or poorly ventilated compartment fire is being controlled or limited by the availability of oxygen at a rate capable of supporting the mass loss rate of the fuel.

The fuel-controlled or well-ventilated compartment fire does not indicate that combustion is efficient. As we know, combustion efficiency is a function of the availability of fuel and oxygen to mix properly in the combustion reaction. Most, if not all, combustion reactions that fire safety professionals deal with have a significant degree of inefficient combustion, especially when considering diffusion flames. Likewise, most, if not all, enclosure fires have substantial inefficient combustion resulting in the production of incomplete combustion products (e.g., carbon monoxide, etc.).

Progression of a Fuel-Controlled Enclosure Fire

A well-ventilated or fuel-controlled enclosure fire is one that is not adversely affected by the availability of oxygen until the fire nears full-room involvement and is limited only by the availability of fuel in a ready state for combustion. This type of enclosure fire, after the initial growth phases, can follow three paths: a slow progression to full-room involvement, a fast transition to full-room involvement (flashover), and no **full-room involvement** due to fuel exhaustion (see Figure 12.10).

full-room involvement
■ Condition in a compartment fire in which the entire volume is involved in fire.

FIGURE 12.10
(1) Fire growth curves
for self-extinguishing
fire. (2) Fire growth
curve for flashover.
(3) Fire growth curve
for enclosure fire with
full-room involvement
but no flashover.

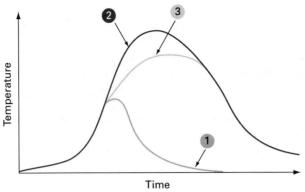

(1) Fire growth curve for self-extinguishing fire
(2) Fire growth curve for flashover (gray line above)
(3) Fire growth curve for enclosure fire with full-room
    involvement but no flashover (curve 3)

CURVE NUMBER 1

A well-ventilated fire is limited by the availability of fuel in a state capable of supporting combustion. The first path that this fire can take is consuming all available fuel and then self-extinguishing. Typically, this happens when only one item is burning and other fuel items are located at a distance too far away for radiant ignition to occur and continue combustion. In this case, the fuel burning continues to burn until it is exhausted.

CURVE NUMBER 2

This curve depicts a fast transition to full-room involvement as a result of the enclosure transitioning through flashover. (This flashover discussion is adapted from Gorbett, Kennedy, and Hopkins, 2007.) The preferred definition for **flashover** given by NFPA is:

> A transitional phase in the development of a compartment fire in which surfaces exposed to thermal radiation reach ignition temperature more or less simultaneously and fire spreads rapidly throughout the space resulting in full room involvement or total involvement of the compartment or enclosed area. (NFPA 921, 2008, p. 13)

This preferred definition by NFPA has been shown by an exhaustive review of the available scientific literature regarding flashover to be the only definition that encompasses all of the pertinent elements for a flashover (Kennedy, 2003). Kennedy showed in his review that there are six common elements found in all previous attempts to define flashover accurately, including the following:

flashover
■ A transition phase in the development of a compartment fire in which surfaces exposed to thermal radiation reach ignition temperature more or less simultaneously and fire spreads rapidly throughout the space, resulting in full-room involvement or total involvement of the compartment or enclosed space.

1. *Flashover represents a transition in fire development.* Flashover is not a discrete event occurring at a single point in time, but a transition in the growth and spread of a fire.
2. *Flashover occurs rapidly.* Although not an instantaneous event, flashover happens rapidly, in a matter of seconds, to spread full fire involvement within the compartment.
3. *Flashover occurs in a confined space or contained fire.* There must be an enclosed space or compartment such as a single room or enclosure.

4. *All exposed surfaces ignite.* Almost all combustible surfaces existing in the lower layer of the enclosed space and exposed to the upper layer radiant heat flux become ignited.
5. *Fire spreads throughout the compartment.* The rapid ignition of combustibles within the lower layer of the compartment spreads the fire.
6. *Flashover results in full-room involvement.* The result of the flashover is that every combustible surface within the room, compartment, or enclosure becomes ignited; the entire volume is involved in fire; and this fire can no longer be contained within the room of origin.

Flashover is a rapidly occurring transitional event in the development of a compartment fire. It represents a significant increase in fire growth from a distinct source of burning or single fuel item to the ignition and burning of almost every other exposed combustible fuel surface in the compartment.

A characterization of flashover is the spread of flaming combustion without any actual flame contact (flame impingement) between the original fuel(s) and the subsequent fuels. While the initial heat transfer mechanism in the early fire stages of a compartment fire is largely by convection, the heat transfer mechanism at and beyond flashover is primarily by radiation (NFPA 921, 2008).

Typically, as a compartment fire begins, a single fuel item is burning. This produces a buoyant fire plume that begins spreading heat energy primarily by convected gases rising in the plume. At this point in the fire, the effect of convective and radiant heat transfer to other fuel packages and the walls, floor, and ceiling of the compartment are relatively minimal. As the buoyant plume's gases and other heated products of combustion begin to collect below the ceiling and spread laterally, the upper layer begins to form. From this point on in the fire, radiant heating is occurring from both the original fuel item's plume and the ever-deepening upper layer. As the upper layer continues to become deeper and contain more heat energy, the radiant portion of the total heat transfer within the compartment increases and the ratio of the convective heat to radiant heat within the compartment decreases. As the fire approaches flashover, radiant heating becomes the dominant heat transfer mechanism. Outside the compartment and in other adjacent spaces, convection remains the predominant heat transfer mechanism until the same process occurs.

Figure 12.11 is based on the reciprocal of the typical compartment fire time/temperature curve. The illustration in the figure shows the dynamically changing

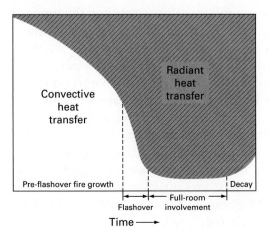

FIGURE 12.11
Relationship of heat transfer mechanisms in a compartment fire.
Source: Adapted from Kennedy, P. (2003).

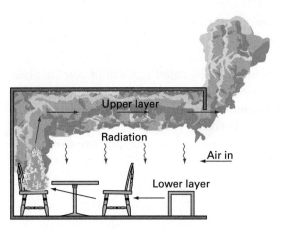

FIGURE 12.12 Radiant heat transfer in enclosure fire nearing flashover.
Source: Adapted from Kennedy, P. (2003).

relationship between convective and radiated heat transfer mechanisms during the course of compartment fire growth.

The relatively uniform upper layer descends as the temperature and heat release rate increases. Thermal radiation from the bottom interface of this hot upper layer heats the surfaces of the various fuels in the lower layer throughout the compartment. These various fuels typically include the compartment furnishings, contents, wall and floor coverings, and the lower walls. As the fire continues to grow, the heat release rate of the original fire plume and temperature of the upper layer is also increasing. As the heat energy of the increasingly deeper and lowering upper layer increases, and the distance between the bottom of the upper layer and the fuels in the lower layer decreases, the radiant flux on the unburned but now pyrolyzing fuels present in the lower layer grows exponentially. Thus, fire growth and the rate of radiant flux increase until nearly simultaneous ignition of the target combustibles in the lower layer of the compartment occurs: This is flashover.

In Figure 12.12, the heavy arrows represent radiant heat energy from the bottom of the upper layer. The narrow arrows in the lower layer represent air entering the room at the bottom of the doorway and entraining into the chair fire. The narrow arrows in the upper layer represent heat and smoke movement from the burning chair at the left.

The dynamics of flashover require a positive imbalance between the input of heat energy into the compartment and the energy leaving the compartment through vents and conduction through the materials lining the room. When or whether flashover occurs at all depends on the excess of heat energy input and the ability of the compartment to retain the heat. Energy input is comprised of the total available heat of combustion of the fuel items, the heat release rate (HRR) of the burning fuel(s), available ventilation to keep the fire growing, and the location of the fire within the compartment. The loss of the energy is through available vents (openings in doors, windows, walls, and ceiling, and active heating, ventilation, and air conditioning [HVAC]), and thermal conduction through the compartment's walls and ceiling. Figure 12.13 illustrates a typical enclosure fire that begins with a single fuel burning and the eventual transition into full-room involvement, indicating that flashover has occurred.

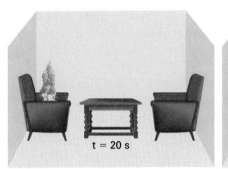

FIGURE 12.13 Enclosure fire transitioning through flashover.
Source: Adapted from NFPA 921 (2008).

Components That Control Flashover

Many varied components of the fire and the compartment control whether and when flashover will occur. Thus, the components of various flashover prediction equations and computer fire models include the following:

Ambient temperature at the beginning of the fire: Heat transfer is a transfer of energy driven by a difference in temperature; thus, the beginning temperatures are important to the quantity and intensity of heat transferred. Buoyancy forces and the development of the fire plume are greatly affected by the ambient temperature.

Geometry of the enclosure (size, shape, area, and volume): The formation of the upper layer and its relationship to the combustibles located in the lower layer is a function of the geometry of the enclosure.

Area, height, width, and soffit (header) height of open doors and windows, or other vents: The source of oxygen for combustion becomes very important as the fire progresses from one that is controlled by the availability of fuel to one that is controlled by the availability of oxygen.

Lining material: The lining materials often have a significant impact on the growth of the fire and may be a determining factor in whether an enclosure transitions through flashover. Lining materials that affect the loss and/or retention (conductance of the surface lining) of energy within the compartment include surface area, material properties, thickness, and thermal inertia.

Heat release rate (kW): The heat release rate of the initial fuel item and subsequent fuels ignited are the power behind the fire. Therefore, the heat release rate is the most important element in assessing the flashover potential for an enclosure fire.

Fire growth rate (kW/s): The heat release rate over the duration of the fire drives the progression of fire growth and affects the time to flashover.

Fuel height: A fuel item located higher in the compartment is more likely to develop a flaming ceiling jet, instead of just heated gases, thereby promoting a greater potential for heat transfer from this upper layer to the combustibles located in the lower layer.

Ceiling jet: The ceiling jet consists of the hot gases from the fire plume intersected by the ceiling. The ceiling jet is the source for radiant heat transfer to the lower layer combustibles. Therefore, the temperature of the gases and particulates in the upper layer has a major impact on the transfer rate of heat to the lower layer combustibles.

Indicators of Flashover

Through years of actual full-scale and scaled model compartment fire testing and the subsequent production and testing of mathematical algorithms, fire researchers have developed sets of physical indicators that can suggest that flashover has probably occurred within a given compartment.

Technical Indicators   Scientists and engineers must have quantitative data to do their studies with anything approximating certitude. When researching flashover, they must have technical indicators that flashover has occurred and that must be measurable (quantitative). The actual definitive elements of flashover—rapidity, transition to full-room involvement, ignition of exposed surfaces, and fire spread—are too subjective and qualitative to be used in any mathematical, scientific, or engineering analysis. The two commonly accepted technical indicators of flashover involve temperature and radiant heat flux, respectively. The technical indicators of flashover include the observations of an average upper layer temperature of approximately 600°C (1,112°F) or radiant flux at floor level of approximately 20 kW/m^2 (Peacock Reneke, Bukowski, and Babrauskas, 1999). Some texts refer to these technical indicators as triggering conditions (Custer, 2008; Drysdale, 1998).

In many early testing scenarios and research burns, in which expensive water-cooled radiometers were unavailable, telltale evidence of crumpled newsprint pages was used by placing them on the floor of the test room and physically observing when they ignited by radiant heat, thereby indicating that flashover had occurred. The critical radiant flux of these so-called telltales was approximate to the 20 kW/m^2 now considered the critical radiant flux for flashover to occur.

Nontechnical Indicators   At or near flashover, several other physical observations are frequently reported. Eyewitnesses commonly report that the fire seemed to explode within the compartment, very rapid flame extension moved laterally throughout the compartment, general floor level burning occurred, external windows broke, flame extension escaped the compartment doors or windows, or culminating full-room involvement itself occurred.

The breaking of external windows is commonly associated with flashover or reported as frequently occurring just after transition to full-room involvement. Thus, this window-breaking phenomenon is a commonly reported observation by eyewitnesses and can, with judicious caution, be used as an indicator of when flashover has occurred. It had been widely believed that the rapid increase of pressure within the flashed-over compartment was the cause of this window breakage. Testing conducted by Fang and Breese (National Bureau of Standards [NBS]) in 1980 and by Skelly (National Institute for Science and Technology [NIST]) in 1990 indicates that it is not the relatively small overpressure that results from flashover (0.013 kPa to 0.028 kPa, or 0.002 psi to 0.004 psi), but rather the temperature differential of approximately 70°C (158°F) between the exposed and unexposed surfaces of the glass (beneath the glazing) that creates the window breakage. The commonly accepted minimum failure pressure of residential windowpanes is 0.689 kPa to 3.447 kPa (0.1 psi. to 0.5 psi.), well above the pressures reported in the NIST tests (Zalosh, 2003). Thus, the rapid increase in the heating of the windowpanes causes this effect to occur at or near flashover.

MISCONCEPTIONS ABOUT FLASHOVER

Unfortunately, the phenomenon of flashover and its proper evaluation is currently misunderstood in much of the professional fire safety community. Listed below are some of the most commonly encountered misconceptions about flashover.

MISCONCEPTION: "FLASHOVER IS DEFINED BY ITS INDICATORS"

The indicators of flashover do not define flashover. Rather, flashover is defined by its nature (rapid transition to a full-room involvement). The presence of one or more indicators of flashover does not a flashover make. The technical indicators of flashover (i.e., approximately 600°C [1,112°F] upper layer temperature, or approximately 20 kW/m^2 radiant flux), and even the other nontechnical indicators, can commonly occur in fires that have never experienced actual flashover. The mere presence of one or more of the indicators does not define flashover. The definition of flashover, as reported previously, does not even contain in its defining elements any of the listed indicators other than the ultimate outcome of flashover, that is, full-room involvement. This is a misconception commonly held even by some well-respected fire researchers. Caution must be taken by the fire safety professional not to make this fundamental mistake of defining "the disease as the symptoms" or "the symptoms as the disease."

MISCONCEPTION: "FULL-ROOM INVOLVEMENT IS FLASHOVER" AND "FULL-ROOM INVOLVEMENT MEANS FLASHOVER OCCURRED"

The fact that a compartment fire ultimately resulted in full-room involvement does not, in and of itself, indicate that flashover occurred. This problem is generally brought about by the indiscriminant interchanging of the word *flashover* with the phrase *full-room involvement* in some texts and lectures. Though full-room involvement is the culminating condition when a flashover occurs, they are separate and distinct fire dynamics phenomena. They are not the same, and though they are frequently closely related, neither is the singular defining element of the other. *Flashover* and *full-room involvement* are not synonymous concepts and care should be taken to use the terms exactly. See the text discussion under Curve 3 for more information.

MISCONCEPTION: "VENTILATION PREVENTS FLASHOVER"

Another common misconception is that ventilation prevents flashover. The reality is that increased ventilation can slow progression to flashover if the fire is fuel-controlled. However, if the fire has progressed to a ventilation-controlled fire, then improper ventilation without coordination with suppression activities may induce flashover due to an increased heat release rate and radiant feedback to fuel packages within the compartment.

FLASHOVER CALCULATIONS

The progression of an enclosure through flashover is an extremely important aspect for life safety. The occurrence of flashover in an enclosure is the ultimate end to tenable conditions within a compartment. The room that has transitioned through flashover has become untenable, and the adjacent compartments are quickly approaching untenable conditions due to the excessive production of incomplete combustion by-products (i.e., carbon monoxide, smoke, soot). A number of experimental studies have been performed over the past 30 years

(continued)

reviewing full-scale fire behavior inside enclosures, and they have provided simple correlations to predict the heat release rate required to cause flashover within an enclosure.

METHOD OF THOMAS

This procedure contains an equation for estimating the amount of energy needed in a room or similar confined space to raise the level of temperature to a point likely to produce flashover. Though there are some limitations on its usefulness, the correlation can give the investigator a quick overview of the minimum heat release rate necessary for flashover to occur in a moderate size compartment. Its input data consist only of the size and height of the room and the size and heights of the vents (e.g., open doors and windows).

The theory behind Thomas's flashover correlation results from simplifications applied to a hot upper layer energy balance in a room. These simplifications result in the equation below. The term A_T within the equation represents heat losses to the total internal surface area of the compartment, and the term $(A_v H_v^{1/2})$, represents energy flow out of the vent opening. The two constants, 7.8 for A_v and 378 for $H_v^{1/2}$, represent values correlated to experimentally tested flashover conditions (Jones and Forney, 1990). Solving Thomas's flashover correlation equation for heat release rate provides the minimum heat release rate needed for flashover to occur in the defined compartment.

$$\dot{Q}_{fo} = 7.8\, A_T + 378\, A_v \sqrt{h_v}$$

Where:

$\dot{Q}_{fo}$ = heat release rate to cause flashover (kW)
A_T = total area of the compartment enclosing surfaces (m²), excluding area of vent opening
A_v = area of ventilation openings (m²)
h_v = height of ventilation opening (m)

The equation does not take into account the location of the vent or whether the vent is a window or a door, though the equation was developed from tests that included window venting. The equation does not consider whether or not the walls are insulated. Use of the equation for compartments with high thermal inertia and high-conductance lining materials, such as thin metal walls, would be inappropriate. The experiments were conducted with compartments not greater than 16 m² (172 ft²), with thermally thick walls and fueled by fires in wooden cribs. Babrauskas (1984) later verified the equation in gypsum wallboard–lined rooms with furniture-fueled fires.

METHOD OF BABRAUSKAS

Building on Thomas and others, Babrauskas gives us a formula for determining the minimum heat release rate of a fire that can cause a flashover in a given room as a function of the ventilation provided through an opening. Known as the ventilation factor, and colloquially referred to within the fire science community as "A root H," it is calculated as the area of the opening (A_v) times the square root of the height of the opening (H_v) (Babrauskas, 1980).

An approximation of the heat release rate required for flashover to occur from the method of Babrauskas is given in the following equation:

$$\dot{Q}_{fo} = 750\, A_v \sqrt{h_v}$$

Where:

$\dot{Q}_{fo}$ = heat release rate to cause flashover (kW)
A_v = area of ventilation openings (m²)
h_v = height of ventilation opening (m)

Heat release rates at flashover from thirty-three actual full-scale tests with a variety of fuels is reported by Barauskas, Peacock, and Reneke (2003) as high as 5.9 MW and as low as just over 1 MW, with the median average at 1.7 MW, which they denote as "probably more characteristic of the data." The majority of these reported heat release rates was between 1 MW and 2 MW (Babrauskas et al., 2003).

METHOD OF McCAFFREY, QUINTIERE, AND HARKLEROAD (MQH)

Upper layer temperature (u-temp) is a fast, mathematical subroutine for predicting pre-flashover upper-layer gas temperatures in a compartment fire with a door and/or window, which is also contained in the original FPEtool FIREFORM computer fire model routines. It was developed from a regressional fit to a large number of experimentally measured fire data. This large database is, in large part, a reason for the procedure's robustness. The authors of this method are McCaffrey, Quintiere, and Harkelroad (1981).

The prediction of upper layer temperature begins with an energy balance about a control volume. This control volume includes the hot pyrolyzates and entrained air that together rise and form the gaseous upper layer within the room. The control volume does not include the barrier surfaces (ceiling and walls); the control volume extends to, but not beyond, the openings from the vents. By applying conservation of energy to this control volume, a general expression for the temperature of the upper layer in the room becomes available, thereby predicting flashover when the technical indicator of approximately 600°C (1,112°F) upper layer temperature $(T - T_\infty)$ is reached.

$$\dot{Q}_{fo} = 610 \sqrt{h_k A_T A_v \sqrt{h_v}}$$

Where:

$\dot{Q}_{fo}$ = heat release rate to cause flashover (kW)

A_T = total area of the compartment enclosing surfaces (m²), excluding area of vent opening

A_v = area of ventilation openings (m²)

h_v = height of ventilation opening (m)

h_k = effective heat transfer coefficient (kW/m²K). The quotient of the thermal conductivity (kW/m K)/ material thickness (m)

Note: These correlations were developed from simplified mass and energy balance on a single compartment with ventilation openings. The data were collected from separate sets of experiments, which were used in the development of each correlation. In other words, the correlation may or may not have predictive capabilities based on the size, geometry, and given conditions for a given fire scenario. It is recommended that the user of these correlations refer to Iqbal and Salley (2004) and/or Icove and DeHaan (2008).

CURVE NUMBER 3

An enclosure fire may achieve full-room involvement, but it may not transition through flashover. Flashover, though quite common, is not a requisite phase of compartment fire growth and does not necessarily occur in every compartment fire that progresses to full-room involvement. Many fully involved compartment fires have never experienced flashover. The transition to full involvement need not always be rapid, as in flashover. It may also be slower, representing different fire spread and heat transfer mechanisms. Issues of the compartment shape, area, ceiling heights, fuel heat release and fire growth rates, and particularly venting and ventilation can affect whether flashover (the rapidity portion of transition to full-room involvement) ever actually occurs.

For example, high rates of ventilation within the compartment, with attendant reduction in heat accumulation, can prevent the effective production of a hot upper layer and flashover. Continued normal fire spread under those conditions can ultimately bring the compartment to full involvement, only more slowly. This concept is extremely important from a life safety standpoint. The fast transition to full-room involvement that is found with flashover may not allow occupants time to escape, but the slower transition to full-room involvement that we just discussed may provide substantially greater time.

Particularly in ignitable liquid fueled fires or flash fires from diffuse gaseous or particulate fuels, full-room involvement can occur nearly from the beginning of the fire event without any initial hot upper layer accumulation.

Progression of a Ventilation-Controlled Enclosure Fire

A ventilation-controlled or poorly ventilated enclosure fire is one where combustion is affected by the lack of oxygen. The fire is limited by the availability of oxygen to support the combustion process. The **ventilation-controlled fire** occurs in compartments where a source of oxygen is not readily present. These fires often occur in smaller compartments, compartments that do not have openings of sufficient size to the outside, compartments that do not have openings of sufficient size to adjacent areas, compartments that begin entraining combustion by-products instead of air, and smaller structures. Most fires begin as fuel-controlled, but they become ventilation-controlled due to the lack of an available oxygen source. Because there is limited air available, the combustion by-products that are now collecting in the compartment are often entrained into the plume and further decrease the combustion efficiency. This type of enclosure fire follows two paths through the initial growth phases: extinguishment of the fire due to lack of oxygen (vitiation) or flashover (see Figure 12.14).

ventilation-controlled fire
■ A fire in which the heat release rate or growth is controlled by the amount of air available to the fire.

CURVE NUMBER 4

The first curve under a ventilation-controlled enclosure fire represents the potential that the fire may extinguish due to inadequate oxygen to support combustion.

FIGURE 12.14
(4) Fire growth curve for ventilation-controlled fire: self-extinguished. (5) Fire growth curve for ventilation-controlled fire transitioning through flashover.

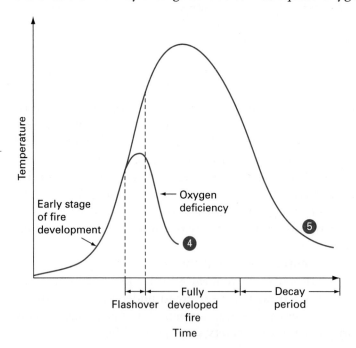

CURVE NUMBER 5

There is a slight possibility that the compartment may transition through flashover. This is rare because combustion is limited due to oxygen; therefore, the heat release rate is also greatly limited and may not be sufficient to cause the compartment to transition through flashover.

Flameover

Another phenomenon often seen during enclosure fires when they are underventilated is flameover or rollover. **Flameover** can be defined as the condition where unburned fuel (pyrolyzate) from the originating fire has accumulated in the ceiling layer to a sufficient concentration (i.e., at or above the lower flammable limit) that it ignites and burns. The initial flame plume serves as the piloted ignition source (see Figure 12.15).

Typically, as a compartment fire begins, there is a single fuel package burning. This produces a buoyant fire plume that begins spreading heat energy primarily by convective gases rising in the plume. At this point in the fire, the effect of convective and radiant heat transfer to other fuel packages and the walls, floor, and ceiling of the compartment are relatively minimal. As the buoyant plume's gases and the other heated products of combustion begin to collect below the ceiling and spread laterally, the upper layer begins to form. From this point on in the fire, radiant heating is occurring from both the original fuel package's fire plume and the ever-deepening upper layer. As the temperature increases in the upper layer, the unburned fuel that has been accumulating may gradually reach its autoignition temperature (AIT), and pockets of partially mixed fire gases begin to ignite. For ignition to occur, those pockets of gases must have already mixed with fresh air to bring the mixture within its flammable limits. This fresh air is entrained from the compartment into the bottom of the upper layer. As this process continues, the pockets of gases merge into a flame front that begins to propagate through the compartment. Typically, this flame front looks like it is rolling across the ceiling. This is termed as a flameover or rollover. The radiation from the flameover begins to increase the temperatures of those items in the lower layer. A flameover often precedes a flashover, but it is not a required event for flashover to occur.

flameover
■ The condition where unburned fuel (pyrolyzate) from the originating fire has accumulated in the ceiling layer to a sufficient concentration (i.e., at or above the lower flammable limit) that it ignites and burns; can occur without ignition of, or prior to, the ignition of other fuels separate from the origin. Also known as *rollover*.

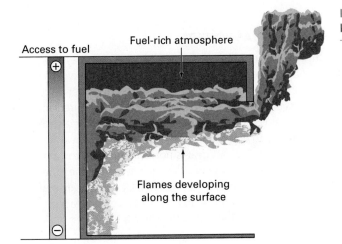

FIGURE 12.15
Illustration of flameover.

Access to fuel

Fuel-rich atmosphere

Flames developing along the surface

The reason that the flameover seems to roll across the ceiling is simply the increase in buoyancy of the now ignited particulates displacing the now cooler smoke from the ceiling layer. This rolling effect is further influenced by the entraining of fresh air into the base of these detached flames.

Components That Control Flameover/Rollover

1. The compartment fire is underventilated.
2. Build-up of unburned oxygen-deficient pyrolysis products form an upper layer within the compartment.
3. One or more of the fuels present in the layer accumulates to within its flammability range.
4. Ignition occurs due to direct flame contact.
5. Oxygen is present in the highest concentrations at the lower portion of the upper layer, forming a flammable region.
6. Ignition occurs at the location of the flammable mixture, and the flame spreads until the local fuel and/or oxygen is exhausted.

Indicators of a Flameover

1. The upper layer begins to thicken (visibility decreases). The amount of incomplete combustion products increases in the upper layer.
2. The upper layer temperature begins to increase, gradually bringing those pyrolyzates to their autoignition temperature. Firefighters may begin to feel the heat from the descending upper layer.
3. Turbulent mixing occurs in the upper layer. If the upper layer begins to mix vigorously, the temperature distribution throughout the upper layer is increasing and is causing the mixing of fresh air at the underside of the upper layer to increase greatly.

 MISCONCEPTIONS REGARDING FLAMEOVER

Many misconceptions were found in the current literature, including many of the current traditional training documents.

MISCONCEPTION: *"FLAMEOVER IS FLASHOVER"*

Many textbooks and articles publish the accounts of firefighters who state that they have lived through a flashover. The likelihood of a person living through this event is very small and most likely those accounts are better related to the flameover concept. A flashover is the near simultaneous ignition of all combustibles within the compartment due to the radiant heat transfer from the upper layer. Flashover is a step event, meaning that once the transition is completed, the compartment continues to quickly increase in the amount of heat flux imposed on materials within that compartment. Turnout gear is not manufactured to withstand the heat flux that would be imposed on a firefighter in a flashover event. On the other hand, a flameover is more of a localized, transient event that occurs relatively quickly and subsides. The intensity may be near the same magnitude as that of the beginning stages of a flashover, but the duration of this exposure is low. The issue here is that a flameover, while dangerous and sometimes a sign of an impending flashover, is not nearly as dangerous as that of a flashover and should be characterized as such.

MISCONCEPTION: "FLASHOVER CHAMBER"

Since the early 1980s the practice of using portable shipping containers, sometimes referred to as flashover chambers, for training firefighters has become popular. These shipping containers are set up with a fire compartment and a viewing compartment (see Figure 12.16). A fire is set in ordinary combustibles within the fire compartment and students sit and watch the fire behavior from the lower viewing compartment. An instructor demonstrates the cooling of the upper layer with the different types of hose streams and also demonstrates the effect of ventilation. The instructor allows the fire compartment to increase in heat to illustrate that flames will begin to form dissociated from the original flame plume. Some instructors then state that this is flashover and illustrate to the students several signs of an impending flashover. The training is top-notch for the firefighter and should not cease. It is the incorrect terminology used by some of these instructors that should be changed. The unattached flames that occur in the upper layer are actually the beginning of a flameover. Therefore, the actual term for this training exercise is a *flameover chamber*, unless those technical indicators for a flashover have been reached in the upper compartment. In that case, the instructor would be correct in his or her description. There are several organizations (i.e., the Compartment Fire Behavior Training [CFBT] in Oregon) that have been using these containers in a very positive way to increase the knowledge and live training for firefighters. These instructors and programs should not be lumped in with those organizations that are using the terminology incorrectly.

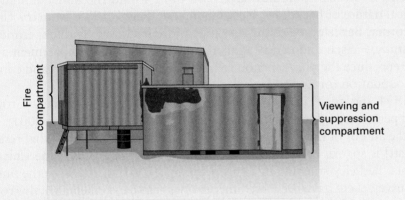

FIGURE 12.16 Flashover chamber in Auburn, Indiana, Fire Department Training Facility.

Impact of Changing Ventilation Conditions

An enclosure fire can follow two additional paths if ventilation conditions are changed during a ventilation-controlled fire. These two additional paths include backdraft and flashover (see Figure 12.17).

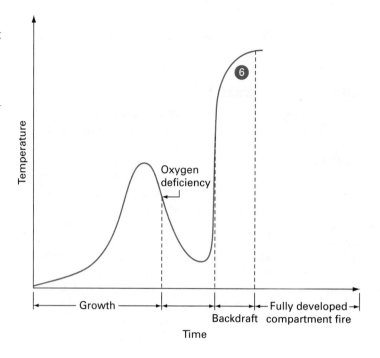

FIGURE 12.17 Fire growth curve: backdraft from ventilation-controlled fire.
Source: Adapted from Bengtsson (2001).

CURVE NUMBER 6

The backdraft phenomenon has not been studied as extensively as that of flashover. Nevertheless, studies have been done to aid the understanding of the theoretical nature of backdraft. Fleischmann and Pagni (1993) from the University of California, Berkeley, were the first to explore backdraft with experimentation. In this study, Fleischmann and Pagni created a small-scale compartment and were able to reproduce the phenomenon reliably. These tests have been and continue to be the foundation for all research of this phenomenon.

In 1994 and again in 2001, the Fire Research and Development Group from England performed a survey of research related to backdraft (Chitty, 2001). Richard Chitty, the author of this survey, states, "[T]he ultimate finding was that research on [backdraft] is sparse, and had identified only one active group at the University of California, Berkeley studying this phenomenon" (2001, p.12). Since the time of the Chitty survey, there have been five other entities that have taken an active role in the study of the backdraft phenomenon, including Essex County Fire and Rescue Services in England; Lund University in Sweden; Eastern Kentucky University in Richmond, Kentucky; University of North Carolina, Charlotte; and Hughes and Associates in the United States. Direct contact has been made with all entities to obtain any recent publication and details of their studies, which are listed in Table 12.1. At present, these studies have presented similar results as those of Pagni and Fleischmann. Therefore, the theoretical basis and general understanding of the backdraft phenomenon is validated and understood. However, the quantitative mechanisms of the backdraft phenomenon still must be researched and delineated.

The first backdraft experiments were conducted by creating a ventilation-limited fire and burning some type of typical fuel (methane or diesel), and once an underventilated fire was reached, the experimenters closed the ventilation openings and then injected more fuel gas. While this produces similar results as that witnessed during a backdraft, it does not replicate an actual backdraft driven by

TABLE 12.1

ID	AUTHOR(S)	TITLE	PUBLISHER	DATE PUBLISHED
1	Babrauskas, V.	*Estimating Room Flashover Potential*	Fire Technology	1980
2	Beyler, Craig	"Major Species Production by Diffusion Flames in a Two-Layer Compartment Fire Environment"	*Fire Safety Journal*	1986
3	Bowen, J.	"Flashover/Backdraft Explosive Situation"	*Western Fire Journal*	1982
4	Caspi, H.	*Illinois Church Backdraft Named Largest Ever Documented*	Firehouse	2002
5	Chen, A., and Francis, J.	Prediction of Pre-Flashover Temperatures and the Likelihood of Flashover in Enclosure Fires		
6	Chitty, Richard	*UK Fire Research and Development Group—A Survey of Backdraught*	UK Fire Research Group	2001
7	Clark, William	"Flashover Symposium, Vent-Point Ignition"	*Fire Engineering*	1995
8	Croft, W.	"Fires Involving Explosions"	*Fire Safety Journal*	1980
9	DeHaan, J.	*Kirk's Fire Investigation*, Fifth Edition	Prentice Hall	2002
10	Drysdale, D.	*An Introduction to Fire Dynamics*	Wiley	1985; 1999
11	Drysdale, D.	"The Flashover Phenomenon"	*Fire Engineers Journal*	1996
12	Dunn, V.	"Beating the Backdraft"	*Fire Engineering*	1988
13	Dunn, V.	"Fire and Explosions"	*Fire Engineering*	–
14	Dunn, V.	"Flameover Fires, Flashover Fires: What Is the Difference?"	*Fire Command and Control*	2002
15	Dunne, T.	"Delayed Backdraft: What We Learned"	*Fire Engineering*	2002
16	Faith, N.	*Blaze*		
17	Fleischmann, C., and Pagni, P.	"Exploratory Backdraft Experiments"	*Fire Technology*	1993
18	Fleischmann, C., and Pagni, P.	*Quantitative Backdraft Experiments*	International Association of Fire Safety Science	1994
19	Fleischmann, C., and McGrattan, K.	"Numerical and Experimental Gravity Currents Related to Backdrafts"	*Fire Safety Journal*	1999

(continued)

TABLE 12.1 (continued)

ID	AUTHOR(S)	TITLE	PUBLISHER	DATE PUBLISHED
20	Friedman, R.	*Principles of Fire Protection: Chemistry and Physics*	NFPA	2002
21	Gojkovic, D.	"Initial Backdraft Experiments," Report 3121	Lund University, Lund, Dept. of Fire Protection Engineering	2000
22	Gojkovic, D., and Bengtsson, L-B	*Some Theoretical and Practical Aspects on Fire Fighting Tactics in a Backdraft Situation*	Interflam	2001
23	Gojkovic, D., and Karlsson, B.	"Describing the Importance of the Mixing Process in a Backdraft Situation Using Experimental Work and CFD"	Third International Seminar on Fire and Explosion Hazards	2000
24	Gottuk, D.	"The Development and Mitigation of Backdraft: A Real-Scale Shipboard Study"	*Fire Technology*	1999
25	Grimwood, P., and Desmet, K.	*Tactical Firefighting: A Comprehensive Guide to Compartment Firefighting and Live Fire Training (CFBT)*	Cemac	2003
26	Grimwood, P.	*Rapid Fire Progress*	Firetactics	2002
27	IFSTA	*Essentials of Firefighting, Fourth Edition*	IFSTA	2000
28	Karlsson, B.	"Flashover, Backdraft and Smoke Gas Explosion—The Fire Service Perspective"	Interflam	1999
29	Karlsson, B., and Quintiere, J.	*Enclosure Fire Dynamics*	CRC	1999
30	Kennedy, J.	*Fire and Arson Investigation*	Investigations Institute	1965
31	Kennedy, J., and Kennedy, P.	*Fires and Explosions: Determining Cause and Origin*	Investigations Institute	1984
32	Kennedy, P.	*Flashover and Fire Analysis: A Discussion of the Practical Use of Flashover in Fire Investigation*	Investigations Institute	2003
33	Kennedy, P., and Kennedy, J.	*Explosion Investigation and Analysis: Kennedy on Explosions*	Investigations Institute	1990
34	Knapp, J.	"Flashover and a Survival Guide"	*Fire Engineering*	1996

ID	AUTHOR(S)	TITLE	PUBLISHER	DATE PUBLISHED
35	Knapp, J.	"Flashover Survival Strategy"	*Fire Engineering*	1996
36	Kraszewski, A.	*Problems with Flashover*	University of Technology—Sydney	1998
37	Lentini, J.	"Flashover: A New Reality for Fire Iinvestigators"	*Journal of Canadian Fire Investigators*	1992
38	McCaffrey, B., and Quintiere, J.	*Estimating Room Temperatures and the Likelihood of Flashover Using Fire Test Data Correlations*	National Engineering Lab	1981
39	NFPA	*Fire Protection Handbook*	NFPA	All editions
40	NFPA	NFPA 921: Guide to Fire and Explosion Investigations	NFPA	1992–2004
41	Novozhilov, V.	"Flashover Control under Fire Suppression Conditions"	*Fire Safety Journal*	2001
42	Peacock, R., and Babrauskas, V.	"Defining Flashover for Fire Hazard Calculations"	*Fire Safety Journal*	1999
43	Peacock, R., and Babrauskas, V.	"Defining Flashover for Hazard Calculations, Part II"	*Fire Safety Journal*	2003
44	Quintiere, J.	*Principles of Fire Behavior*	Delmar	1992
45	Russell, D.	"Seven Fire Fighters Caught in Explosion"	*Fire Engineering*	1983
46	Sutherland, B. J.	*Smoke Explosions*	University of Cantebury, School of Engineering, New Zealand	1999
47	Thomas, P.	"Fires and Flashover in Rooms—A Simplified Theory"	*Fire Safety Journal*	1980
48	Walton, W., and Thomas, P.	"Estimating Temperatures in Compartment Fires"	*Fire Protection Handbook*	1995
49	Weng, W.	"A Model of Backdraft Phenomenon in Building Fires"	*Progress in Natural Science*	2002
50	White, B.	"Firefighting and the High-Pressure Backdraft"	*Fire Engineering*	2000

class A fuels (see Figure 12.18). Recently, the British Fire Service College and Essex County Fire began performing backdrafts in shipping containers with class A fuels. These were the first demonstrations that we are aware of that utilized class A fuels to reproduce a backdraft. Since these initial demonstrations, Eastern Kentucky University's Fire and Safety Engineering Technology program have created several quarter-scale and half-scale compartments and have been able to reproduce backdrafts reliably with class A fuels.

FIGURE 12.18 Half-scale compartment fire with backdraft.
Source: Photos by authors.

FIGURE 12.18
(continued)

Definition of Backdraft

The most accurate, current definition for backdraft is an adaptation from the experimental work done by Pagni and Fleischmann: Limited ventilation during an enclosure fire can lead to the production of large amounts of unburned pyrolysis products. When an opening is suddenly introduced, the inflowing air forms a gravity current and begins to mix with the unburned pyrolysis products, creating a combustible mixture of gases in some part of the enclosure (see Figure 12.19). Any ignition source, such as a glowing ember, can ignite this combustible mixture,

FIGURE 12.19
Gravity current.
Source: Adapted from Fleischmann, C. M., and McGrattan, K. B. (1999), "Numerical and Experimental Gravity Currents Related to Backdrafts," *Fire Safety Journal*, 33:21–24.

which results in an extremely rapid burning of gases and pyrolysis products forced out through the opening and causes a fireball outside the enclosure.

Understanding Backdraft

Similar to the fire described during a flashover event, a fire that can result in a backdraft usually originates from a single item burning. This fire will grow and may spread to other combustibles within the room from direct flame impingement or flaming combustible items dropping down from the upper layer, similar to a typical enclosure fire.

An underventilated fire develops: Sometime during the growth of the fire, the fire becomes underventilated. This underventilated fire can occur in two different ways. First, the fire can become underventilated by utilizing most of the available oxygen within the enclosure and not have any additional sources of ventilation to sustain combustion. Second, the fire can begin having enough ventilation for the fire to transition through flashover but, because of the rapid increase in fuels and combustion, the excess oxygen is consumed, forcing the fire to become ventilation-controlled. An underventilated compartment fire produces excess pyrolyzates (unburned fuel) because of inefficient combustion. In a backdraft situation, temperatures are typically still high enough to sustain pyrolysis of the combustibles located within the compartment, regardless of the lack of oxygen for combustion. This further adds pyrolysis products to the compartment's atmosphere, thus causing the upper layer to descend and nearly fill the entire volume of the compartment (see Step 1 in Figure 12.20).

FIGURE 12.20
Progression of backdraft.
Source: Adapted from Gottuk, D. (1999).

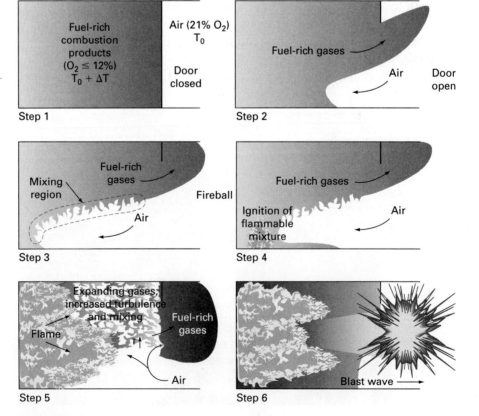

A gravity current is created when a vent is suddenly opened: A gravity current is a current formed when two fluids of differing densities interact so that a vertical interface exists between the fluids. The resulting motion consists of the heavier fluid flowing horizontally beneath the lighter fluid. In other words, when the opening is made, the hot, fuel-rich gases flow out the upper portion of the opening, while the cooler exterior air is being drawn into the compartment at the lower portion of the opening. A gravity current consisting of air is formed when a door or window is suddenly opened either by a building occupant or fire service employee (see Step 2 in Figure 12.20).

An ignitable mixture is formed: An ignitable mixture is formed at the shear interface between the smoke layer and this influx of air created by the gravity current.

An ignition source must be present at the interface for ignition: This ignitable mixture at the shear interface will ignite if presented with a suitable ignition source (i.e., smoldering combustion, hot surfaces, flaming combustion) (see Steps 3 and 4 in Figure 12.20).

A flame front propagates through the compartment: Once this initial ignition takes place, increased mixing of air and the smoke layer commences, thereby allowing more fuel and air to mix and form ignitable mixtures. A flame front forms from this chain reaction of ignitions and propagates through the compartment. An increase in temperature and pressure commences due to the flame front propagation. The increase of pressure inside the compartment forces excess fuel-rich gases through the opened vent. These excess fuel-rich gases suddenly begin mixing with the available oxygen exterior of the compartment and ignite, causing a tremendous fireball upon exit from the compartment (see Steps 5 and 6 in Figure 12.20).

FACTORS THAT CONTROL BACKDRAFT

1. The compartment fire is underventilated.
2. Unburned, oxygen-deficient pyrolysis products (excess pyrolyzates) are created.
3. Air is suddenly introduced (i.e., through a window or door).
4. A gravity current carries fresh air into the compartment.
5. Air mixes with unburned oxygen-deficient pyrolysis products, creating a flammable (combustible) mixture interface.
6. If an ignition source is present at this flammable (combustible) mixture interface, then ignition occurs.
7. Turbulent mixing of air and unburned, oxygen-deficient pyrolysis products results from the ignition of this interface, which results in further flame spread.
8. A deflagration occurs as the flame propagates through the compartment.
9. Excess unburned pyrolyzates are forced through the opening by the positive pressure buildup and heat created by the propagating flame front.
10. The excess pyrolyzates outside the compartment ignite once presented with fresh air and ignited by the following flame front, in turn creating a fireball and blast wave.

Indicators of a Backdraft

The following are indicators that a backdraft may occur:

1. The fire may be pulsating. Windows and doors are closed, but smoke is seeping out around them under pressure and being drawn back into the building.
2. No visible flames appear in the room.
3. Doors and windows are hot.
4. Whistling sounds leak out around doors and windows. If the fire has been burning for a long time in a concealed space, unburned gases may have accumulated.
5. Window glass is discolored and may be cracked from heat (Norman, 1991).
6. Raising and lowering of the hot gas layer is also a newly identified indicator. (See Grimwood, Hartin, McDonough, and Raffel, 2005; these authors identified this indicator in a number of fires in Europe.)

The key indicator that has been witnessed in the past is the in and out movement of the smoke, which makes the building appear to be breathing.

MISCONCEPTIONS REGARDING BACKDRAFT

Many misconceptions were found in the current literature, including many of the current traditional training documents. In this chapter, only the most prevalent misconception about carbon monoxide as a fuel will be addressed.

MISCONCEPTION: "BACKDRAFT IS FUELED BY CARBON MONOXIDE"

Many texts state that carbon monoxide is a major fuel that drives flashover and/or backdraft phenomena. However, there is no scientific proof for these statements; in fact numerous studies have been done to disprove this theory (Gottuk, 1999; Gojkovic, 2000; Fleishchmann and Pagni, 1993; Sutherland, 1999). These studies reveal that the major fuel constituent that drives the backdraft phenomenon is the incomplete solid pyrolysis products within the smoke layer. Many texts incorrectly use the fact that carbon monoxide (CO) has a flammability range as proof that CO is the fuel behind the backdraft phenomenon. However, in reality the lower explosive limit (LEL) of CO is the most important factor when investigating this issue. Carbon monoxide requires a substantial mixture (12 percent) in air before it is flammable or explosive. And studies have shown typical enclosure fires rarely have CO mixtures above 7 percent (Bryner, Johnsson, and Pitts, 1992; Icove and DeHaan, 2009; Babrauskas, 2003; Gottuk, 1999). Therefore, the misconception has no basis or scientific support. Pitts (1994) has performed several small-scale and large-scale studies regarding the production of CO and has found volume percentages on the verge of the lower flammable limit of CO, but no one has related these findings to being a fuel in the backdraft phenomenon.

FIGURE 12.21 Full-room involvement.
Source: Photo by authors.

Full-Room Involvement

The definition of full-room involvement is a condition in a compartment fire in which the entire volume is involved in fire (NFPA 921, 2008) (see Figure 12.21). The production of fuel in a state capable of combustion is overwhelming the compartment and combustion is occurring wherever the appropriate mixture of oxygen and fuel can take place. While the definition states that the entire volume is involved in fire, a better way to describe this state is that every location where an adequate mixture of fuel and oxygen exists within the enclosure is burning. This stage is limited by the availability of oxygen due to the massive production of fuel brought about by the elevated temperatures pyrolyzing all available combustibles within the compartment. In fact, the release of the greatest amount of energy is during this stage of the fire. Because the greatest amount of energy is being released, the greatest amount of combustion is occurring, and the fire is greatly oxygen-deficient as the combustion efficiency becomes much lower. Therefore, the production of carbon monoxide and other incomplete combustion products has also increased.

Full-room involvement is also the stage in the fire where a steady heat release rate typically occurs. The steady-state heat release rate is based on the presence of openings and the incoming air.

COMBUSTION PRODUCTS FOR OCCUPANT SAFETY

One common request for fire safety simulations is, Do the fire protection elements in place for this occupancy provide adequate time for the occupants to escape in the event of a fire? This is especially true for the more progressive performance-based design. Therefore, it is often necessary to know how the combustion by-products are accumulating and possibly affecting the occupants.

The time required for people to move safely out of a building is commonly referred to as the required safe evacuation time (RSET). The RSET concept is compared to the fire environment to evaluate when conditions within the building would become untenable (people would be unable to survive) to evaluate if enough time is provided for people to safely evacuate. The evaluation of the fire environment from ignition to untenable conditions is known as the available safe egress time (ASET). ASET needs to be greater than RSET for safe evacuation. Fire Dynamics Simulator (FDS) provides the user with the option to collect the output data for specific species (i.e., carbon monoxide, carbon dioxide). The user can then use this data to calculate incapacitation levels based on one of the several incapacitation equations, including the Stewart equations or those discussed by Purser in *Handbook of Fire Protection Engineering*, published by the Society of Fire Protection Engineers (SFPE).

FIRE ENVIRONMENT EXPOSURE (TOXICITY)

The ASET concept involves a tenability analysis for the structure by taking into account the fire environment from ignition to untenable conditions. The main mechanisms for incapacitation and death during fires include smoke particles obscuring illumination and vision; irritants attacking sensory nerve endings in the eyes and respiratory tract, causing pain and affecting vision and breathing; and asphyxiant gases (CO, HCN, CO_2, low O_2) causing hypoxic effects on vital organs and systems (Purser, 2008). The main concerns when evaluating ASET are to assess the times when fire by-products affect behavior and delay escape, when escape is prevented by incapacitation, and when exposure is likely to result in permanent injury or death (Purser, 2008). The major mechanisms for incapacitation are summarized below.

Smoke, Irritants, and Visibility

Smoke consists of various particles, gases, and irritants. As people move through the smoke, their vision tends to be obscured by the particles within the smoke, and their eyes, lungs, and nasal cavity are irritated by the acidic gases (i.e., hydrochloride, bromide) and organic irritants (i.e., acrolein, phenol). People attempt to use the exit path most familiar to them, but they may have to search for alternative exits in the event their path is impeded. As visibility in the compartment decreases and irritants affecting the person increase, his or her walking speed, wayfinding ability, and exit choice diminish. The decision to move through the smoke depends on the density of the smoke, yet the research is mixed on the defining line between occupants choosing to move through the smoke or not (SFPE, 2003).

The irritants contained in smoke begin to affect the human body as minor nasal, skin, and eye irritations, but can often become extremely painful and incapacitating. However, irritants are rarely immediately fatal (Purser, 2008). The inflammatory effects caused by these irritants, if they have reached the lungs, can be fatal hours or days after the event (Purser, 2008).

Asphyxiant Gases

Two main asphyxiant gases are common during fire events: carbon monoxide (CO) and hydrogen cyanide (HCN). Carbon dioxide (CO_2) also contributes to incapacitation and death by increasing a person's respiration rate, which in turn increases the uptake of CO and HCN. As oxygen is used in the combustion reaction, low oxygen levels develop within the compartment. Low levels of oxygen can also result in incapacitation and death by hypoxia, but such an outcome has been shown not to be as prominent as with HCN and CO. Thus, CO and HCN will be the primary focus of this discussion.

Carbon monoxide is a colorless, odorless gas that is produced in fires by incomplete combustion (Purser, 2008). It is the best known and greatest toxin produced in a fire (Purser, 2008). If the combustion reaction were complete, CO would not be produced in the fire; only CO_2 and water would be produced. However, diffusion flames and inadequate ventilation for most fires prevents a combustion reaction from being complete. Icove and DeHaan (2009) report that well-ventilated fires produce as little as 0.02 percent (200 ppm) of the total gaseous product, while smoldering or underventilated fires range from 1 to 10 percent. It is evident that the burning regime is an important factor in the production of CO; therefore, it is imperative that the modeler take this into account.

TABLE 12.2	Classical Relationship Between Carboxyhemoglobin Concentration and Signs Exhibited in Humans and Nonhuman Primates		
BLOOD SATURATION %COHb	**AFTER STEWART**	**AFTER SAYERS AND DAVENPORT**	**AFTER PURSER (NONHUMAN PRIMATES)**
0.3–0.7	Normal range due to endogenous production		
5–9	Exercise tolerance	Minimal symptoms in less than 10%	
16–20	Headache, abnormal visual evoked response, may be lethal for patient with compromised cardiac function	Tightness across forehead and headache experienced in 10–20%	
20–30	Throbbing headache; nausea; abnormal fine manual dexterity	Throbbing headache	
30–40	Severe headache; nausea and vomiting; syncope	Severe headache, generalized weakness, visual changes, dizziness, nausea, vomiting, and ultimate collapse	30% caused confusion, collapse and coma in active animals during 30-minute exposures, with nausea after
40–50		Syncope, tachycardia, and tachypnoea	40% caused coma, bradycardia, arrhythmias, and EEG changes in resting animals during 30-minute exposures
50+	Coma, convulsions	Coma and convulsions	
60–70	Lethal if not treated	Death from cardiac depression and respiratory failure	

Source: Data derived from Purser (2008).

When CO is inhaled, it passes into the bloodstream, where it combines with the hemoglobin molecule to form carboxyhemoglobin (COHb). This reduces the percentage of hemoglobin available for carrying oxygen, which reduces the amount of oxygen delivered to the tissues. Eventually, the lack of oxygen to the body causes cells to die and the person to become incapacitated and then later die. CO in the blood is not abnormal. COHb levels of 0.5 to 1.0 percent are found normally within the body. Smokers can have levels between 4 and 10 percent (Icove and DeHaan, 2009). COHb levels between 30 and 50 percent typically result in incapacitation, while levels between 50 and 70 percent typically result in death. Table 12.2 provides a more detailed list of COHb concentration levels and signs exhibited. Another important aspect of

(continued)

time to incapacitation for fire victims is the level of physical activity during the event. If a person is at rest, she or he has a relatively low respiration rate, but while a person performs light work or heavy work, his or her respiration increases. Therefore, the amount of CO entering into the body also increases and should be taken into consideration in the calculations.

Hydrogen cyanide (HCN) is a colorless gas, with a bitter almond taste (Purser, 2008). HCN prevents oxygen metabolism in the mitochondrion by inhibition of cytochrome c reaction with water (Purser, 2008). HCN is dispersed quickly into the tissue and causes incapacitation and possibly death at 150 to 400 ppm (Icove and DeHaan, 2009).

CALCULATIONS

All of the above toxins can be calculated based on the fire environment. The combination of these elements can be calculated and predicted by the fractional effective dose (FED) method (Purser, 2008). Basically, the exposure dose acquired over any period of time during a fire is expressed as a fraction of the dose required to cause incapacitation.

Decay

As the fuel and/or oxygen is exhausted, the heat release rate and the temperatures begin to diminish and the fire begins to die out. This is known as the decay stage of the fire. The fire may also transition from ventilation-controlled back to fuel-controlled at this stage. The decay stage of a fire still poses a great threat to life because collapse can still occur due to smoldering combustion and increased load (water), and carbon monoxide concentrations can still be high.

Summary

Fires inside enclosures can develop in many different ways and not all can be covered in one chapter. This chapter provided a basic knowledge of the common factors that affect fire growth inside an enclosure. These factors include plume formation, ceiling jet, upper layer development, pressure profiles, ventilation, and rapid transition events.

Case Study

On February 20, 2003, a fire broke out in an overcrowded nightclub (the Station Nightclub in West Warwick, Rhode Island) when pyrotechnics were misused near a non-code-compliant combustible wall lining as the band took the stage for their performance that night. Media captured the entire event on video, and the nation watched as 100 people lost their lives and over 200 more were injured. Several things went wrong that led up to this event and allowed this fire to be so devastating. This event will have an everlasting impact on fire safety and will forever be a cornerstone of research regarding code issues, sprinkler ordinances, human behavior in fires, and fire dynamics.

While the courts ruled on who and what agencies were ultimately responsible for the deaths and injuries of the individuals in the nightclub, the by-products of combustion and enclosure fire dynamics were what sealed their fates that evening. The stage was located along the west wall. Directly behind the main stage was a smaller room known as the drummer's alcove. The drummer's alcove was a raised floor stage area with a much smaller volume and height than the rest of the building. The combustible wall lin-ing was ignited by the misuse and close proximity of the pyrotechnics. However, the enclosure fire dynamics occurring inside the drummer's alcove would result in the fast transition through flashover and substantially increase the production of combustion by-products (i.e., CO, HCN) that ultimately resulted in the deaths of many.

Once the fire entered into the much smaller volume of the drummer's alcove, it quickly developed an upper layer that resulted in a nearly uniform heat flux to the remaining combustible wall lining. Shortly after (approximately 90 seconds) the exposure to this elevated heat flux and fire dripping from the foam, the drummer's alcove flashed over. When this small volume transitioned through flashover, then the production of the carbon monoxide, hydrogen cyanide, carbon dioxide, and other products of incomplete combustion rapidly increased. These gases quickly traveled from the drummer's alcove throughout the rest of the nightclub, causing people to become disoriented, incapacitated, and then asphyxiated. For a more detailed discussion regarding this historical fire, please see NIST's investigation report (Grosshandler, Bryner, Madrzykowski, and Kuntz, 2005).

Review Questions

1. In specific, scientific terms, define fire growth within a compartment.
2. Given the geometry of the room on page 264 ($12 \times 12 \times 9$), please determine the minimum heat release rate needed to cause flashover. Assume that the walls are lined with gypsum wallboard ($\frac{3}{4}''$ thickness). Show all three methods. Show all your work.

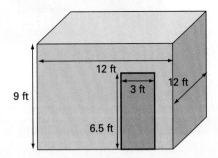

3. What are the six elements that are crucial for a practical flashover definition?

4. What are the driving factors behind backdraft? Specifically, what are the fuel, ignition mechanism, and following effects?

5. Explain in your own words the three modes of heat transfer. Give examples of each and explain their importance in fire dynamics. Diagrams may assist in your discussion.

6. Which is the predominant mode of heat transfer as a compartment nears flashover?
   a. Conduction
   b. Convection
   c. Radiation
   d. All of the above

7. When the plume is intersected by a ceiling or other horizontal obstruction, the hot gases and incomplete combustion products spreading outward is known as a(n) _____.
   a. flashover
   b. ceiling jet
   c. smoke plume
   d. obstructed flow

8. What is the HRR required to flashover a metal building (¼″ steel) that measures 20 yards long by 10 yards wide and is 3 yards high? Assume that there is an open rollup door measuring 10 feet wide and 8 feet tall. Is a scenario as described in this question practical?

9. Calculate the size fire required to generate a flashover in a room that measures 12 feet wide × 16 feet long and has a ceiling 8 feet above the floor. The door measures 32 inches wide × 80 inches tall. Assume the wall is made of gypsum wallboard (0.5″ thick).

10. Draw the room in which you sleep to scale. Calculate the HRR required to generate flashover within this compartment if only the door provides ventilation.

PEARSON

myfirekit™

For additional review and practice tests, visit **www.bradybooks.com** and click on MyBradyKit to access book-specific resources for this text!

Register your access code from the front of your book by going to **www.bradybooks.com** and selecting the mykit links.

References

Babrauskas, V. (1980). "Estimating Room Flashover Potential," *Fire Technology* 16, no. 2:94–103. Quincy, MA: National Fire Protection Association.

Babrauskas, V. (1984). "Upholstered Furniture Room Fires—Measurements, Comparison with Furniture Calorimeter Data Predictions," *Journal of Fire Sciences*, 2, no. 5:25–39.

Babrauskas, V. (2003). *Ignition Handbook,* Issaquah, WA: Fire Science Publishers.

Barauskas, V., Peacock, R., and Reneke, P. (2003). "Defining Flashover for Fire Hazard Calculations: Part II." *Fire Safety Journal*, 38:613–622.

Bengtsson, L. (2001). *Enclosure Fires.* Karlstad, Sweden: Raddnings Verket-Swedish Rescue Services Agency.

Bryner, N., Johnsson, E., and Pitts, W. (1992). *Carbon Monoxide Production in Compartment Fires—Full-Scale Enclosure Burns.* Gaithersburg, MD: NIST.

Chitty, R. (2001). *UK Fire Research and Development Group—A Survey of Backdraught.* London, Eng.: United Kingdom Fire Research and Development Group.

Custer, R. (2008). "Dynamics of Compartment Fire Growth," in *Fire Protection Handbook,* Twentieth Edition. Quincy, MA: NFPA.

Drysdale, D. (1998). *An Introduction to Fire Dynamics.* West Sussex, Eng.: Wiley.

Fang, J., and Breese, N. (1980). "Fire Development in Basement Rooms," NBSIR 80-2120. Washington, DC: U.S. Department of Commerce, National Bureau of Standards.

Fitzgerald, R. (2004). *Building Fire Performance Analysis.* West Sussex, Eng.: Wiley.

Fleischmann, C., and Pagni, P. (1993). "Exploratory Backdraft Experiments." *Fire Technology* 21: 21–34.

Gojkovic, D. (2000). "Initial Backdraft Experiments." Lund University.

Gorbett, G., Kennedy, P., Hopkins, R. (2007). *The Current Training and Education Regarding Flashover, Backdraft, and Other Rapid Fire Progression Phenomena.* Boston, MA: NFPA.

Gottuk, D. (1999). "The Development and Mitigation of Backdraft: A Real-Scale Shipboard Study." *Fire Technology* 33:261–282.

Grimwood, P., Hartin, E., McDonough, J., and Raffel, S. (2005). *3D Fire Fighting: Training, Techniques, and Tactics,* First Edition. Stillwater, OK: Fire Protection Publications.

Grosshandler, W., Bryner, N., Madrzykowski, D., and Kuntz, K. (2005). *Report of the Technical Investigation of the Station Nightclub Fire.* NIST NCSTAR 2: Vols. I & II. Gaithersburg, MD: NIST.

Hall, J., and Cote, A. (2008). "An Overview of the Fire Problem and Fire Protection." *Fire Protection Handbook,* Twentieth Edition, Quincy, MA: National Fire Protection Association.

Icove, D., and DeHaan, J. (2009). *Forensic Fire Scene Reconstruction,* Second Edition. Upper Saddle River, NJ: Brady/Pearson.

Iqbal, N., and Salley, M. (2004). *Fire Dynamics Tools (FDTs): Quantitative Fire Hazard Analysis Methods for the U.S. Nuclear Regulatory Commission Fire Protection Inspection Program.* NUREG-1805. Washington, DC: U.S. Nuclear Regulatory Commission.

Jones, W., and Forney, G. (1990). *A Programmer's Reference Manual for CFAST, the Unified Model of Fire Growth and Smoke Transport.* TN-1283. Gaithersburg, MD: National Institute of Standards and Technology.

Kennedy, P. (2003), *Flashover and Fire Analysis,* Sarasota, FL: Investigations Institute.

McCaffrey, B., Quintiere, J., and Harkelroad, M. (1981). "Estimating Room Temperatures and the Likelihood of Flashover Using Fire Test Data Correlations." *Fire Technology,* 17, no. 2:98–119.

NFPA 921 (2008). *Guide for Fire and Explosion Investigations.* Quincy, MA: NFPA.

Norman, J. (1991). *Fire Officer's Handbook of Tactics.* Saddle Brook, NJ: Fire Engineering.

Peacock, R., Reneke, P., Bukowski, R., and Babrauskas, V. (1999). "Defining Flashover for Fire Hazard Calculations," *Fire Safety Journal* 32:331–345.

Pitts, W. (1994). *Global Equivalence Ratio Concept and the Prediction of Carbon Monoxide Formation in Enclosure Fires.* NIST Monograph 179.

Purser, D. (2008). "Assessment of Hazards to Occupants from Smoke, Toxic Gases, and Heat," Chapters 2–6, in *SFPE Handbook of Fire Protection Engineering,* Fourth Edition. Bethesda, MD: SFPE.

SFPE (2003). *SFPE Engineering Guide to Human Behavior in Fire.* Bethesda, MD: Society of Fire Protection Engineers (SFPE).

Skelly, M. (1990). "An Experimental Investigation of Glass Breakage in Compartment Fires," NIST-GCR-90-578. Gaithersburg, MD: National Institute of Standards and Technology.

Sutherland, B. (1999). "Smoke Explosions." Canterbury, New Zealand: University of Canterbury, Department of Engineering.

Zalosh, R. (2003). *Industrial Fire Protection Engineering.* West Sussex, England: Wiley.

13

Fire Modeling

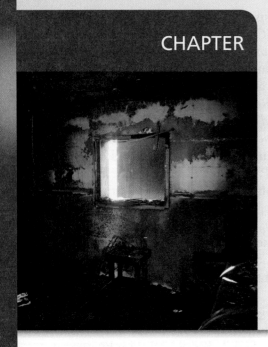

KEY TERMS

field models, *p. 268*

zone models, *p. 270*

OBJECTIVES

After reading this chapter, you should be able to:

- Explain the history and origin of fire modeling.
- Describe the different types of computer fire models available.
- Summarize the general uses and limitations of fire models.
- Identify potential uses of fire modeling for the various fire safety professions.

History and Basics of Fire Testing and Modeling

Many in the fire profession believe that the application of mathematics and science in fire-related dynamics began in the early 1940s (Nelson, 2002). As a result, most scientists would call the science young and relatively undeveloped. However, this chapter proposes that fire is one of the oldest and most studied phenomena. The first studies of fire began at the dawn of humankind, when human beings started to develop insight and understanding of what materials could be used for fuel to continue combustion. Surely, cave dwellers did not quantitatively study the effects of the fuels (i.e., tree bark = 100 J of energy versus wood log = 50,000 J), but it is obvious that they recognized that dry, greater surface area–to–mass fuels were easier to ignite and burned faster than did larger wet fuels. In fact, the study of fire is the basis for all other scientific disciplines. Faraday (1861) summarized this best when he discussed the phenomenon of fire as it related to a candle burning:

> There is no more open door by which you can enter into the study of natural philosophy than by considering the physical phenomena of a candle. There is not a law under which any part of this universe is governed which does not come into play, and is not touched upon, in these phenomena (p. 1).

Therefore, the earliest type of testing and modeling still being utilized today is the actual burning of fuels and examining the results. These studies are the basis for the fire protection profession. Now, standardized tests (i.e., American Society for Testing and Materials [ASTM] D1230, D2859, E603) are utilized to illustrate the hazards associated with different fuels. The first major category of modeling fire dynamics is physical fire testing and modeling, which is the testing and demonstration of fire given various fuels and scenarios. These types of tests and demonstrations fall within two broad categories: full-scale tests and small-scale tests. Full-scale tests are replications of a fire scenario that create a structure or item with similar geometric dimensions and attempt to reproduce fire phenomena. The major benefit of full-scale tests is that they are more representative of the actual scenario. However, these tests are often prohibitively expensive, especially when many variables (known and unknown) must be addressed through a long series of tests. Small-scale tests are replications of a fire scenario that create a structure or item with a scaled-down geometric dimension and/or other variables (i.e., physical scaling laws) when attempting to reproduce fire phenomena. Scale modeling often requires more than just reducing the geometric scale and focuses more on reducing the scale of the physical and/or chemical forces such as those used in Froude modeling. Klote and Milke (2002) provide more information on Froude modeling in their textbook, *Principles of Smoke Management*. The benefits of small-scale tests are that they are relatively inexpensive and may provide mathematical relationships that can be extrapolated for full-scale tests and models (see Figure 13.1).

The physical models lend themselves to the creation of the mathematical models (see Figure 13.2). Mathematical models are sets of mathematical equations that describe the behavior of a physical system (Beyler, Carpenter, and Dinenno, 2008, pp. 3–94). In other words, scientists would observe fire tests and attempt to develop equations based on thermal science fundamentals in order to match the observed physical behavior. These mathematical equations range from simple

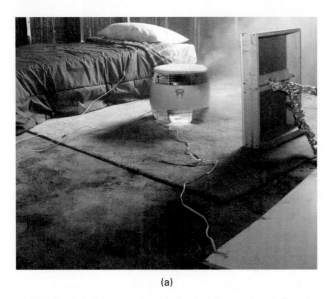 

(a)   (b)

FIGURE 13.1 (a) Examples of physical fire testing: full-scale testing; (b) small-scale testing.
Source: Photo by authors.

field models
- Computer fire models that attempt to predict conditions at every point. Also known as *computational fluid dynamics (CFD) models.*

algebraic equations used for predicting basic fire phenomena (i.e., flame height calculations) to complicated partial differential equations used for predicting enclosure fire phenomena. For purposes of this chapter, mathematical fire models can be placed into three categories based on their use and level of precision and complexity: hand calculations, zone models, and computational fluid dynamics (CFD), also known as **field models**.

Types of Models

HAND CALCULATIONS

Basic hand calculations are typically algebraic equations developed principally on experimental correlations utilized to estimate the effects of simple fire phenomena for simple configurations. Even though these calculations are basic, they can often provide a reliable prediction of the fire phenomena. They can provide the

FIGURE 13.2
Illustration of mathematical computer fire modeling.

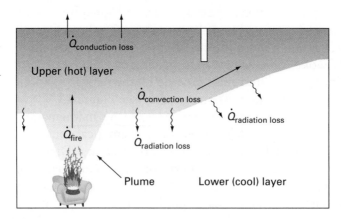

user with a quick calculation or estimate for the given scenario. In fact, the upper level mathematical equations found in the more advanced computer fire models (zone and field) are based on these hand calculations and experimental correlations. These hand calculations are often implemented in spreadsheet software (e.g., Microsoft Excel) as a collection of calculations for ease of use and repetition. The most popular collection is known as fire dynamics tools (FDTs), which were created and are still supported by the U.S. Nuclear Regulatory Commission (NRC). Table 13.1 provides a list of calculations compiled into easy-to-use spreadsheets by the NRC.

Basic hand calculations are based primarily on experimental correlations and are not typically time-dependent. Because of this, they have many limitations and assumptions that must be monitored to ensure that the calculations are appropriate for the given problem. For example, heat detector and sprinkler response times are based on experiments that were performed in compartments with flat, smooth ceilings and the fire located away from walls or other obstructions. Therefore, when using this hand calculation, users must ensure that their scenarios

TABLE 13.1	Fire Dynamics Tools (FDTs) Quantitative Fire Hazard Analysis Methods for the U.S. Nuclear Regulatory Commission Fire Protection Inspection Program*

1. Predicting hot gas layer temperature and smoke layer height in a room fire with door closed, natural ventilation, and/or forced ventilation.

2. Estimating burning characteristics of liquid pool fire, heat release rate, burning duration, and flame height.

3. Estimating wall fire flame height.

4. Estimating radiant heat flux from fire to a target fuel at ground level under wind-free conditions: point source radiation model.

5. Estimating radiant heat flux from fire to a target fuel at ground level in the presence of wind: solid flame radiation model.

6. Estimating thermal radiation from hydrocarbon fireballs.

7. Estimating the ignition time of a target fuel exposed to a constant radiative heat flux.

8. Estimating the full-scale heat release rate of a cable tray fire.

9. Estimating burning duration of solid combustibles.

10. Estimating centerline temperature of a buoyant fire plume.

11. Estimating sprinkler response time.

12. Fire severity calculations.

13. Estimating pressure rise due to a fire in a closed compartment.

14. Estimating pressure increase and explosive energy release associated with explosions.

15. Estimating visibility through smoke.

*Spreadsheets are available at http://www.nrc.gov/reading-rm/doc-collections/nuregs/staff/sr1805/.

have a smooth, flat ceiling and that the fire was located away from any obstructions. If the calculation is run for a beam-pocketed ceiling, then the time to activation will be seriously underestimated.

Most of the math and calculations presented within this text are based on these simple algebraic calculations, which have been shown to be capable of solving many everyday problems associated with fire protection and investigations. Although these hand calculations provide many solutions, they are limited in their ability to account for time-dependent physical and chemical interactions that drive compartment fire dynamics. Consequently, modeling enclosure fires requires computers to synthesize lengthy, time-dependent algorithms through a variety of software and model types (e.g., zone and field models).

ZONE MODELS

The transition from basic hand calculations to the more advanced computer software for fire modeling started in 1975 (Nelson, 2002). Zone fire models are the most common type of modeling utilized for evaluating enclosure fire dynamics. As established in previous chapters, enclosure fires are very complex but behave in a predictable manner due to the governing physics of combustion, buoyancy forces, and pressure profiles of ventilation openings. **Zone models** capitalize on this predictable nature and mathematically simplify various aspects of the enclosure fire to assist in predicting fire conditions. This type of modeling typically separates the compartment into two zones, commonly referred to as the upper (hot) zone and lower (cool) zone. These zones are based on the physics and dynamics of fire inside an enclosure, which include the fire plume, combustion products, and air entrainment. The fire plume and resulting collection of hot gases and combustion products form one zone, typically referred to as the upper zone. The ambient air and entrained air outline the other zone, typically referred to as the lower zone. The interface between the two zones moves vertically in the compartment based on the increasing collection of hot gases in the upper layer.

Zone models solve the conservation equations, including the conservation of mass, conservation of species, and the conservation of energy for each zone. From a mathematical standpoint, zone models are separated into control volumes and calculated accordingly. The upper zone is considered a control volume that receives both mass and energy from the fire and loses energy by convection or mass movement of gases through openings, by radiation to the floor, and to the surfaces in contact with the upper zone by conduction and radiation. The lower zone is considered a control volume that receives mass from air entrained from outside the compartment, and it loses mass to the upper zone by entrainment (see Figure 13.3).

zone models
■ Computer fire models that approximate the fire conditions in a room as two uniform gas layers with a fire energy source.

FIGURE 13.3
(a) Schematic of upper and lower layer separation,
(b) schematic of control volumes and calculation principles for zone models.

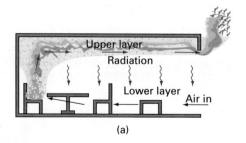

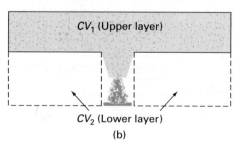

(a)

(b)

A detailed discussion of zone modeling is beyond the scope of this book. Quintiere (1989) and Janssens (2000) provide an excellent introduction to zone modeling. The box below presents a list of the most common zone models.

POPULAR ZONE MODELS

CFAST

Consolidated Model of Fire Growth and Smoke Transport (CFAST) was created and released in the early 1980s by the National Institute of Standards and Technology (NIST) (2005b). NIST continues to support this model and recently released the 6.0.10 version. It is a multiroom model (up to thirty compartments) that predicts conditions within an enclosure resulting from a user-specified fire. The program requires the user to input the geometry of the enclosure; ventilation openings; connections to other compartments; wall, floor, and ceiling lining materials and thermophysical properties; and data regarding the fire (HRR). The program outputs include the temperature, species concentrations, and thickness of the upper and lower layers.

BRANZFIRE

Building Research Association of New Zealand Fire (BRANZFIRE) was created and released in 1997 by the Building Research Association of New Zealand (Wade, 2003). This is a multiroom zone model applicable to room fire scenarios. Model outputs include but are not limited to gas layer temperatures, vent flows, pressure, room surface temperatures, layer height, visibility, and fractional effective dose estimates.

COMPUTATIONAL FLUID DYNAMICS MODELS (FIELD MODELS)

The last type of mathematical computer fire model referenced in this chapter is the computational fluid dynamics (CFD) model, also known as the field model (see Figure 13.4). Field models separate a compartment into hundreds to thousands of

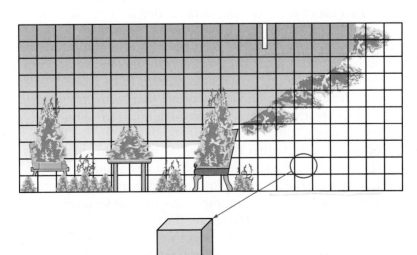

FIGURE 13.4
Computational fluid dynamics (CFD) or field model: illlustration of computational cells.

tiny cubes or calculation cells based on user inputs. Field models are more calculation-intensive than their zone model counterparts. These models calculate each cell using higher-level mathematics to specifically relate energy transfer and flow of fluids to each other. The basic laws of mass, momentum, and energy conservation are applied in each cell and are balanced with all adjacent cells.

From a mathematical standpoint, the field models assume that each cube is a separate control volume and calculate the conservation equations for each cell. Unlike zone models, computational fluid dynamics modeling was developed outside the field of fire dynamics and is used in many other engineering fields. Only recently, these models have been adapted for use in the field of fire dynamics. For a more detailed discussion of field modeling, refer to Galea (1988), who provides an excellent introduction to field modeling.

FIRE DYNAMICS SIMULATOR (FDS)/SMOKEVIEW

Fire Dynamics Simulator (FDS) is the most popular field model. FDS was officially released in 2000 and is another model that was created and is still supported by NIST (2008). Currently, FDS is up to its fifth version. FDS calculates the temperature, pressure, species concentrations, and flow field in relation to the prescribed fire. This model also provides a method for predicting activation of heat detectors and sprinklers.

SMOKEVIEW is the animation software that accompanies FDS. FDS is the mathematical portion of the model; SMOKEVIEW displays the output of FDS and CFAST simulations.

Input Data Needed for Modeling

Computer fire modeling is steadily increasing in use for all fire safety professions. It is important that the fire safety professional obtain the required information to perform computer fire modeling if the need arises. For this reason, this chapter provides a list of general input data needed for computer fire modeling.

1. *Structural dimensions:* The first and most important aspect of re-creating a scenario in computer fire models is to start with an accurate set of blueprints or a scene diagram. This diagram needs to be more detailed than the typical two-dimensional plan view diagram that many fire safety professionals are used to creating. A three-dimensional diagram should be created, including the geometry, soffit(s), sill(s), heights, and widths (see Figure 13.5). The locations of furniture and other fuel items, as well as the thicknesses and heights of those pieces of furniture, should also be obtained.
2. *Lining materials:* In enclosure fire dynamics, the materials lining the walls, ceiling, and floor play an important role in the transferring of energy out of the compartment. Therefore, the type of materials (i.e., carpet, gypsum wallboard) and the thickness of those lining materials need to be accurately documented. During fire reconstruction efforts, it is recommended that fire investigators collect and preserve samples of these lining materials.

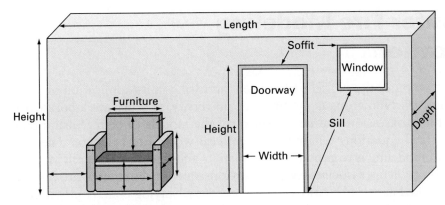

FIGURE 13.5 Schematic illustrating needed dimensions for computer fire modeling.

3. *Fuels and fire growth:* One of the most important variables input into a computer fire model are the types of fuels and their properties. The user of the model needs to input the heat release rate per unit area of the primary and secondary fuels. The computer model also requires a fire growth rate (HRR over time) to be input. Therefore, the more information regarding the type of fuel, the properties of the fuel, size, orientation, and location within the compartment, the less error involved in implementing this variable into the computer fire model. It is also recommended that fuel samples be collected and preserved for further analysis, if warranted.

4. *Ventilation:* The fire safety professional needs to locate and document all the ventilation openings, including heights, widths, soffit, and sill. These include windows; doorways; heating, ventilation, and air conditioning (HVAC); and fire suppression ventilation issues (positive pressure ventilation [PPV]). The fire safety professional needs to determine the positioning of these ventilation openings (i.e., open or closed, on or off). For any mechanical ventilation, the volume and temperature of the air for those vents should also be documented.

5. *Fire protection elements:* The location of all fire protection elements, including heights, should be included on the diagram. One of the biggest problems and failures with fire protection elements is the inaccurate placement of these devices, especially smoke alarms. The investigator and inspector can utilize this information in conjunction with computer fire modeling to evaluate a properly placed fire protection element versus the improperly placed element. Locating automatic sprinklers may also be used for the evaluation of suppression and extinguishment issues.

6. *Changes during the fire:* If the fire is being reconstructed, investigators also need to determine if and what changes occurred during the progression of the fire and when those changes occurred. This may become very important if changes to the ventilation were performed during the progression of the fire. This information can often be obtained through witness interviews and physical evidence.

7. *Photographic survey:* A photographic survey of the structure should be performed to capture other elements that may not be preserved on the diagram.

Computer Fire Modeling Applications

Since 1975, computer fire modeling has been increasing in its application to solving fire problems. Numerous governmental, university, and private laboratories assist with the progression and development of the models to better ensure that the mathematical equations reliably represent real-world fire behavior. The primary focus of modeling is to provide mathematical and scientific research into the behavior and problems associated with fire. Consequently, computer fire modeling can be applied as a tool to any area of fire safety. Staying consistent with the organization of the topics in this book, the application of computer fire modeling is separated into the different areas of the fire safety profession, including fire protection engineering, fire investigation, and fire suppression.

FIRE PROTECTION ENGINEERING

Computer fire models are most commonly used during the design of structures to evaluate fire conditions based on the expected hazards. Typically, a series of possible fire scenarios are predicted based on the hazards and fuel items present within the structure, which are then compared to the protection elements (e.g., smoke/heat detection, sprinklers) in place for the structure. This analysis allows the engineers to evaluate the design and placement of these fire protection elements against the most likely and worst-case fire conditions. The use of computer fire models during the design phase of a structure is often used in lieu of the prescribed, code-compliance approach (prescriptive-based approach). The use of computer fire modeling in conjunction with codes and standards and other engineering guides is known as performance-based design. For more information regarding this approach for fire protection design, refer to the Society of Fire Protection Engineers' (SFPE's) *Engineering Guide to Performance-Based Fire Protection* (2007).

In addition to performance-based design, modeling in the inspection and evaluation of the fire safety of buildings and facilities is currently in widespread use, and the use of modeling is currently being implemented in several U.S. national building codes and fire safety standards, including National Fire Protection Association (NFPA) 1 Uniform Fire Code, NFPA 72 National Fire Alarm Code, NFPA 101 Life Safety Code, NFPA 921 The Guide for Fire and Explosion Investigations, NFPA 5000 Building Construction and Safety Code, and several others (NFPA, 2007).

Businesses and corporations often need to expand, alter, or purchase new building space to accommodate their changing role and economic situation. When economic conditions change, often a building's designed use is no longer applicable, and a change of use must be evaluated. Fire protection systems adequate for the original use may prove unsuccessful against new hazards and/or arrangements. Computational models assist fire protection engineers and code enforcement officials in determining acceptable limits (i.e., storage volume or height limitations) that allow installed system effectiveness. Such analysis may prevent costly retrofitting by minimal changes in arrangement.

The most notable inspection and code enforcement application is the application of modeling by the U.S. Nuclear Regulatory Commission (NRC). The NRC regularly uses computer fire models to assist in the evaluation of enclosures for maximum fire safety protection in nuclear facilities across the United States (NRC, 2006).

FIRE INVESTIGATION

Post-fire reconstruction or fire investigation has also seen an increase in the use of computer fire models. Most notably, the U.S. government, in its analysis of both the World Trade Center fire and the Station Nightclub fire that occurred in West Warwick, Rhode Island, utilized Fire Dynamics Simulator (FDS) and Consolidated Model of Fire Growth and Smoke Transport (CFAST) models in evaluating the reason for the spread, behavior, and impact of the fire. The National Institute for Science and Technology (NIST) was charged with the analysis of both of these national tragedies, as well as many other tragedies around the United States, to evaluate the reasons for the behavior of fire by implementation of computer fire modeling (NIST, 2005a, 2007).

NFPA 921, The Guide for Fire and Explosion Investigations, requires fire investigators to follow a systematic approach in their analysis of the origin of, cause of, and responsibility for a fire. The scientific method has been put forward as this systematic approach. One of the primary steps in the scientific method is to test the hypothesis. Computer fire modeling is a scientifically and generally accepted method for testing a hypothesis. Its primary use falls within the testing of one's hypothesis in the scientific method as it pertains to understanding the fire, timeline analysis, occupant survivability, and fuels analysis, and to analyzing post-fire indicators.

- *Understanding the fire:* Computer fire modeling can assist an investigator in understanding how a fire may have evolved. More specifically, computer fire modeling assesses the relationship of the heat release rate of the burning fuel to other variables (i.e., carbon monoxide production, radiant ignition). Complex scenarios allow for multiple runs to be performed with a range of ventilation variables, which provides the user with a range of outcomes to analyze the effects. Also provided by the models is the ability to calculate the minimum energy required for a compartment to transition through flashover, as well as the timing issues for flashover and full-room involvement. Modeling may assist in evaluating sufficiency of fuels for flashover and damage that exists after the fire due to the heat flux from a burning object.
- *Timeline analysis:* The model can provide a range of timing issues that may assist in understanding eyewitness accounts, the progression of the fire in relationship to other variables, survivability of occupants, possibility of egress, comparison of injuries to fire development, and activation and interaction of fire protection elements, and evaluating ignition and time to ignition issues (see Figure 13.6). The use of computer fire models in a timeline analysis provides an objective way to analyze the progression of events.
- *Survivability Analysis:* People are affected adversely from several different by-products of fire, including temperature, toxic gases, heat and flame, and visibility reduction. These different by-products have tenability limits and can be analyzed with a computer fire model. Investigators can utilize these models to assist with their analysis of egress and escape issues.

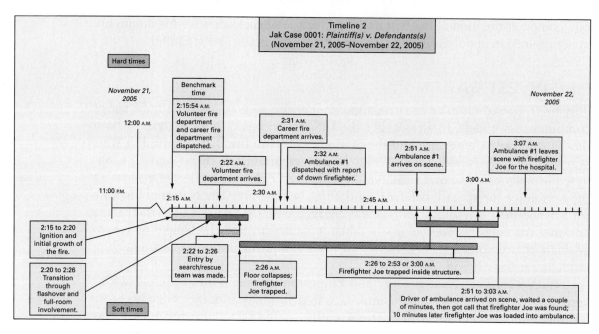

FIGURE 13.6 Sample timeline with range of computer fire modeling soft times.

- *Analyzing post-fire indicators:* Investigators can utilize computer fire models to compare the post-fire damage or physical evidence to the results of the various models (see Figure 13.7). Many of the models can provide insight into the transfer of heat and the subsequent effects of this transfer on materials. More discussion on this topic is presented later in this chapter.
- *Visualization of fire phenomena:* A feature of some of these models is to transfer the mathematical output into three-dimensional computer graphics. FDS and CFAST have companion animation software that provide animation of the fire and can be utilized to visualize fire phenomena.

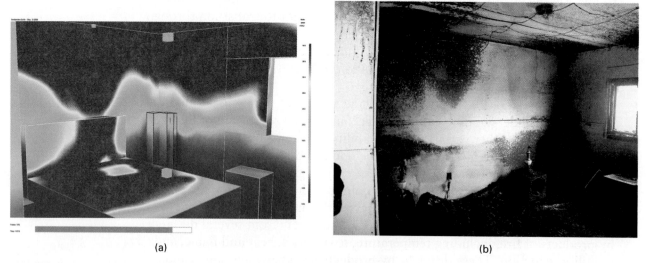

(a)   (b)

FIGURE 13.7 Analyzing fire patterns with computer fire modeling compared to post-fire patterns.
Source: Photo by authors.

- *Multiple hypotheses:* Computer fire modeling is at the heart of the scientific method. Computer fire models may provide an objective means of testing one's hypothesis. It allows an investigator to test her or his hypothesis or others' hypotheses for validation or refutation.

An investigator is cautioned when utilizing any of the models for the above purposes to ensure that the models are appropriately chosen and used within their limitations and assumptions.

Computer fire modeling should be utilized as a *tool* in an investigator's analysis of a fire. The use of computer fire modeling for fire investigations is usually an easier task than for design engineering because other information, such as eyewitness accounts, forensic evidence, and fire department reports, is almost always available. A computer fire model in this case is most often used to supplement the other information in demonstrating that a particular hypothesis is or is not plausible.

Not every case warrants the use of this tool during an investigator's analysis, but all investigators need to begin collecting the data required at the scene to ensure that if the need arises, they have adequately collected the important data. As an analogy, not many investigators can use a gas chromatography/mass spectrometry (GC/MS), but most are aware of the appropriate collection and preservation methods to ensure that samples can be sent to the laboratory for fire debris analysis. The same methodology and knowledge should be passed onto those investigators in the field. The use of modeling should be supported by the investigation community, but it should also be constantly monitored to ensure its proper and objective use.

It is not appropriate for someone to utilize a computer fire model to prove causation. Computer fire models will never replace a good on-scene investigation. Investigators need to be very concerned about those who believe that they can perform an investigation by sitting at a computer. It is important for all investigators to remember that computer fire models should be used as a tool to *supplement* an investigator's on-scene investigation and analysis.

TESTING OF AN ORIGIN HYPOTHESIS WITH COMPUTER FIRE MODELS

One of the primary uses of modeling is to test one's hypothesis. Hypotheses are developed not only for the cause of a fire but, more important, they are also developed for the area of origin (NFPA 921, 2008). The origin hypothesis is more important because the area of origin must first be determined before a cause can be evaluated properly (NFPA 921, 2008). Fire patterns have historically been and continue to be the primary tool used by investigators in determining an area of origin. Use of computer fire modeling is proving to be a tool in validating and/or refuting one's origin hypothesis, based on the resulting fire patterns that remain after a fire, and the boundary heat flux values calculated within the model when testing different areas of origin and fuel packages. This specific application can be implemented when running a field model that calculates boundary heat fluxes and has a companion animation program, similar to FDS/SMOKEVIEW.

Computer fire models are being utilized in conjunction with the Full-Scale Fire Patterns Study that is ongoing at Eastern Kentucky University in cooperation with the National Association of Fire Investigators (Gorbett, Hopkins, Kennedy, and Hicks, 2006; Hopkins, Gorbett, and Kennedy, 2007). To date, there have been thirteen full-scale research burns performed to analyze fire patterns reproducibility and persistence through flashover. Each full-scale test was outfitted with instrumentation, including thermocouples and radiometers. To ensure validation and verification of the models, each has been evaluated against the experimental test results (temperature and heat flux). The first six research burns finalized in March 2006 were intended to analyze reproducibility given similar fuel packages, orientation, and similar origin (i.e., center head of mattress). The two research burns completed in March 2007 were to evaluate an origin in a small (low heat release rate) fuel package (i.e., nightstand) near a larger fuel package (i.e., mattress, with a high heat release rate) to analyze if the larger fuel package's resulting damage would obscure patterns that would assist in determining an accurate area of origin (Hopkins, 2008).

The creation of lines of demarcation has been related to the exposure of a witness surface to varying heat flux intensities (Hopkins et al., 2007). Figures 13.8 through 13.10 illustrates an example of the resulting actual damage from the patterns research burn test compared to a Fire Dynamics Simulator (FDS), SMOKEVIEW heat flux calculation. These tests were performed with the origin of the fire at the head of the mattress. The resulting actual

FIGURE 13.8 (a) East wall actual damage versus (b) FDS heat flux damage. Origin = center of bed.
Source: March 2007 Full-Scale Burn Pattern Study. Photos by authors.

Origin

(a)

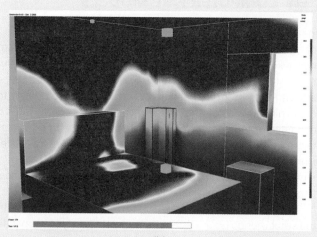

(b)

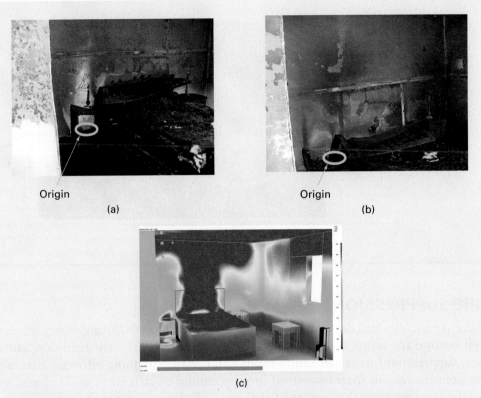

FIGURE 13.9 (a) East wall actual damage, cell 1, (b) cell 2, versus (c) FDS heat flux calculation. Origin = nightstand.
Source: Photos by authors.

damage reveals the effects of the upper layer, flame plume, and ventilation generated patterns. The calculated heat flux shown by FDS/SMOKEVIEW, demonstrated by a color difference (i.e., the darker the color, the higher the calculated heat flux), can be considered anticipated damage calculated by the model. It is evident that the FDS/SMOKEVIEW heat flux calculation (anticipated damage) was consistent with the actual damage.

(a)

FIGURE 13.10
(a) South wall actual damage versus (b) FDS heat flux calculation.
Source: Photos by authors.

FIGURE 13.10
(*continued*)

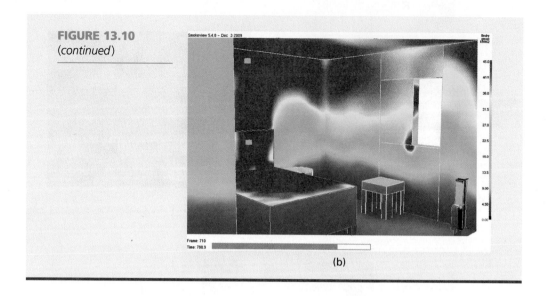

(b)

FIRE SUPPRESSION

Accurate analysis and development of computer fire models during emergencies is well beyond the scope of any fire modeling programs currently available. However, suppression forces can utilize models in pre-fire planning efforts. Currently, the fire suppression forces involved in preplanning of structures utilize basic hydraulics to determine the required fire flow (RFF) for a structure based on several different basic formulas (e.g., the Iowa formula, the National Fire Academy formula, and the Insurance Services Office [ISO] formula). The calculations provide an estimate regarding the amount of water that must be applied to the fire in order to control and/or extinguish it in a reasonable amount of time based on generalized assumptions of fire development. RFF is then evaluated against available water distribution systems' ability to supply the requisite volume of water at the needed rate. Through computer fire modeling, evaluation for specific fire threats and/or areas within the structure is possible; thus, options of suppression strategies can be tested and best practices can be implemented. In fire suppression classes, officers can also visualize changes in fire development, including temperature and smoke density, brought by changes in ventilation efforts.

While these general formulas provide a basis for determining water supply, computer fire modeling may assist in better assessing the hazards and risk of a fire in such an occupancy and result in a better prediction of water supply required. Required flow rates are also important, and not only for manual fire suppression operations. Engineers and designers of fixed fire suppression systems, such as automatic sprinkler systems and standpipe systems, must also know the flow requirements for their systems.

Validation and Verification

Many of the organizations that develop models continue to support these models by performing validation and verification studies. NIST constantly reviews, validates, and verifies these models by comparing real-world fire experiments to the data produced from the model. This assists in ensuring the reliability of the models and their application to fire problems. A community of model users has been organized to provide consistent feedback to the model creators regarding any issue with the models. NIST also reviews and cooperates with independent researchers to utilize their data from experimental tests within the model. Building Research Association of New Zealand (BRANZ), similar to NIST, performs full-scale experiments that are used to continuously validate the mathematical equations contained within the BRANZFIRE model.

Most recently, the U.S. Nuclear Regulatory Commission (NRC) has written a 2,000-page-plus series of validation and verification manuals on their analysis of various computer fire models, including CFAST and FDS (NRC, 2006). The objective of this project was to examine the predictive capabilities of selected fire models. NRC ran a series of full-scale laboratory fire tests and ran the data through the various computer fire models. They were specifically analyzing each model's capability and reliability to reproduce the results from the live fire tests.

The accepted and peer-reviewed methodology for using computer fire models is to provide a range of variables to evaluate both the sensitivity of the model and to ensure that the variables that are not specifically known are accounted for within the series of models (Society of Fire Protection Engineers [SFPE], 2008; Wood, Sheppard, and Custer, 2008). The user must input a range of variables based on the structure, the collected data, and the probable scenarios. This variation of the input variables affects the output or outcomes of the model and provides the user with a range to utilize in her or his analysis.

Summary

The continued improvement and validation and verification of these tools increases their use in all fire safety professions. There are many potential uses for computer fire models, including performance-based design, inspection and code enforcement, development of better standards for fire safety, and reconstruction of fires. It is imperative that all fire safety professions begin to understand these tools better to ensure their proper usage and implementation in their respective disciplines.

Review Questions

1. What is a fire model?
2. What is the difference between a physical model and a mathematical model?
3. How do the physical and mathematical models interrelate?
4. List five applications of mathematical modeling.
5. List the four benefits of mathematical models.
6. When would mathematical models not be useful?
7. Fire models are based on an understanding of:
   a. the chemistry of combustion
   b. the physics of heat transfer
   c. both a and b
   d. neither a nor b
8. Zone models used for computer fire models typically separate information into:
   a. four zones (upper, lower, right and left)
   b. two zones (right and left)
   c. two zones (upper and lower)
   d. a single zone
9. When used to validate fire protection system competency within compartments in performance-based design, modelers should:
   a. validate that input information has been tested and is reliable
   b. use input data that yields desired results
   c. declare the models not usable if accurate inputs are not available
   d. all of the above
   e. a and c
10. **True or false:** Fire models offer a reliable way to determine the origin and cause of fires, which can then be validated by investigation.

PEARSON

myfirekit™

For additional review and practice tests, visit **www.bradybooks.com** and click on MyBradyKit to access book-specific resources for this text!

Register your access code from the front of your book by going to **www.bradybooks.com** and selecting the mykit links.

References

Beyler, C., Carpenter, D., and Dinenno, P. (2008). "Introduction to Fire Modeling," (Section 3, Chapter 3), in *Fire Protection Handbook,* Twentieth Edition. Quincy, MA: National Fire Protection Association.

BRANZ (2003). *A Technical Reference Manual to BRANZFIRE 2003.* Colleen Wade, Building Research Association of New Zealand, Judgeford.

Faraday, M. (1861). *The Chemical History of a Candle.* London: Royal Institute of London.

Galea, E. (1988). "On the Field Modeling Approach to the Simulation of Enclosure Fires." *Journal of Fire Protection Engineering,* 1:11–22.

Gorbett, G., Hopkins, R., Kennedy, P., and Hicks, B. (2006). *Full-Scale Burn Patterns Study.* Cincinnati, OH: International Symposium for Fire Investigations (ISFI).

Hicks, B., Gorbett, G., Hopkins, R., Kennedy, P., and Abney, B. (2006). *Advanced Fire Patterns Study: Single Fuel Packages.* Cincinnati, OH: ISFI.

Hopkins, R. (2008). *Patterns Persistence: Pre- and Post-Flashover.* Cincinnati, OH: International Symposium for Fire Investigations (ISFI).

Hopkins, R., Gorbett, G., and Kennedy, P. (2007). *Patterns Persistence: Pre- and Post-Flashover.* San Francisco, CA: Fire and Materials.

Janssens, M. (2000). *An Introduction to Mathematical Fire Modeling.* Lancaster, PA: Technomic.

Klote, J., and Milke, J. (2002). *Principles of Smoke Management.* Atlanta, GA: American Society of Heating, Refrigerating and Air-Conditioning Engineers.

Nelson, H. (2002). *From Phlogiston to Computational Fluid Dynamics.* Bethesda, MD: Society of Fire Protection Engineers (SFPE).

NFPA (2007). *National Fire Codes.* Quincy, MA: National Fire Protection Association.

NFPA (2008). *Fire Protection Handbook.* Quincy, MA: National Fire Protection Association.

NFPA 921 (2008). *Guide for Fire and Explosion Investigations.* Quincy, MA: National Fire Protection Association.

NIST (2005a). *Analysis of Aircraft Impacts into the World Trade Center Towers.* Chapters 1–8. Federal Building and Fire Safety Investigation of the World Trade Center Disaster. NIST NCSTAR 1-2B; 290 pp.

NIST (2005b). *NIST Special Publication 1041: CFAST—Consolidated Model of Fire Growth and Smoke Transport (Version 6)—User's Guide.* National Institute of Standards and Technology. U.S. Department of Commerce.

NIST (2007). *Reconstructing the Station Nightclub Fire: Computer Modeling of the Fire Growth and Spread.* Volume 2; Interflam 2007. (Interflam '07). International Interflam Conference, 11th Proceedings. vol. 2. September 3–5, 2007, London, England, pp. 1181–1192.

NIST (2008). *NIST Special Publication 1019-5 Fire Dynamics Simulator (Version 5)—User's Guide.* National Institute of Standards and Technology. U.S. Department of Commerce.

NRC (2006). Verification and Validation of Selected Fire Models for Nuclear Power Plant Applications. U.S. Nuclear Regulatory Commission. Office of Nuclear Regulatory Research. Washington, DC 20555-0001

Quintiere, J. (1989). "Fundamentals of Enclosure Fire 'Zone' Models." *Journal of Fire Protection Engineering,* 1:99–119.

Society of Fire Protection Engineers (SFPE) (2007). *SFPE Engineering Guide to Performance-Based Fire Protection.* Bethesda, MD: Society of Fire Protection Engineers.

Wade, C. (2003). BRANZ—*A User's Guide to BRANZFIRE 2003,* Building Research Association of New Zealand, Judgeford.

Wood, C., Sheppard, D., and Custer, R. (2008). "Applying Fire Models to Fire Protection Engineering Problems and Fire Investigations" (Section 3, Chapter 5), in *Fire Protection Handbook,* Twentieth Edition. Quincy, MA: NFPA.

14
Extinguishment

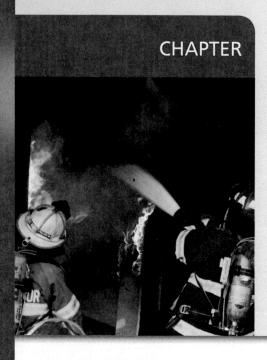

OBJECTIVES

After reading this chapter, you should be able to:

- Understand and cite four methods of extinguishing fires.
- Describe how removal of fuel can extinguish fires.
- Describe how water extinguishes fire by cooling.
- Describe how water reacts when heated above its vaporization point.
- Accurately identify how water removes heat from given situations.
- Understand latent heat.
- List and describe methods to remove oxygen from a fire to extinguish flames.
- Describe how the chemical chain reaction of fire can be interrupted to facilitate extinguishment.

Extinguishment of Fire

When unwanted fires begin, or desired fires are no longer needed, actions to suppress combustion involve scientific principles outlined simply in the fire triangle or the fire tetrahedron. Removal of one component (leg) of the fire triangle or interruption of the chemical chain reaction, as depicted in the fire tetrahedron, is necessary to stop the combustion process. This chapter outlines circumstances and considerations related to extinguishing fires by changing one or more components within the fire tetrahedron. Removal of fuel, removal or reduction of heat, removal or reduction of oxygen, and the interruption of the chemical chain reaction are discussed. Actual extinguishment may combine multiple effects of removing more than a single component. When assessing how to extinguish fires, one must be aware of the benefits and consequences associated with each methodology.

A prime example is the extinguishment of a compartment fire with water. When dispersed into a compartment by a fog nozzle (high surface area–to–mass ratio), water quickly converts to steam. A change in the water's physical state absorbs a tremendous amount of heat from the fire and is therefore believed to be the primary mechanism by which the fire is extinguished. However, when water changes from liquid to a vapor state, it expands 1,700 times. This rapid change from liquid to vapor tends to displace ambient air from the compartment volume, thus reducing oxygen needed for combustion. Simultaneously, the water provides a cooling effect, thus reducing radiant heat imposed on fuel packages, which in turn results in less pyrolysis and fuel gases emitted to the atmosphere. Which of these mechanisms actually extinguishes the fire may not be clear; however, we recognize that combustion ceased. Understanding how the reaction can occur assists in increasing effectiveness of efforts to suppress fires.

REMOVAL OF FUEL

Turn off Fuel Supply

Fires involving fluid fuels may be extinguished by stopping the flow of fuel to a flame. A common example is a cigarette or grill lighter fueled by butane (see Figure 14.1). By clicking a valve open, gas is released in an area where a spark is introduced, resulting in a flame. When the user desires to extinguish the flame, he or she simply releases the control valve (which uses a spring-loaded control to seat the valve), thus extinguishing the fire. The same principle may be a viable option in broken gas mains and gas-fired equipment, and for situations where flowing liquid fuels are involved. More immediate results are probable with gaseous fuels than with liquids, especially liquids discharged outside the

FIGURE 14.1
Suppression by
controlling fuel.
Source: Photo by authors.

normal system for which they are intended. It should be noted that some solid-fuel fires may experience similar extinguishment; however, delayed results should be anticipated.

Separation of Fuel

Separations within fuel storage areas are a common method of preventing catastrophic fire development. These arrangements are common with pile storage of combustible materials in warehouse and industrial settings. Empty aisles of sufficient width in rack storage arrangements prevent radiant heat transfer between fuel packages, and the width of the spaces facilitates water from discharging sprinklers in reaching adjacent fuels.

Separation of fuel can be an effective means in controlling larger fires, especially unconfined fires in large fuel arrays, for example, storage of tires. Occasionally large piles of combustible materials such as tire carcasses are illegally stored, then undergo a fire. Separation of burning fuels from fuel that is not yet burning can enhance fire suppression efforts. A common method of controlling wildfires and forest fires is fuel removal. This is where the phrase "Fight fire with fire" has its genesis. Fuels are removed from the path of flames to prevent continued burning. Often the method of fuel removal is establishing a fire break by physically removing a small path of fuel from the approaching fire, sometimes by igniting a small fire at the break's edge, thus allowing flames to burn back toward the approaching fire. This creates a chasm between the approaching fire and un-burned fuels. When properly handled, this approach is very effective at removing the fuel and serves as a method of fire extinguishment.

Fire Consumes Fuel

Though not often considered a method of extinguishing fires, allowing the fuel to burn out is a method that can be used to control fires. An example is that of liquid fuels burning in an area where little damage could result from heat exposure, but attempts to suppress the fire with traditional methods and tactics may result in spreading flames and creating bigger problems. This method should be considered when burning fuels pose a danger to those who would attempt to extinguish the flames.

EXAMPLE 14.1

A tanker truck caught fire while filling with gasoline. The resulting fire spilled fuel and melted the aluminum tank top away, exposing the fuel surface. An adjacent steel structure was exposed to heat from the tank; however, fuel flow was stopped at the onsite tank. Thus, no additional burning was experienced in the delivery piping. No other exposures were present. Firefighters opted to extinguish the fire with water and *foam*. Water from hose streams settled under the burning fuel surface, due to its higher specific gravity, which caused burning fuel to overflow the vessel. Flames spread, threatening adjacent tanks and properties. Because the fire did not threaten life or property during the original fire, allowing the fuel to simply burn out would have prevented this flame spread and subsequent problems.

REMOVAL OF HEAT

Sufficient heat to initiate chemical change between fuel molecules and oxygen is a necessary requirement for the combustion reaction. Reduction of the heat energy to an intensity below that which facilitates this chemical reaction causes combustion to cease. During the combustion reaction, further heat is typically required for continued pyrolysis, or vaporizing solid and liquid fuels, for the chemical reaction to continue. Reduction of the heat energy to prevent this continued action extinguishes the fire because fuel is not sufficiently being provided in the form necessary for combustion to continue. In a way, removal of heat stops pyrolysis; thus, the fuel is removed from the reaction.

AIR MOVEMENT

Infusion of large volumes of air across a flame can remove the fuel to interrupt the reaction. The most common example is blowing out a candle, wherein movement of air is sufficient to prevent the supply of fuel at a rate to continue the combustion reaction (see Figure 14.2). When fuel is prevented from gasification by heat loss, the fire ceases. Another explanation is diffusion of fuel gas below the lower flammable limit (LFL) when air volume increases. Regardless of the physical reason, the fire is extinguished.

A more extreme example of air movement comes from extinguishing fires with explosives, and oil well fires are the most common example. The method features detonation of an explosive charge near the burning wellhead. Two theories describe how this extinguishes the flame: One is that air movement separates heat from the fuel much as blowing out a small candle. Another theory used to describe this extinguishment method postulates that explosives consume available oxygen within the combustion region; thus, the flames are extinguished.

WATER

The most common method of removing heat from fires, especially turbulent flames, is heat removal with water (see Figure 14.3). Though many propose that heat is removed, actually heat is transferred to all mass within the area equally. Fuel packages continue to receive heat, but water applied in the fire area also takes on heat energy, which prevents that portion of the heat from reaching the fuel package.

When water is applied to fire, heat energy is diverted from pyrolysis by raising the temperature and/or changing the physical state of the water. This action occurs at temperatures lower than the ignition temperature of most fuels; thus, when a sufficient volume of water is introduced, fuel does not receive sufficient heat reflected to support pyrolysis, which in turn results in the fire being extinguished. Water in contact with a fuel must be removed (vaporized) before that fuel can reach its ignition temperature.

A very small quantity of water actually accomplishes extinguishment even in relatively high heat-release-rate fires. Tewarson (2008) reports that 34 g/m^2-s (.034 gpm/ft^2) of water is sufficient to immediately suppress flaming combustion in asymptotically large burning surfaces of polystyrene. This indicates that relatively little of the water applied to the base of a fire actually suppresses the flames; however, we caution against using extremely small quantities or flow rates based on this data. Rather, understand that much of the water is being converted to steam prior to reaching the base of the fire. Therefore, fire suppression by large volume hose streams may occur with brief application periods followed by lower flow rates on smoldering sections of fuels. Sprinklers are incapable of directing water to a specific fire location; thus, they operate to prewet surrounding combustibles. The early termination of sprinkler flow can prove disastrous by allowing flame spread to continue uncontrolled across dry fuels. In polymer fires, deep

FIGURE 14.3 Water being applied to a fire. *Source:* Courtesy of Katherine Steenken, student, Eastern Kentucky University.

smoldering is less likely than in cellulosic fuels and requires different approaches. Deep-seated smoldering may not be extinguished with brief water application. Thus, longer duration application or follow-up water application, known as overhaul, is often required.

Water acts to remove heat from the fire in two ways. First is its high **heat capacity**, which describes the amount of energy required to raise the temperature of a unit of mass 1°. Heat capacity of water is generally designated as 1, indicating the energy required to raise a unit of mass 1°. In the metric system one calorie (1 cal) is needed to raise one gram (1 g) of water one degree Centigrade (1°C). In the English measurement system, that unit is the British thermal unit (BTU), indicating the energy required to raise one pound (1 lb) of water one degree Fahrenheit (1°F). In the SI system, the value is 4.18 Joules (4.18 J), which is what is required to raise one gram (1 g) of water one degree Celsius (1°C). When in the vapor state, water's (H_2O) heat capacity is .48 (48 percent) of that when it is in the liquid state. In the solid state, water's heat capacity is .5 (50 percent) compared with that of its liquid state.

The change in physical state, known as the latent heat, requires a greater amount of energy than does simply raising the temperature of a material and is

heat capacity
■ The amount of heat needed to raise one mass of a substance 1°. In the SI system, the term refers to raising one gram 1°C. In the English system, it refers to raising one pound 1°F.

FIGURE 14.4 Water (SI: Joule).

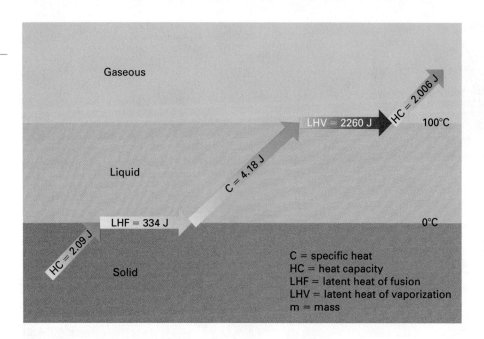

FIGURE 14.4 Water (SI: Joule).

latent heat of fusion (LHF)
■ The amount of heat require to convert a unit of mass of solid substance to a liquid (endothermic). An equal amount of energy is released by the mass to convert from liquid to solid (exothermic). Expressed in J/g, calories, or BTU.

relatively high for water, which makes it a very good extinguishing agent. Significant energy is required to affect the change of state within matter; this heat energy is known as latent heat. The latent heat of vaporization (LHV) is the energy required to change water from its liquid state to its vapor state, or it is the heat released when water changes from a vapor to a liquid. The energy required is 2,260 Joules per gram, 540 calories (per gram), and 970 BTUs (per pound). The **latent heat of fusion (LHF)** is the energy required to change water from a solid (ice) to a liquid and, conversely, it is released when the water moves from the liquid to the solid state (see Figures 14.4 through 14.6).

FIGURE 14.5 Water (SI: calorie).

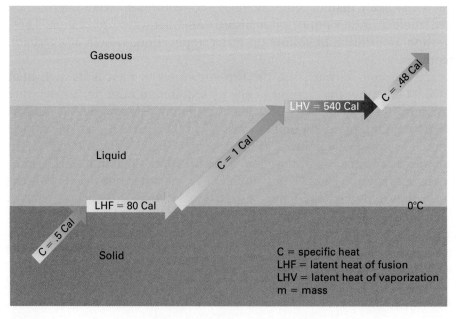

FIGURE 14.5 Water (SI: calorie).

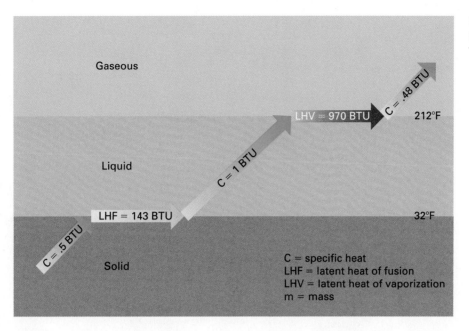

FIGURE 14.6 Water (English: BTU).

Gaseous

LHV = 970 BTU

C = .48 BTU

212°F

Liquid

C = 1 BTU

LHF = 143 BTU

32°F

C = .5 BTU

Solid

C = specific heat
LHF = latent heat of fusion
LHV = latent heat of vaporization
m = mass

EXAMPLE 14.2

How much energy is required to raise the temperature of five gallons (5 gal) of water from 0°F to 240°F?

First, evaluate the energy required to raise one unit of mass from 0°F to 240°F. From 0°F to 32°F is 32° in the solid state. Thus, the heat capacity is .5; 32° multiplied by .5 BTU equals 16 BTUs to bring each unit of mass from the starting temperature to 32°F. For each unit of mass, 143 BTUs are required for the phase change to liquid (LHF). Raising the temperature of each pound from 32°F to 212°F requires 180 BTUs (212 − 32 = 180 * 1 BTU/unit − 180 BTU). The change of state to vapor requires 970 BTUs per pound (LHV). Last, raising the temperature from 212°F to 240°F requires 13.44 BTUs (28 * .48), which is rounded to 13 BTU. Next, determine the mass involved; 5 gallons of water times 8.34 pounds per gallon indicates a mass of 41.7 pounds. Now the energy required for temperature change can be evaluated (see Figure 14.7).

$$1,322 \text{ BTU (per pound)}$$
$$\underline{* \ 41.7 \text{ pounds}}$$
$$55,128 \text{ BTUs required}$$

Another way to solve this example is by calculating the heat required in each phase, including mass:

Vapor	28°F * .48 * 41.7 =	560 BTU
LHV	970 * 41.7 =	40,449 BTU
Liquid	180°F * 1 * 41.7 =	7,506 BTU
LHF	143 * 41.7 =	5,963 BTU
Solid	32°F * .5 * 41.7 =	667 BTU
		55,145 BTU

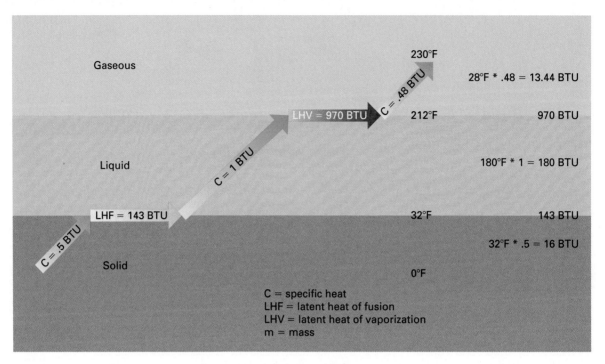

FIGURE 14.7 Temperature and energy diagram of water (English units).

The difference in the results of these methods comes from rounding errors.

EXAMPLE 14.3

EXAMPLE: SI System

Fire suppression forces apply water at 570 lpm (150 gpm). The initial water temperature is 16°C and the ending temperature is 600°C. How much energy is removed from the fire per second?

Between 10°C and 100°C, 4.18 Joules are required to raise each gram 1°C. Thus, 90°C times 4.18 J/g°C is 376.2 joules (376.2 J), which is required for each gram, or 214,434 kJ is required for the total mass (507 kg * 376.2 kJ/kg). Latent heat of vaporization is 2,260 J/g; thus 1,288,200 kJ is required to convert the mass from liquid to steam (570 kg * 2,260 kJ/kg). To raise this mass from 100°C to 600°C requires an additional 571,824 kJ (2.0064 kJ/kg°C * 500°C = 1,003.2 kJ/kg * 570 kg = 571,824 kJ). If the entire mass is raised to 600°, 2,074,458 kJ is removed from the atmosphere (see Figure 14.8).

$$500°C * 2.0064 \text{ kJ/°C/kg} = 1,003.2 \text{ kJ/kg}$$

$$\text{LHV} = 2,260 \text{ kJ/kg}$$

$$\underline{90°C * 4.18 \text{ kJ/°C/kg} = 376.2 \text{ kJ/kg}}$$

$$3,639.4 \text{ kJ/kg}$$

$$3,639.4 \text{ kJ/kg} * 570 \text{ kg} = 2,074,458 \text{ kJ}$$

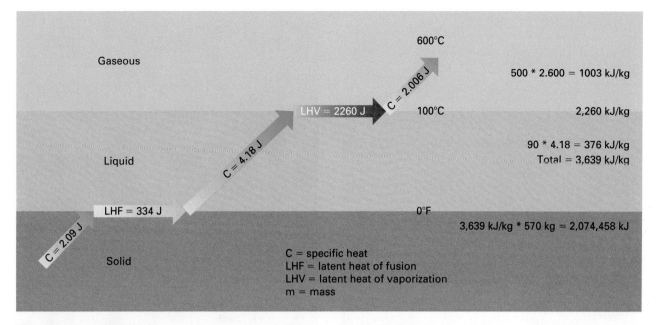

FIGURE 14.8 Temperature and energy diagram of water (SI units)

Rapid Steam Production

The subject of the latent heat of vaporization of water requires special attention. When water is applied as a fog, the droplets in the fog have significantly more surface area and absorb heat more rapidly. Thus, they pass through the latent heat of vaporization much more easily. When converted from liquid to steam, water's volume expands by 1,700 times, which results in dilution of oxygen in the combustion area (see Figure 14.9). The most important concern is that, while converting liquid to steam provides greater efficiency in water usage, persons (firefighters and civilians) within the compartment experience a significant adverse impact from steam reaching their bodies. It should be noted that Lloyd Layman introduced fog application for extinguishing structural fires after learning shipboard suppression methods during his service in the U.S. Coast Guard (Robertson, 2000). Layman touted the application of water in fog form into the upper layer and termed it indirect application. Conversion of liquid to vapor diverts much heat energy while spreading steam to the fuel source. This methodology was applied in the mid- to late 1940s, long before standard methods of extinguishment involved entry into well-developed compartment fires. When inside burning structures, many firefighters in the United States often opt for fog nozzles but use narrow streams that deliver larger droplets in a more concentrated area. These larger droplets do not convert to vapor as readily. Other firefighters maintain that use of solid streams is preferable because water penetrates closer to the fuel package in high heat-release-rate fires.

Similar concerns are posed with fire sprinklers placed to suppress fires. Where fuels have anticipated heat release rate fires within lower ranges, sprinklers are designed to deliver smaller droplets of water. However, when fuel packages are anticipated to produce high heat-release-rate fires, larger droplets may be preferable because smaller droplets tend to vaporize before reaching burning fuels in high heat-release-rate situations.

(a)   (b)

FIGURE 14.9 Fog stream (a) and solid stream (b).
Source: Photos by authors.

WATER VOLUME CALCULATIONS

Understanding that water removes heat and how much heat is removed per unit of mass allows us to calculate the amount of water required to extinguish specific fires. Theoretical volumes of water can be calculated by knowing the starting and ending temperatures for a unit of water mass, then dividing that into the heat release rate for the fire.

EXAMPLE 14.4

For a 2,000,000 W fire (large sofa), you can determine the amount of water required by evaluating the amount of heat removed by each unit of water mass.

First, you must determine the beginning and ending temperatures for the water. For this example, raise the temperature of the water from 20°C to 120°C. Each gram of water diverts 334 J from the fire:

$$20°C \text{ to } 100°C \times 4.18 \text{ J/g°C} = 80°C * 4.18 \text{J/g°C} \quad\quad = \quad 334 \text{ J/g}$$
$$\text{Latent heat of vaporization} \quad\quad\quad\quad\quad\quad\quad\quad\quad = 2{,}260 \text{ J/g}$$
$$100°C \text{ to } 120°C \times 2.0064 \text{ J/g°C} = 20°C * 2.0064 \text{ J/g°C} = \quad\underline{40 \text{ J/g}}$$
$$2{,}964 \text{ J/g}$$

One Joule per second is a watt; therefore, the amount of heat produced from the fire is 2,000,000 Joules per second. If a gram of water removes 2,964 Joules per second, you can determine the point where all energy from the fire is diverted to heating the water:

$$2,000,000 \text{ J/s divided by } 2,964 \text{ J/g} = 675 \text{ g/s}$$

This means that if sufficient water were applied to the fire in 1 second, only .675 kilograms (kg) is needed. The difficult issue is ensuring that all water mass reaches the proper point within the fire to extinguish the fire sufficiently. In reality, not all energy from the fire must be removed to interrupt the flame. At some point insufficient heating of the fuel results in fire extinguishment; however, determination of that point is difficult.

$$675 \text{ g/s} * 60 \text{ sec/min} = 40,500 \text{ g/min} = \text{ or } 40.5 \text{ kg/min}$$

One kg of water is 1 liter of volume. Thus, in this example, 40.5 liters per minute should extinguish the fire. Converting that volume to gallons requires division of 40.5 liters by 3.8 liters per gallon, which reveals that just less than 11 gallons of water per minute should extinguish the fire. If the water were raised to only 100°C rather than absorbing LHV, much more water is required.

$$20°C \text{ to } 100°C \times 4.18 \text{ J/g°C} = 80 * 4.18 = 334 \text{ J/g} \times 100$$
$$2,000,000 \text{ J/s} \div 334 \text{ J/g} = 5,988 \text{ g/s}$$
$$5.988 \text{ kg/s or } 359 \text{ kg/min (359 l/min)}$$
$$359 \text{ lpm/3.8 lpg} = 94 \text{ g/min}$$

FIRE SUPPRESSION CALCULATIONS

Calculation of heat release rate during emergencies is difficult at best. Therefore, the fire service opts for formulas that deliver sufficient water to ensure extinguishment when properly factored. For many years, the fire service used the Iowa formula to determine the volume of water needed to fight a structural fire. The **Iowa formula** required the determination of the structure volume in cubic feet, which is then divided by 100 to determine the volume of water required in gallons per minute that would extinguish the fire.

For a structure that is 30 feet wide, 50 feet long, and 10 feet high, we can calculate the amount of water needed as follows:

$$30' \times 50' \times 10' = 15,000 \text{ cubic feet}$$
$$15,000 \div 100 = 150 \text{ gallons per minute needed for full involvement}$$

In later years, the National Fire Academy (NFA) formula has been promulgated because it calls for increased volume and involves less numbers for calculation. The **NFA formula** factors building (compartment) square footage divided by 3(l * w/3) and is reported in gallons per minute (gpm). The previous example would be calculated as follows, according to the NFA formula:

$$30 \times 50/3 = 500 \text{ gpm}$$

The increased volume provides reserve water for extinguishment when water is misapplied.

Iowa formula
■ A mathematical formula for determining the flow rate of water required for confined fires based on compartment size. Developed at the Fire Service Institute of Iowa State University.

NFA formula
■ A mathematical formula for determining the flow rate of water required for confined fires based on compartment size. Developed at the U.S. National Fire Academy.

SPRINKLER SYSTEMS

Fixed fire protection systems offer opportunities for more accurate determination of water flow needed to control fires. Understanding the peak heat release rate possible for specific fuel packages (mass, volume, and geometry) provides fire protection professionals sufficient information to place nozzles (sprinkler heads) in locations where they can effectively suppress fires before extreme heat release rates are achieved. In addition to fuel package information, sprinkler performance is requisite knowledge, including density (gallons per minute per square foot), operating temperature, and response time index (RTI). Simply, operating temperatures and RTI indicate how much energy must reach the sprinkler before activation results, and the amount of time this temperature must be present before the sprinkler mechanism operates. **Sprinkler density** indicates the amount of water that will reach fuel surfaces every minute. Quantities range from .04 gallons per minute to more than 1 gallon per minute for every square foot protected by the sprinkler head.

Sprinkler operation accomplishes three goals: (1) penetrate the fire plume to cool the burning fire, (2) prewet the surrounding combustibles, and (3) cool the hot gas layer temperature. Generally the cooling of burning slows the reaction, but equally important is preventing flame spread. Water on combustibles must be volatized before they can ignite; thus, a smaller fire (sprinkler water is cooling the flames) is less likcly to initiate pyrolysis because of the wetting action.

The National Fire Protection Association's NFPA 13, Standard for the Installation of Sprinkler Systems, provides area/density curves for sprinkler density over specified areas based on perceived heat release rates (HRRs) of fuels by generalization of occupancy use and/or commodities stored (see Table 14.1). Chapter 11.2 divides typical situations into Light Hazard, Ordinary Hazard Group 1, Ordinary Hazard Group 2, Extra Hazard Group 1, and Extra Hazard Group 2 occupancies. Sprinkler designers choose sprinkler densities over a specified design area to provide protection for the most hydraulically demanding area within the total structure.

> IMPORTANT: *Not all sprinklers within a structure are designed to activate during a fire. Rather only the sprinklers within the design area are expected to control any fire that occurs within the structure. Within this design area, sprinklers cool the fire and prewet the surrounding combustibles, thus preventing fire from spreading beyond the design area. Also, these occupancies do not cover specific higher-challenge conditions such as rack storage of combustibles. Other chapters of NFPA 13 address needed sprinkler density and/or arrangement for those conditions.*

High-challenge fuels and fuel arrangements that have burning characteristics that would overpower standard sprinkler systems should be protected by systems designed to protect those special situations. Pile storage, rack storage, and storage of high heat-release-rate fuels are categories that mandate sprinkler protection beyond the normal density curves.

Performance-based design involves calculation of water delivery specifically for particular compartments and their associated fuels. This provides the opportunity to deliver water more efficiently to specific situations without wasting water or effort. The caution, however, is that performance-based design is for a

---

sprinkler density
■ The rate of water flow delivered through a sprinkler system to an area in gallons per minute per square foot, or liters per minute per square meter.

TABLE 14.1	NFPA 13 Design Area Curves

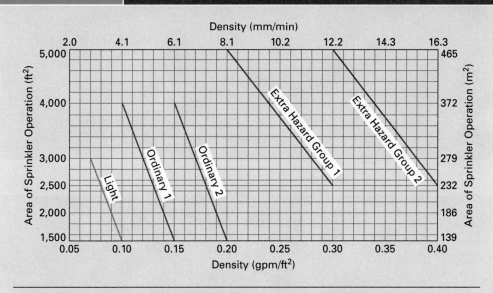

Source: Reprinted with permission from NFPA 13, *Installation of Sprinkler Systems*, Copyright © 2009, National Fire Protection Association, Quincy, MA 02169. This reprinted material is not the complete and official position of the NFPA on the referenced subject, which is represented only by the standard in its entirety.

specific situation. Changes in fuels involved or the arrangement of those fuels can negate sprinkler system effectiveness. Such changes to consider are the change of occupancy use or building occupants.

Where fuels and arrangements are likely to change, such as retail stores commonly known as big box stores, advantage is gained by designing systems for the greatest challenge predicted in the building's use, then providing that protection throughout. This allows managers to move stock freely within the structure without considering if the sprinkler system is designed for the changed situation. Most often these are early suppression fast response (ESFR) systems. ESFR systems have higher flow and pressure requirements for larger quantities of water (see Figure 14.10).

Water Summary

Regardless of application method, it should be noted that small quantities of water can extinguish large fires (see Figure 14.11). This does not mean that we recommend fire attack with small hand lines (booster hoses) or, as some firefighters advocate, with water cans (2½-gal water extinguishers). These smaller lines may work, but little reserve is present to protect those who are taking the attack toward the fire.

It is imperative that firefighters and fire protection designers understand water sufficiently to ensure that the application methodology selected is viable for interrupting the combustion process. Balancing how to prevent additional damage from both fire and water is the measure of understanding.

FIGURE 14.10
Standard spray
sprinkler (a) and ESFR
sprinkler operation (b).
Source: Photos by authors.

(a)

(b)

Sprinkler systems tend to apply more water than the minimum required for extinguishment. Some responders opt to turn sprinklers off before locating the fire and ensuring that it is extinguished. We have not witnessed wisdom in this tactic; rather we strongly recommend shutting sprinklers only after fire suppression forces confirm that the fire has been fully extinguished.

Class A Foam

Surfactants are mixed with water to achieve a mixture that has lower surface tension, which facilitates deeper penetration of water into smoldering Class A fuels, thereby reducing pyrolysis of that area of fuel. The same principle applies to use of soaps and detergents for cleaning to more easily reach deep into surfaces needing to be cleaned. Class A foam has long been widely used to combat forest fires and is now utilized to fight structural fires. Compressed air foam (CAF) is a method by which air is introduced into the foam stream to ensure proper agitation of the foam and thus provide a solution that tends to cling to surfaces, especially vertical surfaces. An additional benefit is a decrease in hose weight and an increase in maneuverability.

REMOVAL OF OXYGEN

Removing oxygen from a fire is a very effective method of extinguishment when properly applied. Smothering a fire involves removal of oxygen by chemical reaction of the available supply but not replenishing the surrounding environment. A common example is shown in Figure 14.12. Let's say that a fire occurs in a pan on a stove: The fuel (medium used to heat the food) ignites the food and/or the pot.

FIGURE 14.12
Cooking fire extinguished with a lid.
Source: Photo by authors.

Simply placing a lid on the fire removes air; thus, the chemical reaction ceases. Oxygen still remains in the pan, but at a concentration not sufficient to support combustion.

Removal of oxygen by closing a fire area is an acceptable option when ventilation openings can be sealed and no civilians are present. The method is not advocated when fuels contain oxidizers or are stored in the same compartment with oxidizers. In the case of the SS *Grandcamp,* a cargo hold containing burning ammonium nitrate was sealed so steam could be introduced for extinguishment. Though oxygen from ventilation was not available, ammonium nitrate contains chemically bound oxygen that is released during decomposition. A massive explosion resulted even when ventilation was restricted.

Foam

foam
- A mixture of chemical surfactants with water to enhance fire suppression characteristics of the water.

Foams are a mix of water and surfactants that reduce surface tension of the water. Thus, they allow water to float on top of typical hydrocarbon liquids in a manner that prevents vapors from mixing with oxygen in sufficient quantities to sustain combustion. Typically this occurs when soap bubbles float on fuels and thus provide a liquid film to create the seal and/or a membrane to occlude the air from the fuel's surface. The foam concentration and the material used to generate the foam is based on the fuel on which the foam will be applied. For example, foams that work on common hydrocarbon fuels may degrade when applied on fuels containing alcohols. This becomes extremely important when addressing gasoline that contains ethanol; foams selected for suppressing fires containing this fuel must be compatible with both gasoline and ethanol. Foam is selected for specific chemicals in industrial applications, while response crews are forced to use a few types of concentrate for multiple applications.

Types of Foam and Predominant Uses

High expansion: Large volumes of surfactant bubbles (soap bubbles) tend to separate air from the fuel source.

Aqueous film forming foam (AFFF): A surfactant film spreads over the fuel surface to occlude air. A thin layer of foam then covers the application to reduce degradation of that film.

Polar solvents (alcohol type): Most hydrocarbon fuels have a different polarity than does water, which is demonstrated by hydrocarbons floating on water. The chemical separation is not the same with alcohols and other solvents that have similar polarity with water. Foams intended for standard hydrocarbons tend to degrade quickly when applied to alcohols or hydrocarbons containing alcohols (gasoline blended with methanol). Alcohol type concentrate (ATC) is blended to reduce this degradation.

Fluoroprotein foams: Consist of protein foam, which is biodegradable and more resilient to degradation, combined with fluoridated hydrocarbons. They are touted as having the characteristics of both synthetic surfactants (i.e., quick knockdown) and the durability of protein foams.

Class A foam: Class A foam (previously discussed in this chapter) acts as a surfactant to reduce surface tension of the water and thus increases water's ability to penetrate fuel surface and char. This in turn reduces or prevents pyrolysis of that fuel.

Kitchen Grease Fires

Though fires in vegetable-based cooking oils (for example, oil in a deep-fat fryer) are ignitable liquid fires, the heat release rate and propensity to retain heat sufficient to reignite poses a greater hazard than that found with typical hydrocarbon fuels. As commercial cooking facilities transitioned from animal-based oils to vegetable-based oils for cooking, many fires continued to be extinguished by dry chemical extinguishing systems, only to reignite, causing significant damage in many locations. This brought about the introduction of wet chemical systems that contain potassium acetate or potassium citrate. When discharged into the burning fuel, these materials react with the vegetable oil to produce saponification, soapy foam that excludes oxygen from the fuel surface. The chemical simultaneously cools the liquid surface to below its ignition temperature. It should be noted, however, that prevention of reignition may depend on ensuring that heat is removed from the cooking oil.

OXYGEN DISPLACEMENT

Removal of oxygen or reducing its concentration in the combustion area can extinguish a fire. Carbon dioxide (CO_2), nitrogen (N_2), and argon (Ar) are gases commonly used to extinguish fires by displacing oxygen. Displacement of *all* oxygen is not the result; rather, the concentration (percentage in the atmosphere) of oxygen is reduced when the **inert** (or chemically inactive) gas is introduced. The National Fire Protection Association's standard NFPA 12, Standard on Carbon Dioxide Extinguishing, indicates that the design criteria for a total flooding system is as low as 34 percent concentration of CO_2, which is sufficient for some fuels (i.e., butane), while 72 percent is needed for other fuels such as carbon disulfide (CS_2) (NFPA 12, 2008) (see Table 14.2).

inert
■ Chemically inactive.

In total flooding systems, the total volume of the protected compartment is determined, then the amount of CO_2 necessary can be calculated to achieve the appropriate design concentration. For example, a compartment that measures 20 feet by 30 feet by 10 feet high occupies a volume of 6,000 cubic feet. If acetone is the hazard and is considered a surface fire hazard, the design concentration is 34 percent. NFPA 12 Table 5.2.2(a) indicates that, for design criteria for specific volumes at 345 concentrations and for 6,000 ft^3, the design is 0.050 lb CO_2/ft^3.

$$6,000 \text{ ft}^3 * 0.050 \text{ lb } CO_2/\text{ft}^3 = 300 \text{ lb of } CO_2 \text{ needed}$$

Local application systems can also be useful for limiting oxygen to fires in limited areas and/or for specific hazards. Systems are designed to create an atmosphere in the protected area that restricts oxygen concentration sufficient to support combustion for the time period required for the heat source to cool, thus preventing reignition when ambient oxygen returns to the vicinity of the fuel.

COOLING EFFECT OF INERT GAS

The introduction of inert gases has the ancillary effect of cooling flame temperatures. Heat transferred to the carbon dioxide or other inert gas is not radiated to other portions of the fuel package; thus, the temperature of combustion diminishes. Though removing heat is not the mechanism of extinguishment, it does occur at least to some limited degree.

	TABLE 14.2	Percentage of Carbon Dioxide for Total Flooding Systems	

MATERIAL	THEORETICAL MINIMUM CO_2 CONCENTRATION (%)	MINIMUM DESIGN CO_2 CONCENTRATION (%)
Acetylene	55	66
Acetone	27*	34
Aviation gas, grades 115/145	30	36
Benzol, benzene	31	37
Butadiene	34	41
Butane	28	34
Butane-I	31	37
Carbon disulfide	60	72
Carbon monoxide	53	64
Coal or natural gas	31*	37
Cyclopropane	31	37
Diethyle either	33	40
Dimethyl ether	33	40
Dowtherm	38*	46
Ethane	33	40
Ethyl alcohol	36	43
Ethyl ether	38*	46
Ethylene	41	49
Ethylene dichloride	21	34
Ethylene oxide	44	53
Gasoline	28	34
Hexane	29	35
Higher paraffin hydrocarbons $C_nH_{2m} + 2m - 5$	28	34
Hydrogen	62	75
Hydrogen sulfide	30	36

*Calculated from accepted residual oxygen values.
Source: Reprinted with permission from NFPA 12, *Carbon Dioxide Extinguishing Systems,* Copyright © 2007, National Fire Protection Association, Quincy, MA 02169. This reprinted material is not the complete and official position of the NFPA on the referenced subject, which is represented only by the standard in its entirety.

MATERIAL	THEORETICAL MINIMUM CO_2 CONCENTRATION (%)	MINIMUM DESIGN CO_2 CONCENTRATION (%)
Isobutane	30*	36
Isobutylene	26	34
Isobutyl formate	26	34
JP-4	30	36
Kerosene	28	34
Methane	25	34
Methyl acetate	29	35
Methyl alcohol	33	40
Methyl butene-I	30	36
Methyl ethyl ketone	33	40
Methyle formate	32	39
Pentatne	29	35
Propane	30	36
Propylene	30	36
Quench, lube oils	28	34

Note: The theoretical minimum extinguishing concentrations in air for the materials in the table were obtained from a compilation of Bureau of Mines, Bulletins 503 and 627.

Class D Fires

Extinguishment of fires in combustible metals poses significant challenges when extinguishing agents tested and listed for that metal are not available. Commonly encountered combustible metals include aluminum, magnesium, titanium, and zirconium (see Figure 14.13). These metals are commonly found in machinery and consumer products. Fortunately, quantities are typically relatively low. Where combustible metals are routinely stored, used or processed, specialized chemicals for each particular metal should be readily accessible. These specialized extinguishing agents work by occluding oxygen and removing heat from the fire. When combustible metals are encountered where specialized extinguishing agents are unavailable, consideration should be given to allowing the fuel to oxidize before initiating other extinguishing activities. One of the more commonly encountered combustible metals is magnesium, which is often found in machine engines. Water flow from typical fire hose streams can be effective in extinguishing small fires involving magnesium, but exercise caution when considering attacking more than very small fires involving this metal.

Water-reactive metals such as lithium and sodium pose an even greater challenge for extinguishment. Water and foam may produce adverse results; thus, research into options is necessary based on the situation encountered. Again, an

FIGURE 14.13
Magnesium fire.
Source: Photo by authors.

option is to allow full reaction of the metal before attempting to control other involved fuels.

Interrupting a Chemical Chain Reaction

Typical fires involve organic materials reacting with oxygen to form water and carbon dioxide. The reaction is not as simple as it is usually expressed; rather, there are intermediary steps of the reaction that facilitate the change. Interruption of the reaction within the intermediary steps alters products and can slow the reaction. When sufficient interruption occurs, radiant heat feedback diminishes and the fire can be extinguished. One of the more common agents used to interrupt the chemical chain reaction is application of halon gases. Bromine, chlorine, fluorine, and iodine comprise the **halon gases**. Halon gases, especially bromine, chemically react with hydroxide radicals in the intermediary steps of combustion to interrupt typical oxidation reactions, thus causing extinguishment of the fire.

Dry chemical extinguishing agents work in much the same manner as halons. However, dry chemical extinguishing agents are more corrosive to electrical equipment in the area, and unreacted chemicals tend to disperse well beyond the intended location of extinguishment. This means that cleaning the area becomes a great problem. Dry chemical agents include traditional sodium bicarbonate (baking soda), potassium bicarbonate (Purple K™), potassium chloride (Super K™), and monoammonium phosphate (multipurpose dry chemical).

halon gas
■ Gases formed with carbon and one or more of the elements identified as halogen gases, including fluorine, chlorine, bromine, and iodine.

Summary

Extinguishment of fire is based on scientific principles involving removal or interruption of components required for combustion. By understanding what can occur, you are better equipped to facilitate positive outcomes during fire suppression efforts. Understanding how water works to extinguish fire is critical to safe and effective fire suppression with hose streams. Simply squirting water may produce the desired results, but those who truly understand how the water works will produce more predictable results, which is what fire safety professionals should strive to attain.

Review Questions

1. Extinguishment of fire can be accomplished by:
   a. adding heat to the atmosphere
   b. increasing the percentage of oxygen in the atmosphere
   c. reducing the oxygen level to that which will not sustain combustion
   d. none of the above
2. What does the term *heat capacity* mean?
3. Why is understanding the concept of latent heat of vaporization important in fire suppression efforts?
4. Water applied on a large heat-release-rate fire in small droplets from a sprinkler system:
   a. probably will be very effective in extinguishing the fire
   b. probably will be effective in controlling the fire but not extinguishing it
   c. may be ineffective because the water may convert to steam before reaching burning fuel
   d. cannot be effective in any fires
5. Carbon dioxide extinguishes fires by removal of oxygen. The percentage of carbon dioxide needed to achieve this:
   a. must be sufficient to displace enough oxygen to prevent continued oxidation reaction with the fuel
   b. must replace all oxygen
   c. can be any concentration to extinguish the fire
   d. is unimportant; carbon dioxide has been outlawed

6. An example of removing fuel to stop a fire is:
   a. turning off the pump when a refueling fire occurs when filling a car
   b. closing a valve feeding a burning machine with propane gas
   c. removing combustible stock from beside a warehouse when an outside fire approaches
   d. all of the above
7. Steam production during fire suppression activities is ideal only when:
   a. the latent heat of vaporization draws the most heat from the fire, which always happens
   b. making an exterior attack
   c. making an interior attack
   d. Steam production is never wanted in fire suppression.
8. Dry powder extinguishing agents are intended for:
   a. specific metals based on their individual properties
   b. all metal fires, regardless of their intended use
   c. use on other fires; they are not effective on metal fires
   d. extinguishing ignitable liquid fires
9. Class K fires involve fuels used in:
   a. painting cars
   b. cleaning metal parts
   c. solvents for paints
   d. cooking

10. Water is the most commonly used extinguishing agent because:
   a. it is inexpensive
   b. it is readily available in most locations
   c. it is very effective
   d. all of the above
   e. none of the above

References

NFPA 12. (2008). *Standard on Carbon Dioxide Extinguishing Systems*. Quincy, MA: NFPA.

Robertson, J. (Fire Chief). (May 2000). Chicago, IL. Retrieved March 27, 2001, from http://firechief.com/mag/firefighting_father_fog_lloyd/.

Tewarson, A. (2008). "Generation of Heat and Gaseous, Liquid, and Solid Products in Fires" (Chapters 3 and 4), in the *SFPE Handbook of Fire Protection Engineering*, Fourth Edition. Bethesda, MD: SFPE.

advection One of the mechanisms that comprises the convection heat transfer mode dealing with the transport due to bulk fluid motion.

aerosol A fine mist or spray containing minute particles.

air entrainment The process of air being drawn into a fire, plume, or jet. (NFPA 921, 2008)

alkanes Hydrocarbon compounds that have multiple carbons with a single bond with other carbons. Also known as *saturated hydrocarbons* because hydrogen atoms are present at every possible location.

alkenes Hydrocarbon compounds that have multiple carbons where at least one carbon-to-carbon bond has a double bond.

alkynes Hydrocarbon compounds that have multiple carbons where at least one carbon-to-carbon bond has a triple bond.

area The quantity of space occupied in a two-dimensional plane (length × width).

atom The smallest particle of an element that can be identified with that element.

atomic number The number of protons possessed by an atomic nucleus.

atomic weight The mass of an atom compared to carbon-12, which is assigned the atomic weight of 12.

autoignition temperature The lowest temperature at which a combustible material ignites in air without a spark or flame. (NFPA 921, 2008)

backdraft Limited ventilation during an enclosure fire can lead to the production of large amounts of unburned pyrolysis products. When an opening is suddenly introduced, the inflowing air forms a gravity current and begins to mix with the unburned pyrolysis products, creating a combustible mixture of gases in some part of the enclosure. Any ignition source, such as a glowing ember, can ignite this combustible mixture, which results in an extremely rapid burning of gases and/or pyrolysis products forced out through the opening and in turn causes a fireball outside the enclosure.

blackbody A perfect absorber and emitter of radiation.

boiling liquid expanding vapor explosion (BLEVE) Failure of a confining vessel due to pressure increase facilitated by heating of liquids within the container. Failure often is associated with heat weakening the metal container and with a sudden fireball when contents are combustible. (NFPA 921, 2008)

boiling point The temperature at which the vapor pressure of a substance equals the average atmospheric pressure.

boilover The phenomenon associated with the production of steam during a crude petroleum fire, which drives the burning petroleum up and over the walls of its confining tank.

British thermal unit (BTU) The quantity of heat required to raise the temperature of one pound of water 1°F at the pressure of 1 atmosphere; a British thermal unit is equal to 1,055 Joules, 1.055 kilojoules, and 252.15 calories. (NFPA 921, 2008)

buoyancy The upward force exerted on a body or volume of fluid by the ambient fluid surrounding it. If the volume of fluid has a positive buoyancy, then it is lighter than the surrounding fluid and tends to rise. Conversely, if it has negative buoyancy, it is heavier and tends to sink. Buoyancy of a fluid depends on both its molecular weight and its temperature. (Drysdale, 2008)

burning rate The mass of fuel consumed per unit time in the fire.

ceiling jet A relatively thin layer of flowing hot gases that develops under a horizontal surface (e.g., ceiling) as a result of plume impingement and the flowing gas being forced to move horizontally. (NFPA 921, 2008)

ceiling layer A buoyant layer of hot gases and smoke produced by a fire in a compartment. (NFPA 921, 2008)

cellulosic fuels Hydrocarbon fuels that contain cellulose molecules, which are chains of molecules with the chemical formula of $C_6H_{10}O_5$.

Celsius A temperature measurement system that designates the freezing point of water at 0° and the boiling point of water at 100°. Absolute zero is −273.16°.

char Carbonaceous material resulting from pyrolysis, often appearing to be blackened blisters on the surface of cellulose and other solid organic fuels.

char development A process of pyrolysis that results in char on or through solid fuels.

circumference The linear distance around the outside of a circle.

Class A fire A fire in ordinary combustible materials, such as wood, cloth, paper, rubber, and many plastics. (NFPA 10, 2007)

Class B fire A fire in flammable liquids, combustible liquids, petroleum greases, tars, oils, oil-based paints, solvents, lacquers, alcohols, and flammable gases. (NFPA 10, 2007)

Class C fire A fire that involves energized electrical equipment. (NFPA 10, 2007)

Class D fire A fire in combustible metals, such as magnesium, titanium, zirconium, sodium, lithium, and potassium. (NFPA 10, 2007)

classification of fire A system that categorizes fires into groups based on the similarities in burning and extinguishment.

Class K fire A fire in cooking appliances that involves combustible cooking media (vegetable and animal oils and fats). (NFPA 10, 2007)

code enforcement personnel Persons who engage in inspections to ensure conditions are within prescribed parameters that provide favorable fire protection conditions.

combustible liquid Any liquid that has a closed-cup flash point at or above 100°F (37.8°C), as determined by the test procedures and apparatus. (NFPA 30, 2008)

combustion efficiency The ratio of the output of a combustible's kinetic energy to the total input (potential energy).

compound A substance comprised of two or more elements in chemical combination.

concurrent flow The flame spread direction is the same as the gas flow or wind direction.

condensed phase Phase in which substances are solids or liquids but not gases. (Babrauskas, 2003)

conduction The transfer of energy in the form of heat by direct contact through the excitation of molecules and/or particles driven by a temperature difference. (NFPA 921, 2008)

conduction heat Transfer of heat to another body or within a body by direct contact.

configuration factor The fraction of the radiation leaving a surface i that is intercepted by surface j. Also commonly referred to as the *view factor*.

convection heat Transfer of heat by circulation within a medium such as a gas or a liquid.

convection heat transfer coefficient A quality that represents the ability of heat to be transformed from a moving fluid to a surface. (Quintiere, 1999)

coordinate system A system that uses incremental measures on three planes to designate the exact location of a point. Planes are commonly designated x, y, and z.

counterflow flame spread The flame spread direction is counter to or opposed to the gas flow. (NFPA 921, 2008)

decay phase The point at which the heat release rate from a fire diminishes. This can be brought on by a decrease in fuel available to oxidize or by the restriction of air.

decomposition reaction A chemical reaction involving the breakup of a compound into two or more elements or compounds.

deflagration Rapid burning, faster than open-air burning of the material but slower than detonation. (Thurman, 2006)

dehydration Removal of water, i.e., removal of water from solid fuel through heating the fuel.

density The property of a substance that measures its compactness. The mass of a substance divided by the volume it occupies.

detonation Instantaneous combustion or conversion of a solid, liquid, or gas into larger quantities of expanding gas accompanied by heat, shock, and noise. Flame propagation in detonations is more than 3,300 feet per second. (Thurman, 2006)

diameter The measure across a circle.

diffusion flame Flame resulting from fire where oxygen mixes with the fuel at the combustion zone. (NFPA 921, 2008)

double replacement reaction The type of chemical reaction in which two different compounds exchange their ions to form two new compounds.

element A substance comprised of only one type of atom.

endothermic A chemical reaction that occurs with the absorption of heat.

endothermic reaction A chemical reaction that requires energy input from an external source for the reaction to occur.

energy The property of matter that enables it to do work.

English system A system for quantifying measurement that uses feet, gallons, pounds, and British thermal units.

entrainment The process of air or gases being drawn into a fire, plume, or jet. Also known as *air entrainment*. (NFPA 921, 2008)

established burning The point in an oxidation reaction when piloted flame is unnecessary to sustain pyrolysis. Radiant heat transfer from the fuel package is sufficient for continued pyrolysis.

exothermic A chemical reaction that releases heat energy.

expansion of matter Change in volume resulting from a change in temperature. Increasing the temperature in a unit of matter results in increased volume or pressure, while lowering the temperature in that matter results in reduced volume or pressure.

expansion of vapor The amount of heat required to convert a unit of mass of liquid substance to a vapor (endothermic). An equal amount of energy is released by the mass to convert from vapor to liquid (exothermic). Expressed in J/g, calories, or BTUs.

Fahrenheit A temperature measurement system that designates the freezing point of water at 32° and the boiling point of water at 212°. Absolute zero is −459.69°.

field models Computer fire models that attempt to predict conditions at every point. Also known as *computational fluid dynamics (CFD) models*. (Quintiere, 1999)

fire A rapid oxidation process that is a chemical reaction resulting in the evolution of light and heat in varying intensities. (NFPA 921, 2008)

fire growth The initiation, growth, and decay of a fire's heat release rate over the duration of the fire.

fire investigators Persons who work to determine the origin, cause, and circumstances of fires.

fire plume The column of hot gases, flames, and smoke rising above a fire. (NFPA 921, 2008)

fire point The lowest temperature at which a liquid will ignite and achieve sustained burning when exposed to a test flame in accordance with the American Society for Testing and Materials (ASTM) D 92, Standard Test Method for Flash and Fire Points by Cleveland Open Cup Tester. (NFPA 30, 2008)

fire protection engineers Persons who seek to use scientific methods to prevent or mitigate fires through study and testing.

fire retardant A liquid, solid, or gas that tends to inhibit combustion when applied on, mixed in, or combined with combustible materials. (NFPA 1, 2006)

fire spread The movement of fire from one place to another. (NFPA 921, 2008)

fire suppression personnel Persons who engage in operations to extinguish fires for public and private entities.

fire tetrahedron A model that describes four components necessary for continued combustion: fuel, heat, oxygen, and an uninterrupted chemical chain reaction.

fire triangle A model that describes three components necessary for ignition: fuel, heat, and oxygen.

flame front The leading edge of burning gases of a combustion reaction. (NFPA 921, 2008)

flame jets Horizontal flame movements below ceilings or other restrictions resulting when convective heat transfer can no longer move upward and thus transfers laterally.

flameover The condition where unburned fuel (pyrolyzate) from the originating fire has accumulated in the ceiling layer to a sufficient concentration (i.e., at or above the lower flammable limit) that it ignites and burns. Can occur without ignition of, or prior to the ignition of, other fuels separate from the origin. Also known as *rollover*. (NFPA 921, 2008)

flame plume A body or stream of gaseous material involved in the combustion process and emitting radiant energy at specific wavelength bands determined by the combustion chemistry of fuel.

flame spread The movement of flaming combustion from one place along a fuel item to another. Also known as *flame propagation*. (NFPA 921, 2008)

flaming combustion The hot, light-emitting portion of a combustion process. The heat released is sufficient to raise the temperature of gases so that they emit energy waves in the visible light range.

flammable limits The maximum or minimum concentration of combustible gas and air mixture that will ignite. The lower (lean) limit is the point of fuel deficiency; the upper (rich) limit is the point of oxidizer deficiency to sustain combustion. Also known as explosive limits. (Kuvshinoff, 1977)

flammable liquid Any liquid that has a closed-cup flash point below 100°F (37.8°C), as determined by the test procedures and apparatus and a Reid vapor pressure that does not exceed an absolute pressure of 40 psi (276 kPa) at 100°F (37.8°C), as determined by Section 4.4 of the American Society for Testing and Materials (ASTM) D 323, Standard Test Method for Vapor Pressure of Petroleum Products (Reid method). (NFPA 30, 2008)

flash fire A fire that spreads rapidly through a diffuse fuel, such as dust, gas, or the vapors of an ignitable liquid, without the production of damaging pressure. (NFPA 921, 2008)

flashover A transition phase in the development of a compartment fire in which surfaces exposed to thermal radiation reach ignition temperature more or less simultaneously and fire spreads rapidly throughout the space, resulting in full-room involvement or total involvement of the compartment or enclosed space. (NFPA 921, 2008)

flash point The minimum temperature of a liquid at which sufficient vapor is given off to form an ignitable mixture with the air, near the surface of the liquid or within the vessel used, as determined by the appropriate test procedure and apparatus. (NFPA 30, 2008)

foam A mixture of chemical surfactants with water to enhance the fire suppression characteristics of the water.

free burning Unrestricted combustion of flammable materials. (Kuvshinoff, 1977)

freezing point The temperature at which the liquid and solid states of a substance coexist at one atmosphere or 101.3 kPa.

fuel A material that maintains combustion under specified environmental conditions. (NFPA 53, 2004)

fuel-controlled fire A fire in which the heat release rate and growth rate are controlled by the characteristics of the fuel, such as quantity and geometry, and in which adequate air for combustion is available. (NFPA 921, 2008)

fuel item Any article that is capable of burning. (NFPA 921, 2008)

fuel load The total quantity of the combustible contents of a building, space, or fire area, including interior finish and trim, expressed in heat units or the equivalent weight in wood. (NFPA 921, 2008)

fuel package A collection or array of fuel items in close proximity with one another such that flames can spread throughout the array of fuel items. Multiple fuel packages may be involved in a single compartment fire. (NFPA 921, 2008)

full-room involvement Condition in a compartment fire in which the entire volume is involved in fire. (NFPA 921, 2008)

fully developed fire Fuel and oxygen are readily available with sufficient heat to pyrolize the fuel. The fire reaches a steady burning state. Also called *free burning state* and *steady-state burning*.

gas The physical state of a substance that has no shape or volume of its own and expands to take the shape and volume of the container or enclosure it occupies. (NFPA 921, 2008)

glowing combustion Luminous burning of solid material without a visible flame. (NFPA 921, 2008)

halon gases Gases formed with carbon and one or more of the elements identified as halogen gases, including fluorine, chlorine, bromine, and iodine.

hazard classification system A system of identifying substances by their hazard characteristics, including threat to humans, flammability, reactivity, and other hazards.

heat The form of energy transferred from one body to another because of the temperature difference between them. Energy arising from atomic or molecular motion. (Cengel and Boles, 2006)

heat capacity The amount of heat needed to raise one mass of a substance 1°. In the SI system, the term refers to raising one gram 1°C. In the English system, the term refers to raising one pound 1°F.

heat flux The measure of the rate of heat transfer to a surface, expressed in kilowatts/m^2, kilojoules/m$^2 \cdot$ s, or BTU/ft$^2 \cdot$ s. (NFPA 921, 2008)

heat of combustion The quantity of heat evolved by the complete combustion of one mole (gram molecule) of a substance. (Kuvshinoff, 1977)

heat of gasification (L_g or L_v) The quantity of heat required to cause a mass unit of solid or liquid to convert to the gaseous state. It is expressed in J/g or BTU/ lb. Also known as *heat of vaporization*. (Kuvshinoff, 1977)

heat release rate The rate at which heat energy is generated by burning. (NFPA 921, 2008)

ignition temperature Minimum temperature that a substance should attain in order to ignite under specific test conditions.

incipient stage fire The initial stage of a fire following ignition, typically involving a single fuel item or fuel package. It does not significantly alter conditions within the compartment.

inert Chemically inactive.

ionic bonds Chemical bonding resulting from one or more electrons leaving the atoms of one element to satisfy the valence of atoms of another element in that compound. Resulting electrical difference brings the oppositely charged particles together.

Iowa formula A mathematical formula for determining the flow rate of water required for confined fires based on compartment size. The formula was developed at the Fire Service Institute of Iowa State University.

isotope Any of a group of nuclei having the same number of protons but different number of neutrons.

Joule The preferred SI unit of heat, energy, or work. A Joule is the heat produced when one ampere is passed through a resistance of one ohm for one second, or it is the work required to move a mass of one meter against a force of one newton. There are 4.184 Joules in a calorie, and 1,055 Joules in a British thermal unit (BTU). A watt is a Joule/second. (NFPA 921, 2008)

Kelvin A temperature measurement system that begins at absolute zero, the freezing point of water is at 273.16 K, and the boiling point of water is at 373.16 K.

laminar flames Flames that are smooth, with negligible fluctuations, and are quiescent in nature.

latent heat The amount of heat required to change phase from a solid to a liquid (latent heat of fusion) or from a liquid to a vapor (latent heat of vaporization).

latent heat of fusion The amount of heat required to convert a unit of mass of solid substance to a liquid (endothermic). An equal amount of energy is released by the mass to convert from liquid to solid (exothermic). Expressed in J/g, calories, or BTUs.

latent heat of vaporization The quantity of heat required to cause a mass unit of solid or liquid to convert to a vapor state. It is expressed in J/g or BTU/ lb. Also known as *heat of gasification*.

liquid Matter that possesses a definite volume but lacks definite shape; it assumes the shape of its containing vessel.

lower explosive limit (LEL) The concentration of gas or vapor in air below which a flame will not propagate upon exposure to an ignition source. Also known as the *lower flammable limit (LFL)*.

mass A measure of matter's inertia, often referred to as weight; however, weight is matter's mass times the gravitational pull.

matter Anything that has mass and occupies space.

melt To change the physical state of matter from a solid to a liquid.

metric system A system for quantifying measurement that uses meters, liters, grams, and calories.

minimum ignition energy (MIE) The amount of heat required to initiate piloted combustion within a mixture of flammable gas and air. (NFPA 921, 2008)

mixture Blending of different elements or compounds where chemical combination does not occur.

moisture content Amount of moisture intrinsic to an object; expressed in percentage of total mass for that object.

mole Based on Avogadro's number: 6.022×10^{23} atoms of an element or molecules of a compound. Mass is the number of grams equal to the element or compound's atomic weight.

monomer One or more single substances that combine to form polymers. Generally hydrocarbons capable of linking and thus forming polymers.

neutral plane The height above which smoke will or can flow out of a compartment; the height of zero pressure difference across a partition. (Quintiere, 1999)

NFA formula A mathematical formula to determine the flow rate of water required for confined fires, based on compartment size. The formula was developed at the U.S. National Fire Academy (NFA).

oxidation A chemical reaction in which a compound called a radical loses electrons. The opposite reaction, in which electrons are gained, is called reduction. (Babrauskas, 2003)

pi (π) The relationship of a circle's circumference to the diameter, which is 3.1415 to 1. Represented by the Greek letter pi (π).

piloted ignition Minimum temperature that a substance should attain in order to ignite in the presence of a pilot source (i.e., pilot flame, spark, hot surface). (NFPA 921, 2008)

polymer High-molecular-weight substances produced by the linkage and cross-linkage of its multiple subunits (monomers).

pool fires Fires involving horizontal fuel surfaces.

premixed flames A flame for which the fuel and oxidizer are mixed prior to combustion, as in a laboratory Bunsen burner or a gas cooking range. Propagation of the flame is governed by the interaction among the flow rate, transport processes, and chemical reaction. (NFPA 921, 2008)

pressure Force applied per unit of area.

pyrolysis The chemical decomposition of a compound into one or more other substances by heat alone. Pyrolysis often precedes combustion. (NFPA 921, 2008)

radiation heat Transfer by way of electromagnetic energy. (NFPA 921, 2008)

radius The measurement from the center of a circle to its circumference.

Rankine A temperature measurement system that begins at absolute zero. The freezing point of water is 491.67 R and the boiling point of water is 671.67 R.

SI The system used for quantifying measurement that is accepted by the scientific community. Also known as *Système International*.

single replacement reaction The type of chemical reaction where one atom replaces another in a compound.

smoke The airborne solid and liquid particulates and gases evolved when a material undergoes pyrolysis or combustion, together with the quantity of air that is entrained or otherwise mixed into the mass. (NFPA 921, 2008)

smoldering combustion Combustion without flame, usually with incandescence and smoke. Also known as *nonflaming combustion*. (NFPA 921, 2008)

solid Matter that possesses definite shape and volume.

solution A homogeneous mixture of two or more substances.

specific gravity The ratio of the average molecular weight of a given volume of liquid or solid to the average molecular weight of an equal volume of water at the same temperature and pressure. (NFPA 921, 2008)

specific heat The ratio of the heat capacity of a substance to the heat capacity of water at the same temperature.

spontaneous ignition Initiation of combustion of a material by an internal chemical or biological reaction that has produced sufficient heat to ignite the material. (NFPA 921, 2008)

sprinkler density The rate of water flow delivered through a sprinkler system to an area in gallons per minute per square foot, or liters per minute per square meter.

steady state A stage within a fire's growth where the heat release rate is neither increasing nor decreasing. Essentially the energy being released by the fire has leveled off and is emitting roughly the same amount of energy for a given duration. This stage of the fire is commonly found in compartment fires that have become ventilation-limited and combustion is being controlled by the amount of oxygen that can be entrained into the volume.

stoichiometric mixture Chemical conditions where the proportion of reactants is such that there is no surplus of any reactant after the chemical reaction is completed. Also known as *stoichiometry*. (Babrauskas, 2003)

surface area–to–mass ratio The ratio of surface area to total mass of an object. Thin objects have more surface per unit of mass than do thick objects comprised of like matter. Also known as *surface-to-mass ratio*.

synthesis reaction The type of chemical reaction involving two or more substances that results in the formation of a single product.

temperature Measurement of average kinetic heat energy of an object, which results from molecular motion.

thermal conductivity The property of matter that represents the ability to transfer heat by conduction. (Quintiere, 1999)

thermal diffusivity The ratio of thermal conductivity to the heat capacity, with the units m^2/s.

thermal inertia The properties of a material that characterize its rate of surface temperature rise when exposed to heat. Thermal inertia is related to the product of the material's thermal conductivity (k), its density (ρ), and its heat capacity (c). (NFPA 921, 2008)

thermally thick solid A solid that, while heated from one face, shows a negligible temperature rise at its opposite face. This characteristic is not simply a property of a material; it also depends on the time of exposure and the heat flux.

thermally thin solid A solid that, while heated from one face, shows a back face temperature that is nearly identical to the temperature of the heated face. This characteristic is not simply a property of a material; it also depends on the time of exposure and the heat flux.

thermal runaway Self-heating that rapidly accelerates to high temperatures. (Babrauskas, 2003)

time-temperature curve A graph that depicts temperature at specific times during fire development. The standard time-temperature curve is used to evaluate components and assemblies to determine their performance during fire exposure.

turbulent flames Flames that are typically larger or with higher velocity gas flows, with many fluctuations in direction.

upper explosive limit (UEL) The concentration of gas or vapor in air, above which a flame will not propagate upon exposure to an ignition source. Also known as the *upper flammable limit (UFL)*.

upper layer A buoyant layer of hot gases and smoke produced by a fire in a compartment. (NFPA 921, 2008)

valence electrons Any of an atom's electrons that participate in chemical bonding to other atoms.

vapor A gaseous form of a substance that exists either as a solid or liquid at normal ambient conditions.

vapor density The ratio of the average molecular weight of a given volume of gas or vapor to the average molecular weight of an equal volume of air at the same temperature and pressure. Also known as *specific gravity of gas*. (NFPA 921, 2008)

vapor pressure Pressure exerted within a confined vessel by the vapor of a substance at equilibrium with its liquid; a measure of a substance's propensity to evaporate.

vent An opening for the passage, or dissipation, of fluids, such as gases, fumes, smoke, and the like. (NFPA 921, 2008)

ventilation Circulation of air in any space by natural wind or convection or by fans blowing air into or exhausting air out of a building.

A firefighting operation of removing smoke and heat from the structure by opening windows and doors or making holes in the roof.

ventilation-controlled fire A fire in which the heat release rate or growth is controlled by the amount of air available to the fire. (NFPA 921, 2008)

volatile The quality of a solid or a liquid when it passes into the vapor state at a given temperature. Solids and liquids that readily pass into the vapor state are said to be volatile or highly volatile.

volume The quantity of space occupied in a three-dimensional arrangement (length × width × height).

watt Unit of power, or rate of work, equal to one Joule per second, or the rate of work represented by a current of one ampere under the potential of one volt. (NFPA 921, 2008)

zone models Computer fire models that approximate the fire conditions in a room as two uniform gas layers with a fire energy source. (Quintiere, 1999)

References

Babrauskas, V. (2003). *Ignition Handbook.* Issaquah, WA: Fire Science Publishers.

Cengel, Y., Boles, M. (2006). *Thermodynamics: An Engineering Approach.* Fourth Edition. New York, NY: McGraw Hill.

Drysdale, D. (1999). *An Introduction to Fire Dynamics.* New York, NY: Wiley.

Drysdale, D. (2008). "Physics and Chemistry of Fire," (Section 2, Chapter 1), in *Fire Protection Handbook,* Twentieth Edition. Quincy, MA: National Fire Protection Association.

Friedman, R. (1998). *Principles of Fire Protection Chemistry and Physics.* Third Edition. Quincy, MA: National Fire Protection Association.

Incropera, F., Dewitt, D., Bergman, T., Lavine, A. (2007). *Introduction to Heat Transfer.* Fifth Edition. Danvers, MA: Wiley.

Kuvshinoff, B. (1977). *Fire Sciences Dictionary.* New York, NY: Wiley and Sons.

Meyer, E. (2010). *Chemistry of Hazardous Materials,* Fifth Edition. Brady Fire, New York, NY: Pearson.

NFPA 10 (2007). *Standard for Portable Fire Extinguishers.* Quincy, MA: National Fire Protection Association.

NFPA 30 (2008). *Flammable and Combustible Liquids Code.* Quincy, MA: National Fire Protection Association.

NFPA 53 (2004). *Recommended Practice on Materials, Equipment, and Systems used in Oxygen-enriched Atmospheres.* Quincy, MA: National Fire Protection Association.

NFPA 921 (2008). Guide for Fire and Explosion Investigations. Quincy, MA: National Fire Protection Association.

Quintiere, J. (1999). *Principles of Fire Behavior.* Clifton Park, NY: Delmar.

Thurman, J. (2006). *Practical Bomb Scene Investigation.* Danvers, MA: CRC Press.

INDEX

C